ACETIC ACID
AND ITS
DERIVATIVES

CHEMICAL INDUSTRIES

A Series of Reference Books and Textbooks

Consulting Editor

HEINZ HEINEMANN
Berkeley, California

ADDITIONAL VOLUMES IN PREPARATION

ACETIC ACID AND ITS DERIVATIVES

edited by
Victor H. Agreda
Joseph R. Zoeller

Eastman Chemical Company
Kingsport, Tennessee

CRC Press
Taylor & Francis Group
Boca Raton London New York

CRC Press is an imprint of the
Taylor & Francis Group, an **informa** business

CRC Press
Taylor & Francis Group
6000 Broken Sound Parkway NW, Suite 300
Boca Raton, FL 33487-2742

First issued in paperback 2019

© 1993 by Taylor & Francis Group, LLC
CRC Press is an imprint of Taylor & Francis Group, an Informa business

No claim to original U.S. Government works

ISBN-13: 978-0-8247-8792-9 (hbk)
ISBN-13: 978-0-367-40255-6 (pbk)

Library of Congress Cataloging-in-Publication Data

Acetic acid and its derivatives / edited by Victor H. Agreda, Joseph
    R. Zoeller.
        p.   cm. -- (Chemical industries ; 49)
        Includes bibliographical references and index.
        ISBN 0-8247-8792-7 (alk. paper)
        1. Acetic acid.   2. Acetic acid--Derivatives.   I. Agreda, V. H.
    II. Zoeller, Joseph R.   III. Series: Chemical industries
    ; v. 49.
    QD305.A2A16   1993
    547'.437--dc20   92-36590
                                                            CIP

Visit the Taylor & Francis Web site at
http://www.taylorandfrancis.com

and the CRC Press Web site at
http://www.crcpress.com

Preface

Acetic acid has been a useful chemical since antiquity. Its availability probably precedes the first written language. The first recorded report of acetic acid as vinegar from alcohol is reported to be dated at earlier than 3000 B.C.* Until late in the 19th century, all acetic acid was derived by the same, age-old process of sugar fermentation to ethyl alcohol and subsequent, often undesirable, oxidation to acetic acid to produce vinegar. For most of human history, this was the sole source of acetic acid. Late in the 19th century, this process was supplemented by the advent of wood distillation, which provided an additional source of acetic acid.

In 1916, the first dedicated plant for the production of acetic acid by chemical rather than biological means became a commercial process.* This method was based on acetylene-derived acetaldehyde, and it marks the advent of inexpensive, commercial acetic acid and, hence, the birth of a viable industry based on its use.

Early in this process, Tennessee Eastman Company, now the largest division of Eastman Chemical Company (a division of Eastman Kodak Co.), and other companies began to use acetic acid to produce acetic anhydride and diketene. Eastman Chemical Company has become one of

---

*LeMonnier, E., in *Kirk-Othmer Encyclopedia of Chemical Technology*, Vol. 8, 2nd ed., Wiley-Interscience, New York, 1965, p. 386.

the centers of expertise in the chemistry of acetic acid, its derivatives, and their end uses as a result of its long-term investment in this chemistry. This expertise continues to grow today and is evidenced in their introduction of a new entirely syngas-based process for the generation of acetic anhydride, their introduction of new, commercial esterification technology for the production of methyl acetate and other acetate esters, and their continued introduction of useful acetoacetylation reagents based on acetic acid–derived diketene.

The purpose of this book is to describe the chemistry, thermophysics, and selected process engineering and economics of acetic acid and its derivatives. The first section of the book is devoted to describing the chemistry and methods of generating acetic acid. This includes the catalyzed carbonylation of methanol, oxidation processes, and other carbonylation-based processes. Our intent is to be as comprehensive as possible in describing processes for the production of acetic acid.

The second section is devoted to the thermophysical properties of acetic acid and its aqueous mixtures. Mathematical models and correlations are described in detail, with a thorough appraisal of the data used in their development or an asessment of their predictive ability. The thermophysical properties of aqueous mixtures of acetic acid are included because of the ubiquitousness of such mixtures in processes involving acetic acid and the economic importance (due to high-energy requirements) of the acetic acid–water separation.

The third section of this book deals with end uses and products derived from acetic acid. Included in this discussion, among other topics, will be the generation of acetic anhydride, vinyl acetate, ketene, diketene, esters, haloacetyls, acetamides, acetonitrile, and chemicals or polymers derived from these materials. No attempt is made to generate a definitive treatise on each of the subjects in this section of the book. Instead, this set of chapters focuses on the chemical means of arriving at the derivatives and their chemically useful applications.

Our intention is to present a multidisciplinary chemistry and chemical engineering text. Thus, each subject area describes the pertinent chemistry of the intermediates and processes involved. When commercial processes are discussed, this book attempts, when pertinent and possible, to also describe what engineering details might be available and discuss associated economic estimates. Extensive referencing of review articles (when available and useful) is made to supplement the information in this text.

These objectives have been accomplished by tapping the pool of expertise in acetyl chemistry and engineering at Eastman Chemical Company. However, in an undertaking of this magnitude, it may be inevitable that some information that someone may regard as important may be omitted at the discretion of the individual authors or simply overlooked. This could be particularly true in the third section of this book. A balance between chemistry and engineering is used where appropriate, and efforts are focused on providing reasonable coverage to the most industrially relevant chemistries and processes. Nevertheless, compromises had to be made, given the enormity of the task of preparing a comprehensible and, hopefully, comprehensive text. We hope a reasonable balance was achieved and that readers find this book useful in their work.

Victor H. Agreda
Joseph R. Zoeller

Contents

Contributors

Victor H. Agreda Research, Eastman Chemical Company, Kingsport, Tennessee

Robert J. Clemens Chemistry Research Division, Research Laboratories, Eastman Chemical Company, Kingsport, Tennessee

Frank Cooke Technical Support Division, Research Laboratories, Eastman Chemical Company, Kingsport, Tennessee

Steven L. Cook Development and Control Department, Acid Division, Eastman Chemical Company, Kingsport, Tennessee

A. Thomas Fanning Research and Development Laboratories, Texas Eastman Division, Eastman Chemical Company, Longview, Texas

Bruce L. Gustafson Catalysis Research Laboratory, Chemistry Research Division, Eastman Chemical Company, Kingsport, Tennessee

J. Adrian Hawkins Eastek, Eastman Chemical Company, Kingsport, Tennessee

Philip C. Heidt   Coatings and Resins Research Laboratory, Chemistry Research Division, Eastman Chemical Company, Kingsport, Tennessee

William H. Heise   Development and Control Department, Polymers Division, Eastman Chemical Company, Kingsport, Tennessee

Gether Irick   Chemistry Research Division, Research Laboratories, Eastman Chemical Company, Kingsport, Tennessee

Griffin I. Johnson   Development and Control Department, Cellulose Ester Division, Eastman Chemical Company, Kingsport, Tennessee

Peter N. Lodal   Engineering and Construction Division, Eastman Chemical Company, Kingsport, Tennessee

Lee R. Partin   Filter Products Research Laboratory, Fibers Research Division, Eastman Chemical Company, Kingsport, Tennessee

Peter W. Raynolds   Cellulose Chemistry Research Laboratory, Chemistry Research Division, Eastman Chemical Company, Kingsport, Tennessee

Ryan C. Schad   Process Engineering Department, Tennessee Eastman Division, Eastman Chemical Company, Kingsport, Tennessee

William H. Seaton*   Research Laboratories, Eastman Chemical Company, Kingsport, Tennessee

Charles E. Sumner   Catalysis Research Laboratory, Chemistry Research Division, Eastman Chemical Company, Kingsport, Tennessee

Lanny C. Treece   Polymer Process Design and Evaluation Research Laboratory, Engineering Research Division, Eastman Chemical Company, Kingsport, Tennessee

J. Stewart Witzeman   Coatings Laboratory, Technical Service and Development, Eastman Chemical Company, Kingsport, Tennessee

Paul R. Worsham   Development and Control Department, Acid Division, Eastman Chemical Company, Kingsport, Tennessee

Joseph R. Zoeller   Industrial Chemicals Research Laboratory, Chemistry Research Division, Eastman Chemical Company, Kingsport, Tennessee

---

*Private consultant, retired from Eastman Chemical Company.

ACETIC ACID AND ITS DERIVATIVES

I

Chemistry and Methods for the Manufacture of Acetic Acid

1

Bioderived Acetic Acid

Lee R. Partin and William H. Heise
Eastman Chemical Company, Kingsport, Tennessee

I. INTRODUCTION

The production of acetic acid via biological means is both of historical significance and the topic of current research. The objective of this chapter is to provide a brief background of bioderived acetic acid, discuss the current trends and directions of ongoing research, and present what we believe to be the optimal process description and economics for bioderived acetic acid. The latter will allow the reader to make comparisons to acetic acid production by oxidation and carbonylation processes as discussed elsewhere in this book.

Although the biotechnology route for acetic acid production does not currently compare favorably from an economic standpoint to carbonylation processes, there are several advantages of the biotechnology process. The principal advantage is that the raw materials may be obtained from renewable resources (agricultural crops or biomass versus coal or petroleum). In addition, a wide range of products is possible through both traditional biological means and genetic engineering.

In general, the production of chemicals using biotechnology may be divided into three areas [1]:

1. Large scale, where the biotechnology process competes with coal and petroleum feedstocks

2. Medium scale, where the biotechnology process competes both with petroleum-based feedstocks to produce commodity chemicals and with agriculture to yield products such as proteins and lipids
3. Small scale, where the biotechnology process yields products for which no other route has been discovered

Obviously, small-scale processes will continue development and growth in which the cost of the products is not of consequence. These products are commonly found in the medical and pharmaceutical fields. However, the medium- and large-scale processes are probably not, in general, economically favorable except in a few cases such as alcoholic drinks and food-grade acetic acid (vinegar).

II. BIOLOGICAL PROCESSES

A. Vinegar Processes

Vinegar, which is an aqueous solution of acetic acid, has been produced as long as wine making has been practiced and therefore dates back to at least 10,000 B.C. [2]. It is assumed that the first vinegar was a result of spoiled wine [3], given that the Latin word *acetum* means sour or sharp wine.

The first known process to manufacture vinegar is known as the "slow process," "French method," or "Orleans process." The process involves partially filling wooden barrels with wine and good-quality vinegar to speed up the reaction. Wine is added once a week for 4 weeks, until after 5 weeks vinegar is withdrawn and replaced with wine, resulting in a slow, continuous process. Air is introduced just above the liquid surface, where a gelatinous mat forms containing a large number of bacteria; this is known as the "mother of vinegar" [4,5].

In the early 1800s, the "quick process" or "German process" was developed in which wine trickles down a packed bed of wood shavings or coke. A large volume of air is sparged through the bottom of the tank. This process, which achieves about 98% conversion in 5 days [6], was used extensively for about a century. The Frings generator was introduced in 1929, utilizing forced aeration and temperature control with a trickle bed [7,8]. Its principal advantages included low cost, ease of control, attainment of higher acetic acid concentrations, and lower evaporation losses.

Hromatka and Ebner [9] in 1949 applied submerged fermentation to the oxidation of ethanol to acetic acid. Since the *Acetobacter* bacteria are sensitive to the level of gas-phase oxygen (fermentation stops at <5% O_2), the submerged culture process is dependent on the efficiency of broth aeration. A schematic of a typical submerged cultivator is shown in Figure

1. The advantages of submerged cultivation over the trickling bed processes are: (1) faster alcohol oxidation rates (up to 30 times faster), (2) less reactor volume required for a given production rate, (3) greater efficiency and conversion, and (4) ease of operability [3]. Complete aeration is required in the submerged cultivator because the depletion of oxygen for only 30 s will kill the *Acetobacter* bacteria [1].

B. Aerobic Process

Traditionally, bioderived acetic acid has been produced by the "two-step vinegar" process. The first step, in which ethanol (and by-product CO_2) is

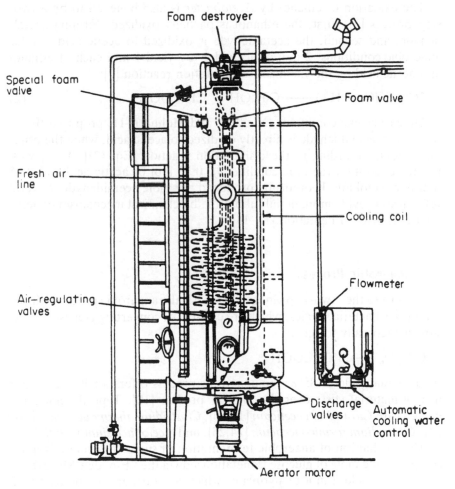

Figure 1 Schematic diagram of a typical submerged cultivator. (From Ref. 3.)

produced from glucose, is carried out normally at 300 K using an anaerobic yeast such as *Saccharomyces cerevisiae* with a yield of about 90%. The second step is the oxidation of the ethanol, typically using the aerobic bacterium *Acetobacter aceti* at 300–310 K, with a yield of about 85% [10].

$$C_6H_{12}O_6 \longrightarrow 2CH_3CH_2OH \ (+ \ 2CO_2) \xrightarrow{+2O_2} 2CH_3COOH + 2H_2O \quad (1)$$

Thus, for an overall average yield of nearly 77%, the two-step process utilizes 1.9 lb of glucose (or 0.9 lb of ethanol) per pound of acetic acid produced. For this process to be considered economically viable at a large or medium scale, one needs to consider the value of ethanol relative to the acetic acid product.

The oxidation of ethanol by *Acetobacter* is also believed to be a two-step process [3]. First, the ethanol is partially oxidized, forming acetaldehyde, and second, the acetaldehyde is oxidized to acetic acid. Under anaerobic conditions, 2 mol of acetaldehyde yield 1 mol each of ethanol and acetic acid by the Cannizaro dismutation reaction [3]:

$$2CH_3CHO + H_2O \longrightarrow C_2H_5OH + CH_3COOH \qquad (2)$$

Under aerobic conditions, Neuberg and Molinari [11] proposed that 1 of 2 mol of acetaldehyde is directly oxidized to acetic acid, while the other mole reacts according to the dismutation reaction [Eq. (2)]. Thus, with adequate aeration (aerobic conditions), 0.75 mol of acetic acid and 0.25 mol of ethanol are theoretically produced per mole acetaldehyde. Several references provide more details about the biochemical mechanism of aerobic acetic acid fermentation [12–18].

C. Anaerobic Process

In contrast to the "two-step vinegar" process, the anaerobic process utilizes acetogenic bacteria species, which are capable of converting glucose almost stoichiometrically to acetate:

$$C_6H_{12}O_6 \longrightarrow 3CH_3COOH \qquad (3)$$

At an actual yield of 85%, the anaerobic process offers a 60% reduction in raw material costs over the vinegar process [10]. Typical acidogenic bacteria are *Clostridium aceticum* [19,20], *Clostridium thermoaceticum* [21–23], *Clostridium formicoaceticum* [24,25], and *Acetobacterium woodii* [26].

The mechanism of anaerobic fermentation of sugars to acetate is quite complex. One mole of glucose is metabolized via the "Embden-Meyerhof" pathway to yield 2 mol of pyruvate, which are further metabolized to 2 mol of acetate and 2 mol of carbon dioxide [27]. The carbon dioxide is further reduced to a mole of acetate. Barker [28] and Wood [29] postulated the two-step mechanism as follows:

$$C_6H_{12}O_6 + 2H_2O \longrightarrow 2CH_3COOH + 2CO_2 + 8H^+ + 8e \qquad (4)$$

$$2CO_2 + 8H^+ + 8e \longrightarrow CH_3COOH + 2H_2O \qquad (5)$$

The net reaction resulting from Eq. (4) and (5) is given by Eq. (3). The synthesis of acetate from carbon dioxide has been extensively studied using *C. thermoaceticum* and was reviewed in 1982 by Ljungdahl and Wood [30]. For more detail on anaerobic fermentation mechanisms, the reader is referred to an excellent review by Wise [27].

D. Comparison of Aerobic and Anaerobic Processes

Aerobic fermentation offers several distinct advantages over acidogenic fermentation. The primary advantage lies in the fact that only acetic acid is produced, whereas acetogenic bacteria also produce significant amounts of propionic and butyric acids. In addition, aerobic fermentation is a faster process, and a higher concentration of acetic acid (up to 11%) can be obtained. Therefore, for commercial vinegar production, the aerobic process is used exclusively.

However, most current research is focused on acidogenesis, driven by its process economic advantages. Process equipment costs are less, operation is relatively simple, and aeration costs are eliminated. A theoretical 3 mol of acetic acid can be obtained from 1 mol of glucose, yet for the aerobic process only 2 mol of acetic acid are produced. The difference in overall conversion lies in the fact that 2 mol of carbon dioxide are liberated in the aerobic process [see Eq. (2)].

A significant problem with acidogenic bacteria is their inherent low tolerance of acidity, thus limiting the ultimate acid concentration. Research in this area has attempted to alleviate the problem by injecting acid into the reactor to adapt the culture to higher acid concentrations [3]. As a result of maintaining a neutral pH, the product is normally found as an acetate salt rather than in the free acid form, as found in the aerobic process. This has direct impact on the product recovery alternatives discussed below.

III. PRODUCT RECOVERY

A. Introduction

Fermentation liquors contain mostly water with a small amount of acetic acid, generally <5 wt%. The task of product separation is to recover the acetic acid in the desired purity, possibly recycle water to the process, and recover other products, such as a tar stream, which can be burned for fuel in a boiler. The selection of the recovery scheme depends on several factors, including:

1. The fermentation selection and its liquor composition since the form of the acetic acid (free or salt) and its concentration are important parameters
2. The product specifications on the acetic acid purity and allowable contaminant levels
3. The tolerance of the fermentation process to the addition of new chemicals that may be used in the separation process
4. The relative costs of steam and electricity
5. The potential for integration into an overall facility

The product recovery costs are quite significant, so they become a major issue in the economic success of the facility.

Normally, the commercial process of choice would be simple distillation. Water and acetic acid can be separated by a single distillation tower. Water has a normal boiling point of 100°C, acetic acid boils at 118.5°C, and there is no azeotrope. The separation is feasible but difficult, requiring many stages and a significant reflux rate owing to the close boiling points and a pinched vapor-liquid equilibria curve. Furthermore, it requires that the water be distilled overhead. For a 5 wt% feed, there are 19 kg of water per kg of acetic acid. This results in an enormous energy cost, which makes simple distillation unattractive for fermentation broths.

The commercial choice for acetic acid purification has been solvent extraction, followed by azeotropic or extractive distillation. A suitable solvent is selected for the extraction of acetic acid from water, considering the affinity toward acetic acid, the mutual solubility with water, distillation purification compatibility, reactivity, special hazards such as flammability, and compatibility with the fermentation process and possible product contamination. A wide selection of potential solvents have been proposed, including low-molecular-weight ketones, ethers, and acetates (methyl propyl ketone, diethyl ether, ethyl acetate, etc.) [31].

B. Traditional Acetic Acid Recovery Process

Busche [10] presented the traditional process (see Fig. 2) along with cost data (see Fig. 3). The cost information is for the product purification system of a 250 million pounds per year glacial acetic acid plant with start-up in 1985. The operating cost data include a 30% pretax return on the investment. The cost plot shows the significant contribution of the recovery costs to the overall cost.

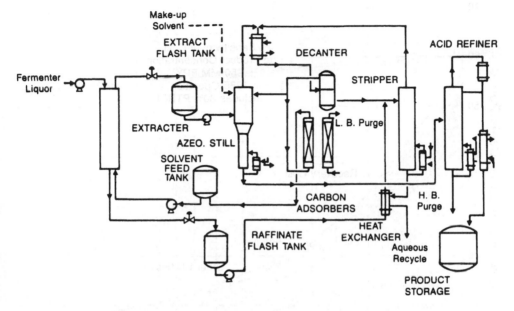

Figure 2 Traditional acetic acid recovery process. (From Ref. 10.)

The first unit of the traditional process is an extractor in which recycled solvent is used to extract acetic acid from the fermentation liquor. The recycled solvent must be nearly free of acetic acid so that the extractor can achieve high efficiency in the acetic acid recovery. This is especially important for the present case with low concentration of acetic acid in the feed. The raffinate stream contains the water, unrecovered acetic acid, and the solubility limit of the solvent. It is further processed in a stripping column for recycle of the solvent and purification of the water for recycle. The extract stream with solvent, acetic acid, and some water is fed to an azeotropic distillation column. Therefore, the extractor has removed most of the water.

Continued purification via azeotropic distillation can apply when the solvent forms a heterogeneous, minimum-boiling azeotrope with water and there is no ternary azeotrope. Solvent and water are distilled overhead with a minimum of acetic acid impurity. The vapors are condensed and decanted to form solvent-rich and water-rich layers. The solvent-rich layer provides reflux to the tower and it is recycled back to the extractor. The flowsheet shows the water-rich layer as feed to the stripping column for removal of the dissolved solvent before the water is recycled. The water-rich layer should also be considered for partial reflux to the azeotropic column with its rate adjusted as an optimization parameter for minimizing

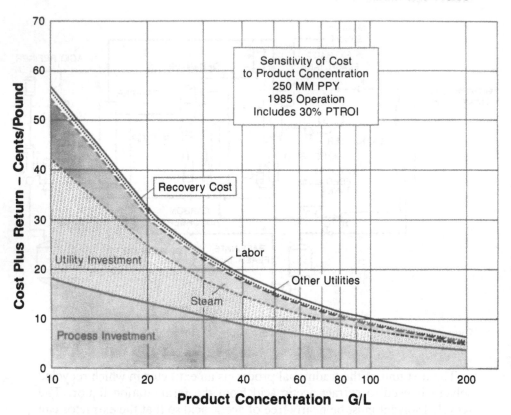

Figure 3 Sensitivity of product recovery costs to acetic acid concentration. (From Ref. 10.)

the cost. The bottoms stream from the azeotropic column has the water removed from the acetic acid, but the stream must be further refined to remove high-boiling contaminants. The flowsheet also shows activated carbon beds for the removal of low-boiling components. This may not be required since many low boilers that may be present could be removed in a vent from the azeotropic column condenser.

The recovery flowsheet depicts the case of a solvent that boils overhead. Extractive distillation is also an option using a high-boiling solvent. The high-boiling solvent is used both in the extractor and as a means to complete the purification within extractive distillation by feeding it near the top of the column. Its presence alters the relative volatilities of water and acetic acid, making it easier to recover the water overhead. The solvent flows under with the acetic acid. Therefore, the second column is then used to separate the acetic acid product from the solvent for recycle. An additional step is then required for removal of high-boiling impurities accumulating in the solvent, which can be performed by sludging a portion of the solvent in an evaporation system.

C. Recommendations and Conclusions

The work of Doherty and co-workers [32–36] has advanced our understanding of the synthesis, design, and optimization of these systems. Residue maps combined with liquid-liquid equilibria plots are an important tool for the synthesis phase. They present the capabilities of extraction and distillation systems to perform the separation, allowing the process engineer to synthesize feasible flowsheets. The diagrams also provide the means for calculating the stream material balances and extractor performance. The process synthesis phase should include the review of how it can be integrated into the overall process. The next step is to apply distillation design equations for the columns within the flowsheet and perform optimization studies on the major parameters, such as the solvent flow rate and distillation column sizes and feed tray locations. Pinch technology and related techniques are applicable for further optimization to ensure the best trade-offs between capital cost and energy usage [37,38].

There are numerous potential schemes and the one of choice will depend on the situation. The cost plot in Figure 3 shows that much work must be done on improving the product separation process. Experimenters must work on supplying the highest possible acetic acid concentration in the fermentation liquor. Also, process engineers must continue in the development of better solvents and flowsheets if the cost is to be reasonable.

The previous discussion has assumed that the product is present in a free state. Anaerobic fermentation processes have the added complexity of converting to acetic acid from the sodium salt form. The conversion adds more to the recovery cost and more salt waste for disposal. Pressurized CO_2 has been proposed as a cost-effective means of performing the conversion [39].

Alternate means of separation have been under development. Continuous melt crystallization in units such as the Brodie Purifier is a potential process, especially if high purity is not required [40]. Also, membrane separation may prove to be commercially feasible as functionality, cost, and durability improve.

In conclusion, technology breakthroughs are required to get commercially and economically viable recovery processes. The breakthroughs are possible.

REFERENCES

1. Atkinson, B., and Mavituna, F., *Biochemical Engineering and Biotechnology Handbook*, Nature Press, New York, 1983.

2. Nickol, G. B., in *Microbial Technology* (H. J. Peppler and D. Perlman, eds.), Academic Press, New York, 1979.
3. Ghose, T. K., and Bhadra, A., in *Comprehensive Biotechnology* (M. Moo-Young, ed.), Pergamon Press, New York, 1985.
4. Herrick, H. T., and May, O. E., *Chem. Metallurg. Eng.*, *42*, 142 (1935).
5. Lai, M. N., and Wang, I. H., U.S./R.O.C. Symposium on Fermentation Engineering Fundamentals, University of Pennsylvania, May 30–June 1, 1978.
6. Conner, H. A., and Allgeier, R. J., *Adv. Appl. Microbiol.*, *20*, 81–133 (1976).
7. Prescott, S. C. and Dunn, C. G., *Industrial Microbiology*, McGraw-Hill, Tokyo, 1959.
8. Owens, C. H., U.S. Patent 2,089,412 (1937).
9. Hromatka, O., and Ebner, H., *Enzymologia*, *14*, 96 (1950).
10. Busche, R. M., in *Biotechnology Applications and Research* (P. N. Cheremisinoff and R. P. Ouellette, eds.), Technomic Publishing, Lancaster, PA (1985).
11. Neuberg, C., and Molinari, E., *Naturwissenschaften*, *14*, 758 (1926).
12. Wieland, H., *Berl. Chem. Ges.*, *46*, 3327–3342 (1913).
13. Lutwak-Mann, C., *Biochem. J.*, *32*, 1364 (1938).
14. King, T. E., and Cheldelin, V. H., *Biol. Chem. J.*, *198*, 127 (1952).
15. King, T. E., and Cheldelin, V. H., *Biol. Chem. J.*, *220*, 177 (1956).
16. Atkinson, D. E., *Bacteriol. J.*, *72*, 195 (1956).
17. Nakayama, T., *Biochem. J.*, *49*, 158 (1961).
18. Nakayama, T., *Biochem. J.*, *49*, 240 (1961).
19. Wieringa, K. T., *Antonie van Leeuwehhoek*, *6*, 251 (1940).
20. Braun, M., Mayer, F., and Gottschalk, G., *Arch. Microbiol.*, *128*, 288 (1981).
21. Fontaine, F. E., Peterson, W. H., McCoy, E., Johnson, M. J., and Ritter, G. J., *J. Bacteriol.*, *43*, 701 (1942).
22. Andreesen, J. R., Schaupp, A., Neuranter, C., Brown, A., and Ljungdahl, L. G., *J. Bacteriol.*, *114*, 743–751 (1973).
23. Schwartz, R. D., and Keller, F. A., Jr., *Appl. Environ. Microbiol.*, *43*, 117 (1982).
24. El Ghazzawi, E., *Arch. Mikrobiol.*, *57*, 1 (1967).
25. Andreesen, J. R., Gottschalk, G., and Schlegel, H. G., *Arch. Microbiol.*, *72*, 154 (1970).
26. Balch, W. E., Schoberth, S., Tanner, R. S., and Wolfe, R. S., *Int. J. Syst. Bacteriol.*, *27*, 355 (1977).
27. Wise, D. L., *Organic Chemicals from Biomass*, Benjamin/Cummings Publishing Co., Menlo Park, CA (1983).
28. Barker, H. A., *Proc. Natl. Acad. Sci. USA*, *31*, 219 (1944).
29. Wood, H. G., *J. Biol. Chem.*, *194*, 905 (1952).
30. Ljungdahl, L. G., and Wood, H. G., in *B12*, Vol. II (D. Dolphin, ed.), Wiley, New York, 1982.
31. Brown, W. V., *Chem. Eng. Prog.*, *59*, 65 (October 1963).
32. Doherty, M. F., *Chem. Eng. Sci.*, *40*, 1885 (1985).
33. Knight, J. R., and Doherty, M. F., *Ind. Eng. Chem. Res.*, *28*, 564 (1989).

34. Doherty, M. F., and Caldarola, G. A., *Ind. Eng. Chem. Fundam.*, *24*, 474 (1985).
35. Pham, H. N., Ryan, P. J., and Doherty, M. F., *Am. Inst. Chem. Eng. J.*, *35*, 1585 (1989).
36. Pham, H. N., and Doherty, M. F., *Chem. Eng. Sci.*, *45*(7), 1845 (1990).
37. Linnhoff, B., and Polley, G. T., *Chem. Eng. Prog.*, 51 (June 1988).
38. Smith, R., and Linnoff, B., *Chem. Eng. Res. Des.*, *66*, 195 (1988).
39. Yates, R. A., U.S. Patent 4,282,323 (Aug. 4, 1981).
40. *The Brodie Purifier—A Breakthrough in Fractional Crystallization*, C. W. Nofsinger, Kansas City, MO (1982).

2

Ethylene- and Acetylene-Based Processes

A. Thomas Fanning

Eastman Chemical Company, Longview, Texas

I. INTRODUCTION

There are three main commercial methods for acetic acid production in the world today: acetaldehyde oxidation, liquid phase hydrocarbon oxidation, and methanol carbonylation. Ethylene- and acetylene-based methods for acetic acid production are concerned with the oldest, but still important, acetaldehyde oxidation route. The incorporation of ethylene and acetylene as fundamental chemicals in this route is shown in Figure 1. Most of the processes shown in Figure 1 are well established in the chemical industry. These processes will be reviewed and newer methods will be discussed where applicable.

II. ETHANOL FROM ETHYLENE

Ethanol is a key intermediate for acetic acid processes based on the ethylene–ethanol–acetaldehyde–acetic acid sequence. There are two industrial methods for ethanol production: the indirect, or sulfuric acid, process and the direct hydration process.

A. Indirect or Sulfuric Acid Process

This is the oldest commercial process for ethanol (used in the United States and Britain in 1930–1940) and is based on the conversion of ethylene to

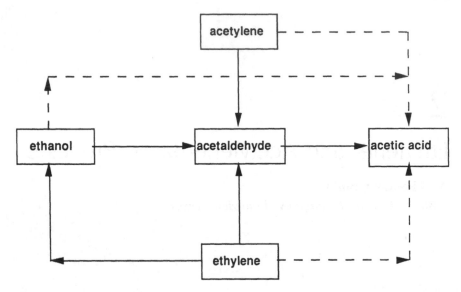

Figure 1 Acetic acid from ethylene and acetylene. (——) established industrial process; (---) nonindustrial methods.

mono- and diethylsulfates by absorption in concentrated sulfuric acid followed by hydrolysis to ethanol and sulfuric acid [1–3] [Eq. (1) and (2)]. Approximately 99% of the ethylene feed is absorbed in concentrated sulfuric acid at 75°C and 200 psig to give a 90% yield of alkylsulfates. Hydrolysis of the alkylsulfates at atmospheric pressure and 70°C affords ethanol in 95% yield. Ethylene conversion to oils, tars, and polymeric material is 1–7%. The efficiency of this process depends on efficient ester hydrolysis, recovery and concentration of sulfuric acid, tar removal from the absorption step, control of by-product ether formation, economical use of steam, and reduced corrosion. These deficiencies resulted in the development of the direct hydration process.

$$C_2H_4 + H_2SO_4 = CH_3CH_2OSO_3H + (CH_3CH_2O)_2SO_2 \qquad (1)$$

$$CH_3CH_2OSO_3H + (CH_3CH_2O)_2SO_2 = CH_3CH_2OH + H_2SO_4 \qquad (2)$$

B. Direct Hydration Process [4–7]

In this process, developed by Shell in the 1940s, ethylene and water heated at 300°C and 1000–1200 psig are passed over a supported (diatomaceous earth) phosphoric acid catalyst [Eq. (3)]. The yield of alcohol is 95%, with a 4.2% ethylene conversion per pass. Diethyl ether is the main by-product, along with trace amounts of acetaldehyde, ethylene oligomers (oils), and hydrocarbons to C-8. The ethylene-to-water mole ratio, reactor temperature, and pressure are important process variables that affect both catalyst activity and selectivity. Concentration of phosphoric acid on the catalyst support is 80–85%, and to maintain catalyst activity, phosphoric acid is injected into the reactor feed stream periodically. Other factors affecting catalyst life are carbon and/or tar formation on the support and catalyst strength. Alternative catalysts for this process are acidic in nature, e.g., acid clays [8], tungstic oxide [4], and heteropolyacids [9]. Improvements in the direct hydration process reported in the literature are generally concerned with catalyst strength [10,11] and activity [11–14].

$$C_2H_4 + H_2O = CH_3CH_2OH \tag{3}$$

C. Other Methods

A three-phase hydration system consisting of water (liquid), tungsten oxide (solid), and ethylene (gas) has been used for the hydration of several olefins [4] but not in an industrial ethanol process; ethylene hydration using dilute acids [1], e.g., 5–80% sulfuric acid, 200°C, 1–40 atm, have been investigated, but corrosion, low crude alcohol concentration (10% v/v), and high heat usages are distinct disadvantages.

III. ACETALDEHYDE PROCESSES

Acetaldehyde is the building block for a host of chemicals in the chemical industry [15], and its use as an intermediate in the ethylene and acetylene route to acetic acid is the heart of this process. Acetylene hydration was the original process for making acetaldehyde in Europe from 1916 to the present, whereas ethanol was used for making acetaldehyde in the United States and Britain. Both processes, since 1960, have taken a back seat to the Wacker acetaldehyde process.

A. Acetylene Hydration

This process has been reviewed thoroughly by Miller [16]. Basically, acetylene is passed into a hot (70–80°C), aqueous solution of 25% sulfuric acid containing 1–2% of mercuric sulfate and 0.5–1% ferric sulfate. This affords a 7% solution of acetaldehyde in 93–95% yield with a 55% acetylene conversion per pass [Eq. (4)]. Ferric sulfate or manganese dioxide is used to retard the reduction of mercury and maintain catalyst activity. This oxidant is maintained in its active, oxidized state by treatment of a catalyst slip stream with 30% nitric acid. The crude acetaldehyde is refined by distillation to give an acetaldehyde purity of 99.9%. Acetone is the major by-product generally recovered; unrecovered by-products are acetic acid, diacetyl, and crotonaldehyde. An alternate process that involves vacuum distillation of the acetaldehyde with no acetylene recycle affords acetaldehyde in 98% yield [16]. Other catalysts in lieu of mercury, most notably $ZnCl_2$ (12–25%) or CuCl in concentrated aqueous $ZnCl_2$, afford acetaldehyde in yields of 85–90%. Although no commercial processes are known, vapor phase hydration experiments using acidic catalysts (e.g., oxides and phosphate salts of Zn, Cd, Hg, Cu, Fe, Co) have been investigated extensively [16]. Catalyst lifetime is apparently a problem with the vapor phase process.

$$HCCH + H_2O = CH_3CHO \tag{4}$$

B. Ethanol Oxidation/Dehydrogenation [17–19]

The industrial ethanol oxidation process is well established and consists in the passage of a vapor stream of ethanol, steam, and air over a silver catalyst at 300–575°C [5]. The reaction is exothermic and affords acetaldehyde in yields of 85–95% with an ethanol conversion per pass of 25–35%. Typical by-products are acetic acid, formic acid, ethyl acetate, methane, and carbon monoxide. Silver is the catalyst of choice; other metals (e.g., V, Mn, Co) are more active, resulting in lower yields of acetaldehyde.

The dehydrogenation process uses a copper catalyst promoted with chromium and requires a reaction temperature of 260–290°C [6]. Ethanol conversion per pass is controlled at 30–50% and provides acetaldehyde in greater than 90% yield. By-products are acetic acid, ethyl acetate, isobutanol, methane, and carbon dioxide. Ethanol dehydration and/or acetaldehyde decomposition are apparent problems with other catalysts (e.g., Ni, Co, Pt, nonreducible metallic oxides). The copper catalyst is reactivated by oxidation in a current of hot air.

Improvements include enhanced silver catalyst activity, which permit operation at lower temperatures [20]; a highly active silica-supported copper catalyst [21]; and a method for reprocessing copper catalyst used in ethanol dehydrogenation [22].

$$CH_3CH_2OH + 1/2\ O_2 = CH_3CHO + H_2O \tag{5}$$
$$CH_3CH_2OH = CH_3CHO + H_2 \tag{6}$$

C. Ethylene Oxidation (Wacker Process) [23–25]

The availability of low-priced petroleum raw material (ethylene) after World War II led to the development of the direct ethylene oxidation process for acetaldehyde in the late 1950s by Hoechst and Wacker-Chemie

[26]. This is a palladium-catalyzed, liquid phase process in which ethylene is passed through an aqueous solution of cupric chloride, cuprous chloride, and hydrochloric acid in the presence of palladium chloride. The overall reaction is shown in Eq. (7). The oxidation of ethylene in the presence of palladium chloride catalyst occurs as shown in Eq. (8). Palladium in the active state ($+2$) is reduced to zero valent palladium. The reduced palladium is oxidized to palladium chloride with cupric chloride [Eq. (9)] and the cuprous chloride formed is oxidized with oxygen [Eq. (10)]. The reactions forming acetaldehyde [Eq. (8)] and the oxidation of cuprous chloride [Eq. (10)] may be conducted simultaneously (one-stage process) or in two steps (two-stage process). Reaction temperatures and pressures are 100–110°C and 130–150 psig, respectively. Acetaldehyde yield is 95%. By-products are acetic acid, carbon dioxide, and chlorinated organics (e.g., ethyl chloride, chlorinated aldehydes). Owing to the highly oxidized and corrosive nature of the catalyst, titanium metal is used in the reactor section of the process. The powerful nature of the catalyst also provides an inherent self-cleansing process, which aids catalyst activity and longevity.

$$C_2H_4 + 1/2\ O_2 = CH_3CHO \tag{7}$$

$$C_2H_4 + PdCl_2 + H_2O = CH_3CHO + Pd + 2HCl \tag{8}$$

$$Pd + CuCl_2 = PdCl_2 + 2CuCl \tag{9}$$

$$2CuCl + 1/2\ O_2 + 2HCl = 2CuCl_2 + H_2O \tag{10}$$

Other noble metals (e.g., Pt, Rh) are effective catalysts in the Wacker process, but do not provide the high reaction rate of Pd salts for an industrial process. Salts other than chloride (e.g., sulfate) have been investigated with the intent to reduce chlorinated by-product formation; however, chloride is necessary for Pd stability and for aiding the rapid oxidation of cuprous chloride. Vapor phase processes have been investigated, but apparently there are problems with catalyst life, corrosion, and adequate heat removal. A summary of direct ethylene oxidation–acetaldehyde methods through 1976, especially the Wacker process, has been reviewed by Ma [27].

Recent developments in Wacker process chemistry have included apparent solutions to the problems of corrosion and chlorination inherent in this process. Mild conditions are reported in a method that uses an oxygen transition metal complex to regenerate active palladium species; high selectivity and good yields of acetaldehyde (98%) are obtained [28,29]; a liquid phase olefin oxidation process with a palladium–heteropoly acid catalyst is claimed to be noncorrosive with no chlorinated by-products due to low chloride or no chloride in the catalyst with an acetaldehyde yield of 97% [30–32]; acetaldehyde synthesis using a fuel cell system, $Pd/H_3PO_4/$

Pd, affords greater than 97% selectivity for acetaldehyde [33]; gas phase oxidation of ethylene in the presence of water vapor and oxygen over a palladium-vanadate catalyst supported on alumina is reported to give long catalyst activity [34,35]; and a silica-supported molten salt catalyst ($PdCl_2$, $CuCl_2$) gives 95% acetaldehyde selectivity [36] via gas phase process.

IV. ACETIC ACID PROCESSES

A. Acetaldehyde Oxidation [37–41]

Acetic acid from acetaldehyde was once the leading method for acetic acid production; as much as 45–75% of the acetaldehyde produced was used for acetic acid and acetic anhydride manufacture. Because of economic reasons, alternate acetic acid routes (hydrocarbon oxidation, methanol carbonylation) have also become important methods for acetic acid production.

Commercial production of acetic acid from acetaldehyde originated in Germany in 1911 and in the United States in 1920. In the batch processes, oxygen or air is sparged through a solution of acetaldehyde containing 0.5% manganese acetate catalyst at 55–80°C and 75–85 psig [Eq. (11)]. If oxygen is used, then dilution with nitrogen or air is recommended to prevent overoxidation and excessive by-products (e.g., methyl acetate, acetone, formaldehyde, formic acid, carbon oxides). Oxygen concentration in the off-gas is maintained at <10% for safety reasons. Acetaldehyde is recovered from the crude product mixture in a separate distillation column, and the crude acetic acid (95%) is further purified by distillation to give 99% acetic acid. Yield is 88–95%.

$$CH_3CHO + 1/2 O_2 = CH_3COOH \tag{11}$$

In newer plants, a continuous process is used that has the particular advantage of increased safety due to steady temperature and pressure operating conditions. The use of metal salts as catalysts makes it possible to oxidize acetaldehyde to acetic acid on a large scale safely. Acetaldehyde conversion is greater than 90%; selectivity to acetic acid is greater than 94%.

In both continuous and batch processes, acetaldehyde is converted to acetaldehyde monoperacetate, which decomposes to peracetic acid and acetaldehyde [Eq. (12)]. This decomposition reaction is promoted by metal salts, especially manganese acetate. Peracetic acid in the presence of acetic acid at temperatures >40°C decomposes to acetic acid and oxygen [Eq. (13)]. Use of cobalt acetate and copper acetate as catalysts promotes the formation of acetic anhydride.

Recent literature references concerning acetaldehyde oxidation show that acetic anhydride and formic acid formation are reduced by inclusion of an alkali metal acetate in the presence of an aqueous manganese, cobalt, or nickel acetate catalyst [42,43]; mild acetaldehyde oxidation conditions are employed using oxygen complexes of cuprous halide and a phosphine oxide in liquid phase [44]; use of a ferrous cyanide complex with oxygen provides mild oxidizing conditions and reduces by-product formation [45]; acetaldehyde oxidation with oxygen over a molybdenum–phosphorus–copper–vanadium–metal oxide catalyst gives high acetaldehyde conversions and acid selectivity without production of a peroxy acid intermediate [46]; and an improved acetic acid process is obtained by the use of a reoxidation stage to eliminate acetaldehyde recycle and allow reuse of manganese acetate catalyst in the process [47].

$$2CH_3CHO + O_2 = [CH_3CHOHOOCOCH_3]$$

$$= CH_3CO_3H + CH_3CHO \quad (12)$$

$$CH_3CO_3H = CH_3CO_2H + 1/2\ O_2 \quad (13)$$

B. Direct Ethanol Oxidation [37,48]

Oxidation of ethanol directly to acetic acid [Eq. (14)] is not an industrial-recognized process, possibly because of yield losses associated with reactions of acetaldehyde intermediate (acetal formation) or esterification of acetic acid. However, there are several references to the production of acetic acid from ethanol in the liquid or gas phase.

Direct ethanol oxidation (liquid phase) is reported to give acetic acid (90% yield) in the presence of acetaldehyde, oxygen (air), acetic acid solvent, and cobalt acetate catalyst at 60–115°C and 15–50 psig [48]. A modification of this process reports increased amounts of ethanol can be used with the incorporation of methyl ethyl ketone in the reaction mixture; higher temperatures and pressures are also apparently involved [49].

Gas phase [300–400°C] ethanol oxidation, usually in the presence of steam, has been investigated over a variety of catalysts (e.g., copper chromite containing Zn, Mn, or Ag; Ag; Cu, Co, and Zn) to give acetic acid [37]. Poor catalyst selectivity and/or low yield are inherent problems associated with the gas phase process. Metals such as V, Mn, Co, Ni, and Fe tend to promote overoxidation. Recent catalyst modifications, e.g., Mo-V-Nb-Sb-Ca catalyst [50], zeolite containing Cu, Mn, and Pd [51], claim improvement in selectivity and acetic acid yield. Modifications in the preparation of a silica-supported platinum catalyst were found to affect the product distribution for the oxidation of ethanol and acetaldehyde [68].

$$CH_3CH_2OH + O_2 = CH_3CO_2H + H_2O \quad (14)$$

C. Direct Oxidation of Ethylene and Acetylene

There are no commercial processes for acetic acid production based on the direct oxidation of ethylene or acetylene. However, there are several patent and literature references concerning the oxidation of ethylene to acetic acid or mixtures of acetic acid and acetaldehyde. In general, these methods utilize oxygen or air in combination with a palladium salt in the presence of vanadium (e.g., vanadium pentoxide). Both liquid and vapor phase processes are known.

Some of the direct ethylene oxidation (liquid phase) reactions consist of the following: contacting ethylene with oxygen in the presence of an aldehyde, palladium salt, and iron or cobalt [52]; reaction of ethylene with oxygen in the presence of palladium chloride, vanadyl sulfate, and sulfuric acid at 100°C and 300 psig [53]; and reaction of cuprous chloride–hexamethylphosphoroamide complex with palladium chloride–benzonitrile complex with air and then ethylene to give 98% yield of acetic acid [54].

Catalysts and reaction conditions described for direct ethylene oxidation to acetic acid in the gas phase consist of the following: reaction of ethylene, oxygen, and water vapor over a palladium and vanadium oxide catalyst at 225°C and 150 psig gave 100% ethylene conversion with 60–85% selectivity to acetic acid, but no acetaldehyde was detected [55–57]; ethylene, oxygen, and steam passed over an alumina support impregnated with palladium and chromium at 200°C gave 60% acetic acid selectivity [58]; oxidation of ethylene to acetic acid with oxygen (100°C, 1 atm, 12% yield) over iridium on silica or alumina [59]; a supported palladium and phosphoric acid catalyst afforded acetic acid in 80% yield from ethylene and oxygen (150°C, 4.5 atm) [60]; a supported palladium and gold catalyst containing a sulfur compound gave acetic acid selectivities of 50–80% from ethylene, oxygen, and water vapor (150°C, 3.4 atm) [61]; vapor phase reaction of ethylene and nitric acid over a silica-supported vanadia catalyst at 350°C afforded acetic acid in 72% selectivity at 92% ethylene conversion [62]; a heteropoly acid catalyst containing palladium afforded acetic acid (75% selectivity at 33% ethylene conversion) from air and ethylene at 237°C [63]; and several catalysts (e.g., phosphoric acid, molybdic oxide, palladium compounds) are described in an article by Sampat and Chandalia [64].

There are few reports concerning the direct oxidation of acetylene to acetic acid. These are gas phase hydration processes with acetylene-water vapor at 300–400°C over a catalyst bed such as zinc fluoride [65], cobalt–molybdenum oxide [66], or cadmium-calcium-phosphate [67]. Mixtures of acetaldehyde and acetic acid are obtained in yields of 20–70% for acetaldehyde and 30–50% for acetic acid. Acetaldehyde may be converted to acetic acid by one of the oxidation methods described previously.

V. SUMMARY

Ethylene- and acetylene-based industrial processes for acetic acid involve acetaldehyde as a key intermediate. Although the acetaldehyde oxidation process accounts for 27% of the world's capacity for acetic acid, future processes will be based on methanol carbonylation technology, which accounts for 47% of the world's capacity for acetic acid, because this route has the lowest overall production costs on a newly constructed plant basis. Technology is currently inadequate for an industrial acetic acid process based on direct oxidation of ethylene or acetylene.

REFERENCES

1. John, J. A., in *Ethylene and Its Industrial Derivatives* (S. A. Miller, ed.), Ernest Benn, London, 1969, pp. 692–709.
2. Brownstein, A. M., *U.S. Petrochemicals*, Petroleum Publishing Co., Tulsa, OK, 1972, pp. 296–297.
3. Lowenheim, F. A., and Moran, M. K., *Faith, Keyes, and Clark's Industrial Chemicals*, 4th ed., Wiley-Interscience, New York, 1975, pp. 357–359.
4. Millidge, A. F., in *Ethylene and Its Industrial Derivatives* (S. A. Miller, ed.), Ernest Benn, London, 1969, pp. 709–731.
5. Lowenheim, F. A., and Moran, M. K., *Faith, Keyes, and Clark's Industrial Chemicals*, 4th ed., Wiley-Interscience, New York, 1975, pp. 355–356.
6. Nelson, C. R., and Courter, M. L., *Chem. Eng. Prog.*, *50*, 526 (1954).
7. Carle, T. C., and Stewart, D. M., *Chem. Ind.*, p. 830 (1962).
8. Sommer, A., and Bruecker, R., Ger. Offen. 2,908,491 (1980).
9. Onoe, Y., *Kagaku Kogyo*, *26*, 355 (1975).
10. De Goederen, C. W. H., and Spek, T. G., Ger. Offen. 2,719,055 (1977).
11. Kadlec, V., Grosser, V., and Rosenthal, J., Belg. BE 888,479 (1981).
12. Sommer, A., and Schlueter, D., Ger. Offen. DE 3,709,401 (1988).
13. Ursu, G., Gaber, D., and Petruc, F., Rom. RO 85,367 (1984).
14. Tanabe, K., in *Heterogeneous Catalysis—2nd Symposium* (B. Sharpiro, ed.), Texas A&M, College Station, TX, 1984, p. 71.
15. Brownstein, A. M., *U.S. Petrochemicals*, Petroleum Publishing Co., Tulsa, OK, 1972, p. 300.
16. Miller, S. A., *Acetylene, Its Properties, Manufacture, and Uses*, Vol. 2, Academic Press, New York, 1966, p. 134.
17. Lowenheim, F. A., and Moran, M. K., *Faith, Keyes, and Clark's Industrial Chemicals*, 4th ed., Wiley-Interscience, New York, 1975, pp. 2–3.
18. Miller, S. A., *Acetylene, Its Properties, Manufacture, and Uses*, Vol. 2, Academic Press, New York, 1966, p. 148.
19. Page, R., in *Ethylene and Its Industrial Derivatives* (S. A. Miller, ed.), Ernest Benn, London, 1969, pp. 767–770.
20. Webster, D. E., and Player, J. P., Eur. Pat. Appl. EP 272,846 (1988).

21. Horyna, J., Slosar, P., and Kral, T., Czech. CS 249,378 (1988).
22. Chhabra, M. S., Yadav, B. L., Nath, D., Ghosh, S. K., Bhattacharyya, N. B., and Sen, S. P., in *Adv. Catal., Proc. 7th Natl. Symp. Catal.* (P. Rao, ed.), Wiley, New York, 1985, p. 83.
23. Jira, R., in *Ethylene and Its Industrial Derivatives* (S. A. Miller, ed.), Ernest Benn, London, 1969, p. 639.
24. Miller, S. A., *Acetylene, Its Properties, Manufacture, and Uses*, Vol. 2, Academic Press, New York, 1966, p. 150.
25. Lowenheim, F. A., and Moran, M. K., *Faith, Keyes, and Clark's Industrial Chemicals*, 4th ed., Wiley-Interscience, New York, 175, p. 1.
26. Jira, R., Blau, W., and Grimm, D., *Hydrocarbon Proc.*, *55*, 97 (1976).
27. Ma, J. J. L., *Acetaldehyde, Report No. 24A2*, Stanford Research Institute, Menlo Park, CA, 1976.
28. Yamada, M., Kamiguchi, T., Tanimoto, H., Arikawa, Y., Kaku, H., and Takamoto, S., U.S. Patent 4,806,692 (1989); EP 189312 (1986).
29. Nishimura, Y., Yamada, M., Kamiguchi, T., Tanimoto, H., Arikawa, Y., and Kuwahara, T., U.S. Patent 4,521,631 (1985).
30. Matveev, K. I., et al., Brit. GB 1,508,331 (1978).
31. Murtha, T. P., U.S. Patent 4,507,507 (1985).
32. Vasilevski, J., Dedeken, J. C., Saxton, R. J., Wentrcek, P. R., Fellman, J. D., and Kipnis, L., WO 8,701,615 (1987).
33. Otsuka, K., Shimizu, Y., and Yamanaka, I., *J. Chem. Soc., Chem. Comm.*, (18) p. 1272 (1988).
34. Scholten, J. J., and Vanderstee, P. J., EP 210705 (1987).
35. Forni, L., and Terzoni, G., *I. E. C. Proc. Des. Dev.*, *16*, 288 (1977).
36. Rao, V., and Datta, R., *J. Catal.*, *114*, 377 (1988).
37. Sieber, R. H., in *Ethylene and Its Industrial Derivatives* (S. A. Miller, ed.), Ernest Benn, London, 1969, pp. 668, 669.
38. Takaoka, S., *Acetic Acid, Report No. 37*, Stanford Research Institute, Menlo Park, CA, 1968, pp. 21–31.
39. Wagner, F. S., in *Encyclopedia of Chemical Technology*, 3rd ed. (Kirk-Othmer, ed.), Wiley, New York, 1978, pp. 129–136.
40. Lowenheim, F. A., and Moran, M. K., *Faith, Keyes, and Clark's Industrial Chemicals*, 4th ed., Wiley-Interscience, New York, 1975, pp. 8–15.
41. Sittig, M., *Organic Chemical Process Encyclopedia*, Noyes Development Corp., Park Ridge, New Jersey, 1967, p. 4.
42. Roscher, G., Schaum, H., and Schmitz, H., Ger. Offen. 2,513,678 (1976); U.S. Patent 4,380,663 (1983).
43. Sikorai, S., Wardzin, W., Spacek, F., Stieranka, J., and Barta, J., Czech. CS 240,479 (1987).
44. Yamada, M., Nishimura, Y., Arikawa, Y., Kuwahara, T., Kamahara, T., Kamiguchi, T., and Tanimoto, H., U.S. Patent 4,691,053 (1984).
45. Cornils, B., De Win, W., and Weber, J., U.S. Patent 4,285,875 (1981).
46. Pedersen, S. E., Hardman, H. F., and Wagner, L. F., U.S. Patent 4,408,071 (1983).

47. Schaum, H., Schenk, F., Voight, H., and Sartorius, R., U.S. Patent 4,094,901 (1978).
48. Sittig, M., *Organic Chemical Process Encyclopedia*, Noyes Development Corp., Park Ridge, New Jersey, 1967, p. 6.
49. Hobbs, C. C., and Van't, H. A., U.S. Patent 3,914,296 (1975).
50. McCain, J. H., and Kaiser, S. W., European Patent Appl. EP 294,846 (1988).
51. Shakhtakhtinskii, T. N., Aliev, A. M., Kuliev, A. R., Mamedov, F. A., Babaeva, A. R., Naraevskaya, S. N., and Medzhidova, S. M., U.S.S.R. SU 1,549,945 (1990).
52. Duncanson, L. A., and Ehrlich, H. W. W., U.S. Patent 3,459,796 (1969).
53. Horike, S., Fujii, C., Ihara, A., and Nedachi, H., Japan JP 45/21490; JP 45/21491 (1970).
54. Kamiguchi, T., Yamada, M., Arikawa, Y., Tanimoto, H., and Nishimura, Y., European Patent Appl. EP 156,498 (1985).
55. Naglieri, A. N., U.S. Patent 3,240,805 (1966).
56. Boutry, P., and Montarnal, R., FR 1,568,742, (1969); FR 2,058,817 (1971).
57. Nakanishi, Y., Kurata, N., and Okuda, Y., JP 46/6763; 71/6763 (1971); JP 47/11,050; 72/11,050 (1972).
58. Capp, C. W., and Harris, B. W., U.S. Patent 3,574,730 (1971).
59. Cant, N. W., and Hall, W. K., U.S. Patent 3,641,139 (1972).
60. McClain, D. M., Heller, C. A., and Mador, I. L., U.S. Patent 3,792,087 (1974).
61. Scheben, J. A., Hinnenkamp, J. A., and Mador, I. L., U.S. Patent 3,970,697 (1976).
62. Matsuda, F., and Kato, K., JP 61/280444 (1986).
63. Kondo, T., JP 54/57488 (1979).
64. Sampat, B. G., and Chandalia, S. B., *Chem. Ind. Dev.*, 7, 25 (1979).
65. Petrushova, N., SU 387963 (1973).
66. Albanesi, G., Moggi, P., and Costa, M., *Chim. Ind. (Milan)*, 63, 325 (1981).
67. Govorov, V. G., SU 1068416 (1984).
68. Gonzalez, R. D., and Nagol, M., *Ind. and Eng. Chem., Prod. Res. and Dev.*, 24, 525–531 (1985).

3

Manufacture via Hydrocarbon Oxidation

Gether Irick
Eastman Chemical Company, Kingsport, Tennessee

I. INTRODUCTION

Acetic acid is the end product of many oxidation processes. Two of the more important, ethanol and acetaldehyde oxidations, have been considered in earlier chapters. Oxidations of paraffinic and olefinic hydrocarbons have been done under thermal, photolytic, enzymatic, electrochemical, and possibly other conditions, resulting in the production of acetic acid in yields varying from barely detectable to nearly quantitative. Thus, the challenge in producing acetic acid from hydrocarbons is not just to produce it, but to do so in a manner that is economically attractive. This requires proper selection of starting material, high yield and high selectivity processes, or, in certain cases, high by-product credit. This chapter will provide a selective review of hydrocarbon oxidation routes to acetic acid.

II. OLEFIN OXIDATION

A. Ethylene

Conversion of ethylene to acetic acid has been practiced commercially by coupling the Wacker oxidation of ethylene to acetaldehyde, followed by air oxidation of the aldehyde to acetic acid. The attractiveness of this two-stage approach has been due to the high yields of each of the two steps and to the value of the intermediate acetaldehyde for other applications.

The direct oxidation of ethylene to acetic acid may, in fact, involve acetaldehyde as an intermediate.

$$CH_2{=}CH_2 + O_2 + H_2O \longrightarrow [CH_3CHO] \longrightarrow CH_3COOH + H_2O$$

$$(1)$$

Halcon vapor phase oxidation of ethylene used palladium chloride and vanadium oxide catalysts supported on alumina [1–3]. Reactions were run in the 215–250°C range at 1035–2070 kPa (150–300 psi) pressures. Ethylene concentrations were kept below about 3%, conversions of 100% were typical, and selectivities to acetic acid of ca. 60% were obtained. Selectivities were lower at higher ethylene concentrations, and combustion of ethylene occurred if water levels were too low.

Other reported vapor phase oxidations include palladium/vanadium oxide [4,5], palladium/vanadium/antimony oxides on alumina [6,7], and palladium/phosphoric acid on silica [8,9]; in the latter case, National Distillers claimed acetic acid selectivities of 90% at conversions of ca. 15%.

Use of a Wacker-type catalyst (CuCl/PdCl_2/acetonitrile solvent) at 101 kPa (1 atm) and 60°C produced mixtures of acetaldehyde (67%) and acetic acid (32%) [10].

One-stage, direct oxidation of ethylene to acetic acid has been demonstrated and appears to offer the potential for a commercial process, but there is no indication that commercialization efforts are underway. The two-stage (Wacker + oxidation) process is too well established and, although non-cost-competitive with methanol carbonylation, would provide the best alternative route if ethylene is the starting material of choice.

B. Propylene

Stamicarbon patents [11–15] describe the use of molybdenum in combination with tin or iron at temperatures of 225–330°C, space velocities of 1000–3000/h, and conversions of 14–80% to provide about 40–45% selectivities to acetic acid. As with ethylene, steam improved selectivity.

$$CH_3CH{=}CH_2 + O_2 \longrightarrow CH_3CO_2H + CO_2 + H_2O \qquad (2)$$

Ruhrchemie [16,17] found that a mixed catalyst of molybdenum oxide with phosphoric and boric acids operated at 250–400°C with 80% conversion of propylene, 5% initial concentration, minimized acrolein production, and gave high selectivity to C1 and C2 oxygenates, including acetic acid.

Vanadium oxide supported on silica, alumina, and titania was moderately selective for the vapor-phase oxidation of propylene to mixtures of oxygenates. One of the best acetic acid catalysts, 7.16 E14 molecules V_2O_5/cm^2 titania, when operated at 250°C/101 kPa, in the presence of steam, gave acetone, acetaldehyde, and acetic acid in selectivities of 11, 1, and 40%, respectively [18].

Rare earths, especially praseodymium, at 10–25 mol% levels enhance the selectivity of molybdenum oxide for acetic acid [19]. Selectivities of 20–30% were obtained in vapor phase reactions at 400°C.

A two-stage liquid phase process has been described [20] for the air oxidation of a mixture of pentane and propylene using a molybdenum acetylacetonate catalyst to produce propylene oxide; the by-products were oxidized at 160°C with manganese catalysis to a mixture of acetic acid (33%), propionic acid (20%), and formic acid (15%).

Although the partial oxidation of propylene to acrolein has achieved commercial status, the oxidation of propylene to acetic acid has been of only limited interest. As with ethylene, propylene is a relatively expensive feedstock for acetic acid production, and alternative, higher-value uses for propylene exist. Molybdenum and vanadium appear to be the catalytic metals of choice for propylene oxidation, with selectivities to acetic acid generally below 50%.

C. Butenes and Higher Olefins

Both vapor and liquid phase direct oxidation processes have been developed for the conversion of butenes to acetic acid; very little work has been done on the oxidation of C5 and higher olefins to acetic acid.

$$CH_3CH_2CH_2CH=CH_2 \quad \text{or} \quad CH_3CH=CHCH_3 + O_2 \longrightarrow CH_3CO_2H$$

$$(3)$$

Arco described the gas phase oxidation of butenes at 166°C over a mixed vanadium–titanum–tin oxide catalyst [21]; low butene concentrations (<2 mol%) were required, along with large amounts of steam, but acetic acid selectivity approached 60% at conversions of 45%.

Liquid-phase, cobalt/acetaldehyde–catalyzed oxidation of butenes in acetic acid at 90°C and 3500 kPa gave acetic acid and mixtures of acetoxylated C4 products [22].

$$CH_3CH=CHCH_3 + O_2 + \xrightarrow[CH_3CHO]{Co}$$

$$CH_3COOH + CH_3CH(OAc)CH(OH/OAC)CH_3 \quad (4)$$

Huls demonstrated high (65–75%) selectivity for acetic acid by the oxidation of butenes over tin and titanium vanadate catalysts [23,24]. Temperatures of 225–275°C and 75% conversions of the ca. 2 vol% butene in the feed were used. This process was piloted, but it did not achieve commercialization, presumably due in part to the large amount of steam used in the optimum process and the concomitant high cost of acetic acid isolation [25].

Bayer devised a unique solution to minimize combustion of butenes and production of tars during their oxidation to acetic acid [26,27]. In their two-stage process, mixed butenes were reacted with acetic acid over an acid catalyst (100–110°C/1500–2000 kPa) to produce sec-butyl acetate; this ester was then oxidized at 180–220°C and 5050 kPa in the absence of a catalyst to produce acetic acid in about 80% selectivity.

$$1\text{- and } 2\text{-Butenes} + CH_3CO_2H \longrightarrow CH_3CH(OCOCH_3)CH_2CH_3 \quad (5)$$

$$CH_3CH(OCOCH_3)CH_2CH_3 + O_2 \longrightarrow CH_3CO_2H \quad (6)$$

As with the other olefins, processes have been developed that could serve as the basis for commercial production of acetic acid, but unfavorable economics have precluded their commercialization.

III. PARAFFIN OXIDATION

A. Ethane

Ethane oxidation to acetic acid has not received a great deal of attention. The usual route from ethane involves dehydrogenation to ethylene, followed by Wacker conversion to acetaldehyde and its subsequent oxidation to acetic acid. Alternatively, ethane is oxidized directly to acetaldehyde, which is then converted to acetic acid by conventional oxidation (see Ch. 2).

ICI [28] reported the gas-phase oxidation of ethane at 350°C (13% conv.) over a silver/manganese oxide catalyst to give 42% selectivity to acetaldehyde and 15% to acetic acid.

Union Carbide [29] described the gas-phase oxidation of a 10/1 mixture of ethane and ethylene at 255°C/6900 kPa (3% ethane conversion) over a vanadium catalyst containing lesser amounts of molybdenum, niobium, antimony, and calcium supported on an LZ-105 molecular sieve to yield 63% selectivity to acetic acid and 14% selectivity to ethylene.

$$CH_3CH_3 + CH_2{=}CH_2 + O_2 \xrightarrow[\text{Zeolite}]{\text{V/Mo}}$$

$$CH_3COOH \ (63\%) + CH_2{=}CH_2 \ (14\%) \quad (7)$$

B. Propane

Gas-phase oxidation of propane/butane mixtures to acetaldehyde represented 11% of all acetaldehyde production as late as 1973 [30]. Celanese pioneered this route in 1943; reactions were typically run at 425–460°C and 700–2000 kPa. Direct oxidation of propane to acetic acid has received little attention. As with ethane, this represents a not-very-attractive two-step route to acetic acid.

C. Butane

Of all the olefins and paraffins, butane has received the most attention as an oxidation feedstock for acetic acid. Celanese initiated production of acetic acid at its Pampa, Texas, location in 1952 based on the liquid-phase oxidation of butane, and by 1966, commercial plants were in operation by Union Carbide, AKZO Zout Chemie, Chemische Werke Huls, and Russian Refinery [31]. By 1973, 40% of total acetic acid capacity was based on this technology [30]. Because of the superior economics of methanol carbonylation processes, little of this capacity survives today. The UCC Brownsville, Texas, plant coproduced methyl ethyl ketone and other oxygenated by-products of value; it is currently shut down.

A typical process uses a cobalt catalyst, although manganese, chromium, vanadium, bismuth, nickel, titanium, and tin are also operable [30–33], with oxidations run in acetic acid solvent at 100–200°C/1000–5000 kPa. Separation of the mixtures of acetic, formic, propionic, acrylic, and butyric acids, methyl ethyl ketone, ethyl acetate, methyl vinyl ketone, and gamma-butyrolactone is done by combinations of extraction, distillation, and extractive distillation. Reaction conditions and catalyst choice vary the product mix substantially; bismuth acetate catalyst provides 97% selectivity for acetic acid with low formic acid production, whereas cobalt acetate can give selectivities for ethyl acetate and methyl ethyl ketone of 46 and 23%, respectively [33,34], or for acetic acid alone of 79% [35]. Acetic acid yields of 79% have been claimed for chromium acetate catalyst, with acetate esters and methyl ethyl ketone accounting for most of the remaining products [36]. The process can be run in the absence of catalyst, with 30% butane conversion to give a 93% yield of acetic acid [37]. The commercial Huls plant operated catalyst-free at 7100 kPa and 170–200°C, 2% butane conversion to give 60% selectivity to acetic acid [30].

$$C_4H_{10} + O_2 \xrightarrow{\text{Co}}$$
$$CH_3CO_2H + HCO_2H, CH_3COCH_2CH_3, \text{ other oxygenates} \quad (8)$$

The mechanism accounting for all the products is complex and has not been fully established; elements have been summarized by Adolfo et al. [31]. It is clear that the reaction is a classical radical chain process, involving propagation by hydrogen abstraction from a secondary carbon on butane by a radical species, followed by addition of oxygen to yield the corresponding hydroperoxide; subsequent disproportionation and cleavage reactions are then on the path to the end products.

D. C5 and Higher Paraffins

Oxidation of light naphthas can also provide acetic acid, but the selectivities to acetic acid are lower than with butane. The Distillers naphtha oxidation process [38–43] was commercialized at several locations and was claimed to produce up to 70% selectivity to acetic acid [30]. Manganese was an effective catalyst, but the process operated well in the absence of catalyst. By-product production was typically 35–75% w/w of the acetic acid.

IV. CONCLUSIONS

Hydrocarbon oxidation, especially the liquid phase oxidation of butane, has proven capable of producing acetic acid on a commercial scale. Although this technology has been superseded by methanol carbonylation for grounds-up construction, existing plants are likely to continue to operate where there is a need for the oxygenated by-products. New grounds-up construction based on this technology is unlikely unless there is a major breakthrough in selectivity coupled with high production rate. It is interesting to note that in the 1940–1970 period, butane (and naphtha) oxidation work was directed toward maximizing acetic acid and minimizing by-product selectivity; in the 1980–1991 period, this work was directed toward minimizing acetic acid and maximizing maleic anhydride selectivity [44–47].

REFERENCES

1. British Patent 1,020,068 to Halcon International (Nov. 17,1965).
2. Naglieri, A. N. U.S. Patent 3,240,805 to Halcon International (March 15, 1966).
3. Japanese Patent 41-17456 to Halcon International (Oct. 4, 1966).
4. French Patent 1,568,742 to IFP (May 30, 1969).
5. French Patent 2,073,229 to IFP (Nov. 5, 1971).
6. Japanese Kokai 7,106,763 to Japan Catalytic (Feb. 20, 1971); *Chem. Abstr.*, 75, 63163.
7. Japanese Kokai 7,211,050 to Japan Catalytic (April 4, 1972); *Chem Abstr.*, 77, 33958.
8. European Patent Application 162,263 to National Distillers (Nov. 27, 1985).
9. Belgian Patent 776,752 to National Distillers (June 15, 1972).
10. Kamiguchi et al., Eur. Pat. Appl EP 156,498 to Babcock-Hitachi K.K. (Oct. 2, 1985).
11. British Patent 977,496 to Stamicarbon (Dec. 9, 1964).
12. British Patent 1,019,426 to Stamicarbon (Feb. 9, 1966).
13. German Patent 1,188,072 to Stamicarbon (Oct. 28, 1965).
14. Japanese Patent 40-8523 to Stamicarbon (May 4, 1965).
15. Japanese Patent 40-10569 to Stamicarbon (May 28, 1965).

16. British Patent 1,016,101 to Ruhrchemie (Jan. 5, 1966).
17. German Patent 1,216,864 to Ruhrchemie (May 18, 1966).
18. Nieto, J. M. L., Kremenic, G., and Fierro, J. L. G., *Appl. Catal.*, *61*, 235–251 (1990).
19. Nieto, J. M. L., Bielsa, R., Kremenic, G., and Fierro, J. L. G., *New Developments in Selective Oxidation*, Elsevier, Amsterdam, 1990, pp. 295–304.
20. Prokopchuk, S. P., Abadzhev, S. S., and Shevchuk, V. U., *Khim Promo-st*, *11*, 666–668 (1988); C.A. 110(10), 76099.
21. H. N. Sun, U.S. Patent 4,448,898 to Atlantic Richfield (May 15, 1984).
22. Codignola, F., Gronchi, P., and Centola, P., European Patent Appl. EP 41,726 to Societa Italiana Serie Acetica Sintetica S.p.A. (Dec. 16, 1981).
23. French Patent 1,470,474 to Huls (Feb.24, 1967).
24. Brockhaus, R., *Chem. Eng.-Tech.*, *38*(10), 1939–1941 (1966).
25. Weissermel, K., and Arpe, H.-J., *Industrial Organic Chemistry*, Verlag Chemie, New York, 1978, pp. 153–156.
26. Kroenig, W., Japanese Patent 41-9173 to Bayer (May 16, 1966).
27. Kroenig, W., et al., British Patent 1,072,399 to Bayer (June 14, 1967).
28. Pyke, D. R., and Reid, R. F., Ger Offen DE 3,209,961 to ICI (Oct. 14, 1982).
29. McCain, J. H., Jr., Kaiser, S. W., and O'Connor, G. L., European Patent Appl. EP 294,845 (Dec. 14, 1988).
30. Weissermel, K., and Arpe H.-J., *Industrial Organic Chemistry*, Verlag Chemie, New York, 1978, pp. 144–156.
31. Aguilo, A., Hobbs, C. C., and Zey, E. G., in *Ullmann's Encyclopedia of Industrial Chemistry*, Vol. A1, 5th ed., VCH, Weinheim (FRG), 1985, pp. 45–64.
32. Wagner, F. S., Jr., in *Kirk-Othmer Encyclopedia of Chemical Technology*, Vol. 1, 3rd ed., Wiley, NY, 1978, pp. 124–147.
33. Morgan, C. S., Jr., et al., U.S. Patent 2,659,746 to Celanese (Nov. 17, 1953).
34. Morgan, C. S., Jr., et al., U.S. Patent 2,704,294 to Celanese (March 15, 1955).
35. Logsdon, J. E., and Kiff, B. W., Fr. Demande FR 2,478,626 to Union Carbide (Sept. 25, 1981).
36. Mitchell, R. L., et al., U.S. Patent 2,653,962 to Celanese (Sept. 29, 1953).
37. French Patent 1,370,545 to Institute of Chemical Physics (Aug. 21, 1964).
38. Elce, A., et al., British Patent 743,990 to Distillers (Jan. 25, 1956).
39. Armstong, G. P., et al., British Patent 766,544 to Distillers (Jan. 23, 1957).
40. Lawson-Hall, G., et al. British Patent 805,110 to Distillers (Nov. 26, 1958).
41. Millidge, A. F., et al., British Patent 767,290 to Distillers (Jan, 30, 1957).
42. Elce, A., et al., U.S. Patent 2,800,504 to Distillers (July 23, 1957).
43. Millidge, A. F., et al., U.S. Patent 2,800,506 to Distillers (July 23, 1957).
44. Centi, G., and Trifiro, F., *Catal. Today*, *3*, 151–162 (1988).
45. Serra, A., Poch, M., and Sola, C., *Process Eng.*, *68*(12), 19 (1987).
46. Hodnett, B. K., Ed., *Catal. Today*, *1*(5), 475–629 (1987).
47. Collection of Papers in *New Developments in Selective Oxidation*, Studies in Surface Science and Catalysis, V55 (G. Centi and F. Trifiro, eds.), Elsevier, New York, 1990, pp. 537–634.

4
Manufacture via Methanol Carbonylation

Joseph R. Zoeller
Eastman Chemical Company, Kingsport, Tennessee

I. INTRODUCTION

Although oxidation and biologically (i.e., wood distillation and fermentation) based routes still contribute significantly to the quantities of acetic acid produced worldwide, the carbonylation of methanol [Eq. (1)] has become generally accepted as the method of choice for the commercial generation of acetic acid and accounts for most new production capacity over the last two decades. The first commercially successful venture was a cobalt iodide–catalyzed carbonylation built by BASF in the mid-1960s [1,2].

$$MeOH + CO \longrightarrow AcOH \tag{1}$$

Although this technology was novel and commercially viable, its reign as the method of choice was very short by industrial standards. In 1968, investigators at Monsanto disclosed a rhodium iodide–based catalyst system that operated under milder conditions with greater selectivity [3]. This renowned process, generally referred to simply as the Monsanto acetic acid process, was rapidly commercialized and now accounts for most newly added capacity.

Two excellent reviews of metal-catalyzed carbonylations of alcohols, which obviously focus closely on the carbonylation of methanol, have appeared recently [4]. These reviews mention several other metals that have also been found to catalyze the reaction. including especially iridium, nickel, and copper. Since the publication of these review articles, several technological advances have occurred, and palladium and ruthenium have been added to the list of potentially useful metals in the conversion of methanol to acetic acid by carbonylation. Although it is our intent to review the potential utility of each metal in this application, we will logically initiate discussion of this technology with the commercial rhodium-catalyzed process.

II. RHODIUM-CATALYZED CARBONYLATION OF METHANOL

At present, the carbonylation of methanol with a rhodium iodide–based catalyst is the method of choice for the commercial generation of acetic acid. This process, which was described by Monsanto in 1968 [3], has been well documented, and the reaction mechanism is well defined [3–19].

Operationally, the reaction is performed by adding practically any source of Rh to a methanolic solution of methyl iodide (or methyl iodide precursor) and water. Carbon monoxide is introduced at a partial pressure of between 10 and 25 bar while the temperature is maintained at ca. 175°C [3,5]. In this region, the reaction is first order in methyl iodide and Rh, but zero order with respect to methanol and carbon monoxide [7–15].

The predominent Rh species has also been identified. Under active catalytic conditions, high-pressure infrared spectroscopy indicates that the active species is cis-[Rh(CO)$_2$I$_2$]$^-$ [8]. Based on the combination of the kinetic and infrared data, several groups have proposed virtually identical mechanisms involving a rate-limiting oxidative addition of methyl iodide to cis-[Rh(CO)$_2$I$_2$]$^-$ as the initial step [7–19]. This mechanistic proposal, shown in Figure 1, has been subjected to further scrutiny, and each step involving Rh has been modeled in the laboratory by the Monsanto group [12].

The addition of the catalytically active cis-[Rh(CO)$_2$I$_2$]$^-$ species to an excess of methyl iodide resulted in the formation of an acetyl Rh species that has been formulated as [Rh(Ac)(CO)I$_3$]$^-$. The monomeric complex was stable in solution, but upon attempted isolation, the dimeric acetyl complex (1) (shown in Fig. 2) was generated instead. (The structure of the dimeric acetyl complex (1) was determined by X-ray diffraction [20].) The acetyl complex formulated as [Rh(Ac)(CO)I$_3$]$^-$ has been shown to spontaneously eliminate acetyl iodide in solution upon the introduction of carbon monoxide. (This rhodium chemistry has since been verified in other laboratories [21].) All the remaining steps in the catalytic cycle involved straightforward organic iodine chemistry.

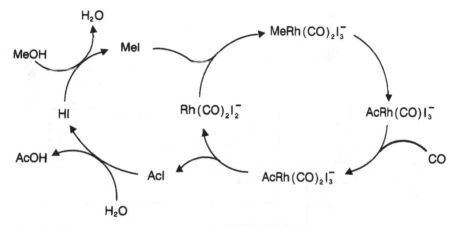

Figure 1 Mechanism for the rhodium-catalyzed methanol carbonylation.

A schematic plant design has already been published [7]. This same report included a detailed study of the effect of the reactor variables on the process. A more detailed plant schematic than that in the earlier reference, displaying approximate flow compositions, is shown in Figure 3. (The iodine removal method, a key operation in this plant, is omitted from the flowsheet because the individual corporate knowledge bases pertaining to acceptable methods for iodine removal are regarded as proprietary intellectual property by all carbonylation practitioners.)

One repeated claim in the Monsanto publications is that alkali metals are ineffective as cocatalysts in this process [7,17]. The reason for this may well be that the alkali metal iodides alone were poor agents for the regeneration of methyl iodide from methanol.

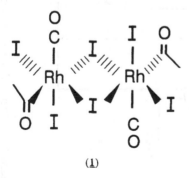

(**1**)

Figure 2 Structure of $[Rh(Ac)CO_2I_2]_2^{2-}$.

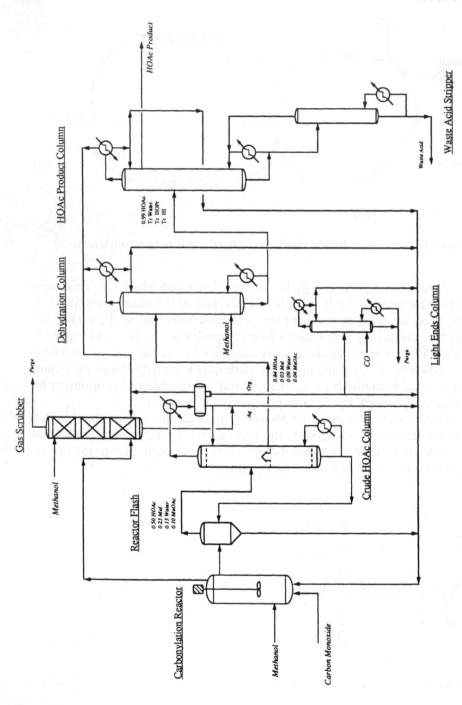

Figure 3 Conceptual flowsheet for the carbonylation of methanol to acetic acid.

However, a group of investigators from Celanese, possibly acting on reports of enhanced carbonylation activity when these salts are used in combination with methyl iodide in the carbonylation of methyl acetate to acetic anhydride [22], described a process improvement involving the addition of lithium iodide [23,24]. The addition of lithium iodide in combination with methyl iodide permits a dramatic reduction in the levels of water required to run the reaction while maintaining high rates. This subsequently reduces the separation costs involved in this process. (Prior to this report, water levels as high as 10 M have been reported to be necessary.)

The Celanese group rationalized the differences in the two systems by invoking a speculative mechanism involving the generation of a $[Rh(CO)_2I_3]^{2-}$ species that was expected to enhance the oxidative addition of methyl iodide. Unfortunately, the proposal was based entirely on kinetic data that was just as easily explained by alternative means. However, some additional background knowledge is required if the rationale behind this alternative is to be understood.

The generally presented rationale for including large excesses of water has been to suppress the formation of methyl acetate. However, a second, often underappreciated, role is in the generation and maintenance of the catalyst in its active form, cis-$[Rh(CO)_2I_2]^-$.

In the presence of water, $[Rh^{3+}]$ species are reduced to cis-$[Rh(CO)_2I_2]^-$ via the water gas shift [25], thus maintaining an active catalyst system. Although this reaction likely consumes only small amounts of water, low water levels are likely to affect the $[Rh^+]$ species in solution. In anhydrous systems, the level of H^+ (the counterion in the Monsanto system) will be severely depressed. The loss of the necessary counterion results in the formation of neutral $[Rh(CO)_3I]$, as observed by high-pressure infrared spectroscopy [22,26]. This species, which was first observed and studied by Morris and Tinker [26], has relatively little activity as a carbonylation catalyst. (These observations have been verified in the laboratories of Eastman Chemical Company.)

A second role for lithium iodide may also be operative. Based on investigations made in the course of developing the methyl acetate carbonylation process to generate acetic anhydride, it was found that lithium iodide is an effective means of converting the methanol equivalent present in methyl acetate to methyl iodide via Eq. (2). However, under low water conditions, HI cannot effectively perform Reaction (2) [22]. (This reaction will be discussed in more detail when we discuss the manufacture of acetic anhydride later in this book.)

$$MI + MeOAc \rightleftharpoons MOAc + MeI \qquad (2)$$
$$M = Li^+, H^+$$

In light of this information, it is likely that the actual role of lithium iodide is twofold. First, lithium iodide would stabilize the rhodium as the catalytically active cis-[Rh(CO)$_2$I$_2$]$^-$ as opposed to the inactive neutral species. Second, lithium iodide effectively transforms methyl acetate to lithium acetate and methyl iodide. The apparent discrepancy between the claims made by the Monsanto group and the Celanese group regarding the addition of alkali metals is likely due to differing water levels, and the beneficial effects of alkali metals are likely limited to low water systems.

In industrial catalysis, it is often preferable from an engineering perspective to utilize hetereogeneous catalysts in continuous processes where possible. Many attempts have been made to convert this homogeneous rhodium system to a hetereogeneous process [27–46]. However, none of these processes has displaced the existing homogeneous system. The general advantages of hetereogeneous systems, namely ease of catalyst separation and product purification, do not appear to materialize in this case.

Specifically, unlike some other homogeneous systems, separation and recycling of the Rh catalyst is simple in the Monsanto acetic acid process. The active cis-[Rh(CO)$_2$I$_2$]$^-$ is readily regenerated from any Rh source. Therefore, the catalyst can generally be simply recycled from the distillation bottoms of the initial distillation, and there is no apparent significant advantage for a hetereogeneous system with regard to catalyst separation and recycling.

In addition, the hetereogeneous systems generally still require a methyl iodide cocatalyst. Separation of the resultant iodine species is critical, and the associated distillation train is essentially unaltered upon switching to a heterogenous process. This essentially negates any separation and purification advantage hetereogeneous systems are generally assumed to exhibit.

The hetereogeneous system has other disadvantages. The process, although generally very clean, still produces a small amount of heavier byproducts. Separation of the catalyst from these materials is potentially more difficult, particularly if the process is to be operated in the vapor phase.

Further, catalyst leaching is a potential problem. This can occur by volatilization as any of several rhodium (0) carbonyl species or as a netural rhodium iodocarbonyl species, such as Rh(CO)$_3$I or [Rh(CO)$_2$I]$_2$, in vapor phase processes. In a liquid phase process using supports to heterogenize the catalyst, the rhodium is leached directly as Rh(CO)$_2$I$_2$$^-$ or Rh(CO)$_2$I$_4$$^-$ in any case known to the author. For all these reasons, it is highly unlikely that a hetereogeneous catalyst will replace the existing homogeneous version of this catalyst.

III. COBALT-CATALYZED CARBONYLATION OF METHANOL

The cobalt-catalyzed carbonylation of methanol, which was discovered and developed by BASF [1,2,47–49], is significant because it was the first carbonylation-based route to acetic acid to be commercialized. Compared to the rhodium-catalyzed process, the cobalt-catalyzed carbonylation of methanol operates under much more severe operating conditions and is less selective with respect to both CO and methanol, yielding acetaldehyde and ethanol (or derivatives thereof) as significant by-products.

The commercialization of the cobalt-catalyzed methanol carbonylation in the early 1960s was quickly followed by the introduction of the rhodium-catalyzed process at the end of the same decade. For this reason, only a few commercial plants were known to have operated processes based on this technology.

The volume of literature describing this reaction is much smaller than for the corresponding rhodium-based process. This is also probably due to the rapid displacement of the cobalt-catalyzed process by the rhodium-based process, which undoubtedly led to a waning academic interest in this process. Although the cobalt-based process is often mentioned in collective works [4,16,17,50–52], only one group outside of BASF has conducted basic research into the commercial iodide-promoted process [16,53,54].

Unfortunately, the reported works are all insufficient to permit as precise a mechanistic understanding as has been attained in either the rhodium or iridium systems (vide infra). The chemistry of cobalt and iodine in this system is undoubtedly complex. CoI_2, CoI_4^{2-}, $Co_2(CO)_8$, $HCo(CO)_4$, $Co(CO)_4^-$, and mixed iodocarbonyl complexes are all likely to be present, and the iodine can play important roles in both the chemistry of the cobalt and the organic reaction pathways.

Probably the most thorough discussion of the mechanism has been presented by Dekleva and Forster, who depended very heavily on the related and better-documented cobalt-catalyzed homologation reactions of methanol to ethanol and acetadehyde to derive their mechanism [4a]. This mechanism, shown in Figure 4, explains most of the observations, is a very thorough summary, and attempts to interpret the myriad observations regarding the role played by cobalt and iodine.

Although the discussion by Dekleva and Forster was very thorough, it may have neglected one possible contribution to the rate acceleration observed with iodide salts. One inevitable intermediate in the carbonylation of methanol to acetic acid is methyl acetate. During studies on the carbonylation of methyl acetate to acetic anhydride, it has been established that iodide salts, either added as the salt or generated from amines or

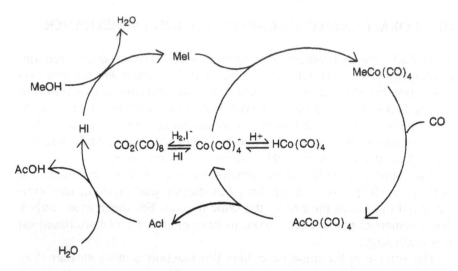

Figure 4 Proposed mechanism for the cobalt-iodine–catalyzed carbonylation of methanol.

phosphines, via quarternization with methyl iodide, accelerate the reaction of methyl acetate to generate methyl iodide and an acetate salt, as per Eq. (2) [22]. [The equilibrium shown in Eq. (2) strongly favors MeI.] Thus, one additional role of iodide salts might be to accelerate the consumption of the methyl acetate intermediate. The acetate salt would be expected to rapidly react with HI or $HCo(CO)_4$ to generate I^- or $Co(CO)_4^-$, respectively. This would likely accelerate the reaction further by accelerating the rate at which the most likely reactive cobalt intermediate, $Co(CO)_4^-$, is cycled through the reaction.

A schematic design of the cobalt-based carbonylation plant operated by BASF has been published [1] but will not be reproduced here because it represents antiquated technology. Interest in this technology would probably only be resurrected in the event of a dramatic increase in the cost of rhodium required for the preferred Monsanto process and a failure to commercialize the alternative processes, which will be discussed in the following sections. Few additional references [55–57] have appeared since the review of Dekleva and Forster [4a].

IV.   NICKEL-CATALYZED CARBONYLATION OF METHANOL

The nickel-catalyzed carbonylation of methanol has been known for several decades [47,58–60] but, until recently, there has been little promise of commercial utility because the reaction requires high temperatures and pressures to attain even moderate rates. Therefore, with the well-established hazards of nickel tetracarbonyl as an additional deterrent to work in this area, this potential process was initially abandoned by industry as a whole.

However, a significant incentive has always existed for the displacement of the rhodium-based catalyst in the Monsanto acetic acid process with a less expensive metal. Responding to this need, investigators at several chemical firms, including Halcon SD [61–67], Eastman Chemical [68,69], Rhone-Poulenc [70–74], and Mitsubishi Chemical [75], reexamined the nickel-based process. More recently, the National Chemical Laboratory of India has also begun to publish work in the area [76].

Only one of these companies, Halcon SD, is known to have developed and successfully operated a pilot plant for the production of acetic acid using a nickel-based catalyst. Halcon SD sold all its rights to this process to Eastman Chemical in 1987.

Several critical observations made by Halcon finally allowed them to identify and pilot a practical, low-pressure, nickel-catalyzed process with commercially acceptable rates. The highest rates are attained using a complex catalyst composition, which entails an added salt or salt precursor, a second additive such as Sn, Cr, Mo, or W compound, and hydrogen. The role of these metals, which are clearly necessary, is not apparent.

Only one detailed study exists in the nonpatent literature [61]. This reference gives considerable guidance about the role of the various components and demonstrates that, at least for the Mo-Li–promoted system, the reaction is first order with respect to nickel and iodide. However, the mechanism proposed is highly unlikely in that it employs phosphine-substituted nickel species. As already noted by Gauthier-Lafaye and Perron [4b], the phosphines are alkylated rapidly and completely in the presence of the excess methyl iodide, as would be present in these systems. Therefore, it is highly unlikely that any phosphine complexes proposed in this report can exist in this reaction system. Further demonstrating this point is the observation that there is no change in the reaction when a quarternary phosphine is substituted for the free phosphine [4b].

The kinetic picture of this process is still incomplete, and considerable work needs to be done before a complete mechanistic understanding of the process is attained. However, two mechanistic proposals have been proposed. They are differentiated solely by the species of nickel believed to initiate catalysis. In their brief overview of the process, Gauthier-Lafaye and Perron [4b] proposed a mechanism for the carbonylation that involved the oxidative addition of methyl iodide to a $Ni(CO)_3I^-$ species [78]. However, Dekleva and Forster evoked the original Reppe proposal involving the oxidative addition of methyl iodide to $Ni(CO)_4$ [4a]. Both authors lacked sufficient data to distinguish between the two proposals, although the dependence on iodide would certainly have favored $Ni(CO)_3I^-$. However, work reported by Nelson and co-workers [77] on the mechanism of the corresponding nickel-catalyzed acetic anhydride process lends additional support to a mechanism involving $Ni(CO)_3I^-$.

In their study, Nelson and co-workers examined the infrared spectroscopic behavior of the nickel-catalyzed carbonylation of methyl acetate to acetic anhydride in N-methyl pyrrolidinone at high pressure using lithium as a promotor. They noted infrared peaks at 1955 cm^{-1} [assigned to $Ni(CO)_3I^-$], 2009 cm^{-1} (unassigned), and 2044 cm^{-1} [assigned to $Ni(CO)_4$]. The relative intensities of the peaks was a complex function of nickel, lithium, methyl iodide, hydrogen pressure, and carbon monoxide pressure. However, the rate of the reaction displayed a first-order correlation with the intensity of the peak assigned to $Ni(CO)_3I^-$, but was independent of the peak assigned to $Ni(CO)_4$. This implies that the $Ni(CO)_3I^-$ was the active species in these methyl acetate carbonylations.

The extrapolation of the results of Nelson and co-workers to a methanol carbonylation is not unreasonable but must be proven in the future. Based on the results of Nelson and co-workers, a mechanistic proposal has been advanced in Figure 5. The proposal attempts to display the complex behavior of the nickel species and the kinetic behavior described by Rizkalla.

This mechanism can be used to rationalize the behavior of several of the additional factors that contribute to the rate. The role of hydrogen is likely to reduce Ni(II) to Ni(0), and higher hydrogen pressures should obviously favor greater levels of Ni(0). The observation of a maximum in the activity of the CO can also be explained. Carbon monoxide is required

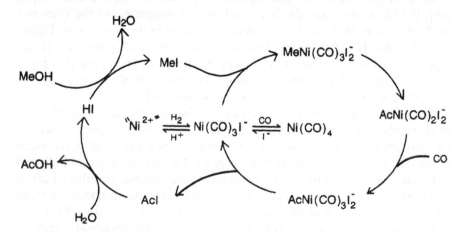

Figure 5 Proposed mechanism for the nickel-catalyzed methanol carbonylation.

to generate the catalyst; however, higher pressures of CO favor the generation of $Ni(CO)_4$ at the expense of $Ni(CO)_3I^-$. [The increased level of $Ni(CO)_4$ was observed by Nelson and co-workers in the high-pressure infrared experiments].

As with the rhodium catalyst, attempts at converting this homogeneous process to a hetereogeneous process have been pursued [79–84]. The problems of hetereogeneous catalysts in these iodine-assisted carbonylations have already been outlined in the discussion of the rhodium-catalyzed carbonylation of methanol, and the problems with a heterogeneous nickel process are completely analogous. However, in the nickel system, metal depletion from the catalyst is likely to be a more severe problem because the formation of the extremely volatile nickel carbonyl would be expected to be more facile.

V. IRIDIUM-CATALYZED CARBONYLATION OF METHANOL

Even though its potential utility is negligible owing to the enormous cost of the catalyst, the iridium-catalyzed carbonylation of methanol to acetic acid has been well documented [12–18,85,86] compared to most other carbonylation systems. Like the rhodium-based process, iridium demonstrates extremely fast rates and high selectivities at moderate temperatures and pressures, and its only apparent deficiency compared to rhodium is the excessive catalyst cost.

Mechanistically, the iridium-catalyzed carbonylation of methanol to acetic acid would be expected to be analogous to the rhodium-catalyzed acetic acid process. However, the iridium system has actually proven to be much more complicated. Unlike the rhodium-catalyzed process, which displayed comparatively simple kinetic and catalyst behavior over a reasonably large range of conditions, the iridium system demonstrated extremely complex kinetic and catalyst behavior, which varies with conditions.

To understand the iridium system, Forster and co-workers [12–15,17,86] recognized and defined three distinct reaction regions. (Obviously, overlapping regions existed that demonstrated mixed behavior.) Each reaction region exhibited distinctly different kinetics and contained a different predominant iridium species. These regions were defined as follows:

Region 1: A region in which the solution contained low concentrations of methyl iodide, water, and anionic iodide

Region 2: A region that contained increased levels of anionic iodide

Region 3: A region characterized by the presence of high concentrations of either methanol or water

In region 1, the form of the iridium was $Ir(CO)_3I$. In this region, the reaction was inhibited by increasing carbon monoxide pressures. This observation implied that, in the rate-controlling step, a carbon monoxide ligand dissociates prior to the initiation of catalysis. The unstable $Ir(CO)_2I$ complex was inferred as the species that actually reacted with methyl iodide. Presumably this generated another unstable iridium species, $MeIr(CO)_2I_2$. Evidence obtained from region 2 (vide infra) suggested that this complex adds an additional carbon monoxide ligand yielding $MeIr(CO)_3I_2$. This species could undergo the sucessive insertions and eliminations necessary to utlimately generate products.

The predominant form of iridium observed in region 2 was $MeIr(CO)_2I_3^-$. The formation of a stable methyl iridium species contrasted sharply with the rhodium system in which carbon monoxide insertion was spontaneous. Forster has suggested that the generally higher carbon-metal bond strengths observed in methyl iridium species (compared to methyl-rhodium species) might account for the difference between the two systems.

Kinetically, in region 2 the reaction was accelerated by increasing carbon monoxide pressures, but was inhibited by anionic iodide. The interpretation of these observations was that the observed complex, $MeIr(CO)_2I_3^-$, must first dissociate an iodide ligand, generating an unstable $MeIr(CO)_2I_2$, and then add an additional carbon monoxide to generate $MeIr(CO)_3I_2$. The connection between regions 1 and 2 is readily apparent at this point as the $MeIr(CO)_3I_2$ is common to both reaction regions.

In the third region, $HIr(CO)_2I_3^-$ was the predominant Ir species. The rate was independent of CO but displayed an apparent dependence on methanol concentration. Of the carbonylations mentioned to date, this represented the first process to exhibit a positive dependence on methanol.

Forster and co-workers have presented a rationalization for the $HIr(CO)_2I_3^-$ and its entry into the catalytic cycle. These authors demonstrated that the addition of HI to $Ir(CO)_2I_2^-$ (the Ir analog of the active species in the rhodium-catalyzed process) was demonstrated to rapidly establish an equilibrium with $HIr(CO)_2I_3^-$, as shown in Eq. (3). Forster and co-workers postulated that the reaction of methyl iodide with $Ir(CO)_2I_2^-$ [Eq. (4)] is irreversible, removing the $Ir(CO)_2I_2^-$ species and generating $MeIr(CO)_2I_3^-$ in the process, and proceeded to demonstrate the stoichiometric reaction. The chemistry of $MeIr(CO)_2I_3^-$ has been discussed previously, and this species forms the connection between region 2 and region 3. The dependence on methanol has never been completely rationalized.

$$HI + Ir(CO)_2I_2^- \rightleftharpoons HIr(CO)_2I_3^- \tag{3}$$

$$Ir(CO)_2I_2^- + MeI \longrightarrow MeIr(CO)_2I_3^- \tag{4}$$

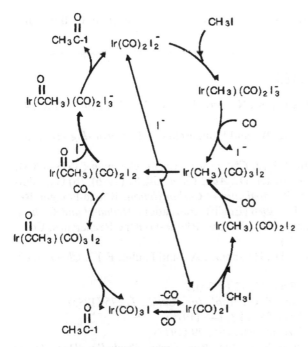

Figure 6 Mechanism of the iridium-catalyzed carbonylation of methanol. (Reprinted, with permission, from Ref. 4a.)

Although the presence of the three reaction regions made the mechanistic interpretation very difficult, the ingenious work represented above allowed Forster and co-workers to present a coherent mechanism that integrated all three regions into a single comprehensive mechanism. This mechanism, shown in Figure 6, is consistent with all the spectroscopic data, the kinetic data, and the extensive list of known stoichiometric reactions assembled for this study.

VI. OTHER METHANOL CARBONYLATION CATALYSTS

Although all the catalysts known to be of practical laboratory or commercial interest have been covered, several additional catalysts have been found to catalyze the carbonylation of methanol to acetic acid. These include iron [2], ruthenium [87], palladium [88,89], and copper [90]. In addition, two mixed-metal systems have been described. A heterogeneous nickel-palladium catalyst [89,91] and a mixed homogeneous ruthenium-cobalt catalyst [92], both requiring the addition of a halide promotor, have been effective in the carbonylation of methanol to acetic acid. None of these processes have been extensively explored and none appear to have any potential at this time.

REFERENCES AND NOTES

1. Hohenschutz, H., von Kutepow, N., and Himmle, W., *Hydrocarb. Proc.*, *45*, 141 (1966).
2. von Kutepow, N., Himmle, W., and Hohenschutz, H., *Chem.-Ing.-Tech*, *37*, 383 (1965).
3. Paulik, F. E., and Roth, J. F., *J. Chem. Soc., Chem. Commun.*, 1578 (1968).
4. (a) Dekleva, T. W., and Forster, D., *Adv. Catal.*, *34*, 81 (1986). (b) Gauthier-Lafaye, J., and Perron, R., *Methanol et Carbonylation*, Rhone-Poulenc Recherches, Courbevoie, Fr., 1986 (English translation: *Methanol and Carbonylation*, Editions Technip, Paris, Fr. and Rhone-Poulenc Recherches, Courbevoie, Fr., 1987, p. 117.)
5. Roth, J. F., Craddock, J. H., Hershman, A., and Paulik, F. E., *Chem. Technol.*, 600 (1971).
6. Roth, J. F., *Plat. Met. Rev.*, *17*, 12 (1975).
7. Eby, R. T., and Singleton, T. C., *Appl. Ind. Catal.*, *1*, 275 (1983).
8. Forster, D., *J. Am. Chem. Soc.*, *98*, 846 (1976).
9. Forster, D., *Ann. N.Y. Acad. Sci.*, *295*, 79 (1977).
10. Hjortkjaer, J., and Jensen, V.W., *Ind. Eng. Chem. Prod. Res. Dev.*, *15*, 46 (1976).
11. Brodzki, D., Leclere, C., Denise, B., and Pannetier, G., *Bull. Chim. Soc. Fr.*, 61 (1976).
12. Forster, D., *Adv. Organomet. Chem.*, *17*, 255 (1979).
13. Dekleva, T. W., and Forster, D., *Adv. Catal.*, *34*, 81 (1986).
14. Forster, D., and Dekleva, T. W., *J. Chem. Ed.*, *63*, 204 (1986).
15. Forster, D., Hershman, A., and Morris, D. E., *Catal. Rev.-Sci. Eng.*, *23*, (1&2), 89 (1981).
16. Mizoroki, T., Matsumoto, T., and Ozaki, A., *Bull. Chem. Soc. Jpn.*, *52*, 479 (1979).
17. Forster, D., and Singleton, T. C., *J. Mol. Catal.*, *17*, 299 (1982).
18. Brodzki, D., Denise, B., and Pannetier, G., *J. Mol. Catal.*, *2*, 149 (1977).
19. Dekleva, T. W., and Forster, D., *J. Am. Chem. Soc.*, *107*, 3565 (1985).
20. Adamson, G. W., Daly, J. J., and Forster, D. *J. Organomet. Chem.*, *71*, C17 (1974).
21. Kent, A. G., Mann, B. E., and Manuel, C. P. *J. Chem. Soc., Chem. Commun.*, 728 (1985).
22. (a) Polichnowski, S. W., *J. Chem. Ed.*, *63*, 206 (1986). (b) See Ref. 4b and Chapter 9 of this book.
23. (a) Smith, B. L., Torrence, G. P., Murphy, M. A., and Aquilo, A., *J. Mol. Catal.*, *39*, 115 (1987). (b) Murphy, M. A., Smith, B. L., Torrence, G. P., and Aquilo, A., *J. Organomet. Chem.*, *303*, 257 (1987). (c) Murphy, M. A.,

Smith, B. L., Torrence, G. P., and Aquilo, A., *Inorg. Chim. Acta*, *101*, L47 (1985).

24. Torrence, G. P., Hendricks, J. D., and Aquilo, A., U.S. Patent 4,994,608 (1991), U.S. Patent 5,026,908 (1991).
25. Singleton, T. C., Park, L. J., Price, J. L., and Forster, D., *Prepr. Div. Pet. Chem.*, *Am. Chem. Soc.*, *24*, 329 (1987).
26. Morris, D. E., and Tinker, H. B., *J. Organomet. Chem.*, *49*, C53 (1973).
27. Scurell, M. E., *Plat. Met. Rev.*, *21*, 92 (1977).
28. Schultz, R. G., and Montgomery, P. D., *J. Catal.*, *13*, 105 (1969).
29. Robinson, K. K., Hershman, A., Craddock, J. H., and Roth, J. F., *J. Catal.*, *27*, 389 (1972).
30. Scurell, M. S., and Howe, R. F., *J. Mol. Catal.*, *7*, 535 (1980).
31. Krzywicki, A., and Martczewski, M., *J. Mol. Catal.*, *6*, 431 (1979).
32. Takahashi, N., Orikasa, Y., and Yashima, T., *J. Catal.*, *59*, 61 (1979).
33. Christensen, B., and Scurell, M. S., *J. Chem. Soc.*, *Faraday Trans. 1*, *73*, 2036 (1977).
34. Christensen, B., and Scurell, M. S., *J. Chem. Soc.*, *Faraday Trans. 1*, *74*, 2313 (1977).
35. Krzywicki, A., and Pannetier, G., *Bull. Chem. Soc. Fr.*, *Pt. 1*, 64 (1977).
36. Fujimoto, K. Tanemura, S., and Kunugi, T., *Nippon Kagaku Kaishi*, 167 (1977).
37. Huang, N. T., Schwartz, J., and Kitajima, N., *J. Mol. Catal.*, *22*, 389 (1984).
38. Anderson, S. L., and Scurell, M. S., *Zeolites*, *3*, 261 (1983).
39. Nefedov, B. K., Dzhaparidze, R. V., and Eidus, Y. T., *Izv. Akad. Nauk. SSR*, *Ser. Khim.*, 1422 (1977).
40. Nefedov, B. K., Sergeeva, N. S., and Eidus, Y. T., *Izv. Akad. Nauk. SSR*, *Ser. Khim.*, 2271 (1976).
41. Nefedov, B. K., Dzhaparidze, R. V., Mamaev, O. G., and Sergeeva, N. S., *Izv. Akad. Nauk. SSR*, *Ser. Khim.*, 376 (1979).
42. Dzhaparidze, R. V., and Krasnova, L. L., *Issled. Obl. Khim. Vysokomol. Soedin. Neftekhim.*, 15 (1977).
43. (a) Shostakovskii, M. F., Nefedov, B. K., Dzhaparidze, R. V., and Zasorina, N. M., *Soobschch. Akad. Nauk. Guz. SSR*, *89*, 93 (1978). (b) Nefedov, B. K., Dzhaparidze, R. V., and Sorokina, A. K., *Izv. Akad. Nauk. Gruz. SSR*, *Ser. Khim.*, *3*, 235 (1977).
44. Shimazu, S., Ishibashi, Y., Miura, M., and Uematsu, T., *Appl. Catal.*, *35*, 279 (1987).
45. Yashima, T., Orikasa, Y., Takahashi, N., and Hara, N., *J. Catal.*, *59*, 53 (1979).
46. Maneck, H. E., Gutschick, D., Burkhart, I., Luecke, B., Miessner, H., and Wolf, U., *Catal. Today*, *3*, 421 (1988).
47. Reppe, W. von Kutepow, N., and Bille, H., U.S. Patent 3,014,962 (1961).
48. Reppe, W., and Friederich, H., Ger. Offen. 902,495 (1951).
49. Reppe, W., and Friederich, H., Ger. Offen. 933,148 (1953).

50. Falbe, J., *New Syntheses with Carbon Monoxide*, Springer-Verlag, New York, 1980.
51. Falbe, J., *Carbon Monoxide in Organic Synthesis*, Springer-Verlag, New York, 1970.
52. Piancenti, F., and Bianachi, M., in *Organic Syntheses via Metal Carbonyls*, *Vol. 2*. (I. Wender and P. Pino, eds.), Wiley, New York, 1977.
53. Mizoroki, T., and Nakayama, M., *Bull. Chem. Soc. Jpn.*, *38*, 1876 (1965).
54. Mizoroki, T., and Nakayama, M., *Bull. Chem. Soc. Jpn.*, *41*, 1628 (1968).
55. Chauvin, Y., Commereuc, D., and Hugues, F., Fr. Demande 2,587,027 (1987).
56. Hoelderich, W., Fouquet, G., Harder, W., and Caesar, F., Ger. Offen. 3,606,169 (1987).
57. Current, S. P., U.S. Patent 4,628,113 (1986).
58. Reppe, W., Freiderich, H., von Kutepow, N., and Morsch, W., U.S. Patent 2,729,651 (1956).
59. Thomas, E. B., and Alcock, E. H., U.S. Patent 2,650,245 (1953).
60. Reppe, J. W., in *Acetylene Chemistry*, Charles A. Meyer and Co., New York, 1949, p. 171.
61. Rizkalla, N., *ACS Symp. Ser.*, *328*, 61 (1987).
62. Rizkalla, N., G.B. Patent 2,128,609 (1984).
63. Naglieri, A. N., and Rizkalla, N., U.S. Patent 4,134,912 (1979).
64. Naglieri, A. N., and Rizkalla, N., U.S. Patent 4,356,320 (1982).
65. Rizkalla, N., Fr. Demande 2,496,643 (1982); U.S. Patent 4,659,518 (1987).
66. Rizkalla, N., U.S. Patent 4,482,497 (1984).
67. Becker, M., and Sachs, H. M., Ger. Offen. 3,319,361 (1983); U.S. Patent 4,661,631 (1987).
68. Holmes, J. D., U.S. Patent 4,133,963 (1979).
69. Holmes, J. D., U.S. Patent 4,218,340 (1980).
70. Gauthier-Lafaye, J., and Perron, R., European Patent Appl. 37,354 (1981); U.S. Patent 4,351,953 (1982).
71. Gauthier-Lafaye, J., and Perron, R., European Patent Appl. 35,458 (1981).
72. Gauthier-Lafaye, J., and Perron, R., European Patent Appl. 39,652 (1981); U.S. Patent 4,436,889 (1984).
73. Gauthier-Lafaye, J., and Perron, R., European Patent Appl. 39,653 (1981).
74. Gauthier-Lafaye, J., and Perron, R., U.S. Patent 4,426,537 (1984).
75. (a) Isshiki, T., Kijima, Y., and Miyauchi, Y., U.S. Patent 4,336,399 (1982).
    (b) Isshiki, T., Kijima, Y., Miyauchi, Y., and Kondo, T., U.S. Patent 4,620,033 (1986).
76. Kelkar, A. A., Jaganathan, R., Kolhe, D. S., and Chaudhari, R. V., U.S. Patent 4,902,659 (1990).
77. Nelson, G. O., Middlemas, E. C., and Polichnowski, S. W., presented to the N.Y. Acad. Sci., March 13, 1986.
78. Ni(CO)$_3$I$^-$ was a known species. See Cassar, L., and Foa, M., *Inorg. Nucl. Chem. Lett.*, *6*, 291 (1970).
79. Tominaga, H., and Fujimoto, K., Jpn. Kok. JP 59/139330 (1984).

80. Fujimoto, K., Mazaki, H., Omata, K., and Tominaga, H., *Chem. Lett.*, 895 (1987).
81. Inui, T., Matsuda, II., and Takegami, Y., *Nippon Kagaku Kaishi*, 313 (1982).
82. Fujimoto, K., Omata, K., Shikada, T., and Tominaga, H., *Ind. Eng. Chem. Prod. Res. Dev.*, *22*, 436 (1983).
83. Wang, R., Ku, C., Chihow, C., Ching, Y., and Huang, Z., *Huadong Huagong Xueyuan Xuebao*, *12*, 211 (1986).
84. Current, S. P., U.S. Patent 4,625,049 (1986).
85. Matsumoto, T., Mizoroki, T., and Ozaki, A., *J. Catal.*, *51*, 96 (1978).
86. Forster, D., *J. Chem. Soc.*, *Dalton Trans.*, 1639 (1979).
87. Jenner, G., and Bitsi, G., *J. Mol. Catal.*, *40*, 71 (1987).
88. Van Leeuwen, P. W. N., European Patent Appl. 90,443 (1983).
89. Van Leeuwen, P. W. N., European Patent Appl. 133,331 (1985).
90. Souma, Y., and Sano, H., *Bull. Chem. Soc. Jpn.*, *46*, 3237 (1973).
91. Mueller, F. J., and Mott, D., U.S. Patent 4,918,218 (1990).
92. Vanderpool, S. H., Lin, J. J. and Duranleau, R. G., U.S. Patent 4,629,809 (1986).

5

Other Synthesis Gas-Based Acetic Acid Processes

Bruce L. Gustafson and Joseph R. Zoeller
Eastman Chemical Company, Kingsport, Tennessee

I. INTRODUCTION

Since the discovery in 1902 by Sabetier and Senderens that carbon mon oxide could be converted to methane [1], extensive research efforts have been maintained on other processes based on synthesis gas. These efforts have been generally spurred by political necessity, represented by surges in the technology in Germany during World War II, in modern-day South Africa, and during the oil crises of the 1970s. As a result of these synthesis gas programs, it is now apparent that most petrochemical feedstocks can be emulated using synthesis gas. Therefore, in principle, acetic acid can be derived from synthesis gas using any of the known oxidation technologies by indirect, presently cost-ineffective, generation of the appropriate hydrocarbon feedstock. Clearly, the generation of acetic acid from feedstocks ordinarily regarded as being of petroleum based origin is not presently economical or part of this book chapter. However, several approaches to acetic acid using synthesis gas or synthesis gas-based feedstocks alone have been developed.

The carbonylation of methanol, which is generally derived from synthesis gas, clearly dominates the synthesis gas-based routes to acetic acid. This commercially important process has been thoroughly described in the previous chapter and represents one of the most successful chemical processes based entirely on synthesis gas.

However, other systems based on synthesis gas have also been described in the literature and clearly deserve mention. As an alternative to the existing two-step synthesis gas route represented by the carbonylation of methanol, the generation of acetic acid directly from synthesis gas is conceptually attractive. However, the technical hurdles have been significant. A second approach, which has been extensively explored, is the rearrangement of methyl formate, which is generated from methanol and carbon monoxide. These two approaches have yet to demonstrate even marginal advantages over the existing rhodium-catalyzed methanol carbonylation and may never be commercial processes but warrant examination.

II. DIRECT GENERATION OF ACETIC ACID FROM SYNTHESIS GAS

The generation of acetic acid directly from synthesis gas may clearly be divided into two systems, a homogeneous Ru-Co system and heterogeneous Rh systems, particularly Rh on silica with any number of promotors. Unfortunately, both systems have the common drawback of poor selectivity and give mixtures of acetic acid containing large levels of ethanol (including its acetate esters) and, in the case of the hetereogeneous system, acetaldehyde. This poor selectivity means that, for a practical application, either a reoxidation step, and possibly a hydrolysis step to eliminate the esters, or a complicated separation system intended to sell the various fractions needs to be included in the process.

Thermodynamically, the formation of the further reduction products is favored over acetic acid. This thermodynamic predilection toward further reduction products will likely be very difficult to overcome because any catalyst used for this process must display significant hydrogenation capacity to activate carbon monoxide. The selectivity problem persists despite enormous efforts in Japan as part of their national C1 program. Until the selectivity problem is overcome, practical application of this technology will not occur.

A. Homogeneous Catalyst Systems

The homogeneous Ru-Co system has been described in detail by Texaco [2-4] and Mitsui-Toatsu Chemicals [5]. These processes use molten quaternary phosphonium salts as solvent, with the best-performing (highest acetic acid selectivity) salts having bromides as the counterion. The systems inevitably produce copious quantities of methane, carbon dioxide, ethyl acetate, methyl acetate, and propyl acetate as by-products.

The principle behind the mixed Ru-Co system is to use ruthenium to induce methanol formation and to use the added cobalt to induce selectivity to carbonylation processes. This principle allows the generation of methanol, but the homologation of methanol and its esters by ruthenium and ruthenium-cobalt catalysts in these molten phoshonium salts has been well documented by the same authors [6]. Therefore, selectivity will be a persistent problem. The cited authors have done an admirable job of optimizing all parameters, but selectivities remain unacceptable for commercial applications.

B. Hetereogeneous Systems

Recent work on the formation of acetic acid from synthesis gas using hetereogeneous catalysts has focused on the use of rhodium as the catalytically active metal. In the mid-1970s, Bhasin and O'Connor at Union Carbide reported that Rh/SiO_2 catalysts were more selective for the formation of C_2 oxygenates from synthesis gas than comparable catalysts containing Ru, Ir, Pd, Pt, Cu, or Co [7]. In subsequent work, it was reported that addition of Mn to the Rh/SiO_2 catalyst resulted in a significant improvement in the rate of C_2 oxygenate formation and a shift in selectivity from ethanol to acetaldehyde [8]. (See Table 1.) The selectivity to acetic acid was not altered by the addition of Mn. Other promoters, such as Fe, result in a shift in selectivity to ethanol but do not increase the rate of oxygenate production. Alkali has also been shown to be effective in lowering the rate of methane by-product formation in the Rh system [11].

The promotional effects of Mn on supported Rh catalysts has been the subject of several studies [12–14]. The general conclusion appears to be

Table 1 Rh/SiO_2-Based Catalysts for Production of C_2 Oxygenates from Synthesis Gas[a]

Catalyst composition	Carbon efficiency %[b]					C_2 rate[d]	Ref.
	MeOH	EtOH	AcH	AcOH	HCs[c]		
1.56% Rh, 0.31% Mn	NR	49.	107.	40.	178.	67.	8
2.5% Rh	3.07	94.	58.	120.	151.	9.4	9
2.5% Rh, 0.8% Mn	0	30.3	138.	116.	138.	85.	9
2.5% Rh, 0.5% Fe	155.	107.	0.9	8.0	169.	NR	10

[a]300°C, 6.89 bar, $H_2/CO = 1$.
[b]Carbon efficiency is based on CO converted to products other than CO_2.
[c]Hydrocarbons.
[d]Rate in $g/m^3/h$.

that addition of a poorly reducible metal oxide, such as MnO, results in a stabilization of partially oxidized Rh sites on the catalyst surface. Presumably the stabilization of Rh(I) surface sites is required for the formation of Rh-CO$_x$ species analogous to those identified in homogeneous carbonylations with Rh. Arakawa et al. recently reported that the rate of acetic acid formation can be correlated with the presence of sites that bind CO in a bridging fashion, and that sites that bind CO in a linear fashion result in ethanol formation [14].

Following the initial reports by Union Carbide, a substantial research effort was begun in Japan under the "C$_1$ Chemistry Project," a National Research and Development Program of the Agency of Industrial Science and Technology, Ministry of International Trade and Industry [15]. Most of this work has focused on adding additional promotors to the Rh-Mn/SiO$_2$ system to enhance selectivity. For example, LiCl and KCl have been shown to reduce the rate of methane formation when added to a Rh/SiO$_2$ catalyst, while increasing the selectivity to higher hydrocarbons [8]. On the other hand, addition of CrCl$_3$ or ZrCl$_4$ to Rh/SiO$_2$ is reported to decrease the formation of higher hydrocarbons. When Zr and Li were both added to a Rh-Mn/SiO$_2$ catalyst, the resulting catalyst produced acetic acid at >63% selectivity when operated at 300°C, 100 kg/cm^2, and a 9/1 CO/H$_2$ feed ratio. A significant effect of CO/H$_2$ feed ratio on acetic acid selectivity was noted in this study, with high CO/H$_2$ ratios being preferred.

Despite the great effort directed toward development of the supported Rh-based catalyst systems, selectivity to acetic acid (or for that matter C$_2$ oxygenates as a group) has remained much too low for practical applications. Coupled with the high cost of Rh (ca. $4000–5000/tr. oz.) and the enormous success of the homogeneous methanol carbonylation process, it is almost certain that this process will not be commercialized in the foreseeable future.

III. METHYL FORMATE REARRANGEMENT

Methanol can also be converted to acetic acid by a two-step process in which methanol is carbonylated in the presence of a basic catalyst, preferably a methoxide salt, with subsequent rearrangement of the methyl formate to acetic acid [16]. This sequence of events is shown in the following equations.

$$\text{MeOH} + \text{CO} \xrightarrow{\text{base}} \text{HCO}_2\text{Me} \tag{1}$$

$$\text{HCO}_2\text{Me} \longrightarrow \text{AcOH} \tag{2}$$

The key rearrangement step was first described in 1929 [17]. However, the reaction generated very little interest until nearly 50 years later, when a rapid expansion of the technology occurred. The reaction has been reported to be catalyzed by Rh [18–28], Ir [18,29–32], Pd [18,29,33], Ru [18,29, 34,35], Ni [36–39], Co [18,40], Hg [40,41], Fe [40], and Re on carbon [42]. In addition, the reaction has been reported to be catalyzed by HF-treated carbon [43] and a mixed Pd-Ni system [44].

Although there are extensive reports involving the observation of this reaction, most are recorded in the patent literature, and details regarding the process are limited. With the exception of the Re on carbon and HF-treated carbon catalysts, and possibly the Hg catalysts, which likely form HgI in situ, all the catalysts are homogeneous catalysts. All require the addition of methyl iodide (or its precursor) and are preferably performed in the presence of carbon monoxide.

The catalysts appear to function best with the addition of salts or salt precursors, such as amines, phosphines, or sulfides, which rapidly form quaternary salts under reaction conditions. The iodide is likely required for efficient methyl group activation. The addition of iodide salts or their precursors probably accelerates this activation in the same manner as it accelerates the activation of the methyl group of methyl acetate in the much better studied acetic anhydride process, which will be described later in this book.

Most of the catalytic systems are similar to those described for direct methanol carbonylation and are probably related to the acetic acid process. Mechanisms for the Rh [19,20] and Ir [30] catalyzed systems have been discussed and tend to reflect this similarity.

The first mechanistic proposal, involving the Rh-MeI catalyst system, proposes the initial decomposition of methyl formate to methanol and carbon monoxide. These components then would be expected to undergo a normal Rh-catalyzed methanol carbonylation, as described in some detail earlier. The evolution of carbon monoxide and its consumption are in balance, and no net carbon monoxide consumption ensues.

A similar mechanism for the very active Li-Rh-MeI system has been proposed by Shreck et al. [20], which differs in the point at which the formyl group dissociates to generate carbon monoxide. The mechanism is similar to the mechanism described by Eastman for the carbonylation of methyl acetate to acetic anhydride, which will be described in complete detail later in this book, except lithium formate is invoked in place of lithium acetate. Acetic acid is generated by the decomposition of an acetic-formic anhydride. The mechanism is recreated in somewhat more detail below (see Figure 1).

It appears that the formyl decomposition to carbon monoxide and either methanol or product acetic acid are the most likely routes for this rearrangement. However, Pruett and Kacmarcik [30] proposed a different

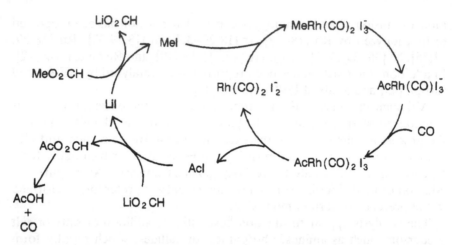

Figure 1 Proposed mechanism for the LiI-Rh–catalyzed methyl formate rearrangement.

mechanism for their Ir system involving a rather complex rearrangement of a hypothetical MeIr(O₂CH)I(ligand) complex to a hypothetical (H)(Me)Ir(CO₂)I(ligand) complex. This mechanism now appears unlikely, and the Ir process likely undergoes decomposition to carbon monoxide and decomposition to either methanol or acetic acid, as proposed for the Rh system.

Roper and co-workers (18) compared several of the transition metal catalysts and found the following reactivity order:

Ir > Rh > Ru > Pd > Co

This study did not incorporate the accelerating effect demonstrated by the addition of iodide salts or their precursors, and the presence of these salts might well alter the relative reactivities of these systems. The relative rate of the remaining catalysts is unknown under similar conditions.

Industrially, it is unlikely that the methyl formate rearrangement will ever be a significant contributor to the acetic acid supply because it converts the same starting material, methanol, to acetic acid in two steps, rather than one. Other circumstances, such as easier processing, particularly with respect to the removal of iodine cocatalysts, or the use of a cheaper catalyst, would need to intervene to justify the significant economic penalty incurred by converting to two steps.

IV. CONCLUSIONS

At this time, it appears that the methanol carbonylation process, which represents a two-step process [Eq. (3) and (4)] for the generation of acetic acid from synthesis gas, is firmly entrenched as the method of choice. For the direct generation of acetic acid from synthesis gas to be competitive, significant developments in selectivity must occur. The difficulty of achieving this development is becoming increasingly apparent, and the achievement appears to be far in the future, if it is to occur at all. The longer methyl formate rearrangement process must demonstrate some significant processing or catalyst pricing advantage if it is even to assume a role in the commercial generation of acetic acid.

$$CO + 2H_2 \longrightarrow MeOH \tag{3}$$

$$MeOH + CO \longrightarrow AcOH \tag{4}$$

REFERENCES AND FOOTNOTES

1. (a) Sabatier, P., and Senderens, J. D., *C. D. Hebd. Seance, Acad. Sci.*, *134*, 514 (1902). (b) Sabatier, P., and Senderens, J. D., *C. D. Hebd. Seance, Acad. Sci.*, *134*, 680 (1902).
2. Knifton, J. F., *J. Catal*, *96*, 439 (1985), and references cited therein.
3. Knifton, J. F., Grigsby, R. A., Jr., and Lin, J. J., *Organometallics*, *3*, 62 (1984).
4. Knifton, J. F., *ACS Symp. Ser.*, *328*, 98 (1987).
5. Ono, H., Hashimoto, M., Fujiwara, K., Sugiyama, E., and Yoshida, K., *J. Organomet. Chem.*, *331*, 387 (1987).
6. See Ref. 2 and references cited therein.
7. Bhasin, M. M., and O'Connor, G. L., G.B. Patent 1,501,892 (1978).
8. Ellgen, P. C., and Bhasin, M. M., U.S. Patent 4,014,913 (1977).
9. Ellgen, P. C., and Bhasin, M. M., U.S. Patent 4,096,164 (1978).
10. Bhasin, M. M., Bartley, W. J., Ellgen, P. C., and Wilson, T. P., *J. Catal.*, *54*, 120 (1978).
11. Bartley, W. J., Wilson, T. P., and Ellgen, P. C., U.S. Patent 4,235,798 (1980).
12. Wilson, T. P., Kasai, P. H., and Ellgen, P. C., *J. Catal.*, *69*, 193 (1981).
13. Van Den Berg, F. G. A., Glezer, J. H. E., and Sachtler, W. H. M., *J. Catal.*, *93*, 340 (1985).
14. Arakawa, H., Hanaoka, T., Takeuchi, K., Matsuzaki, T., and Sugi, Y., *Proc. Int. Congr. Catal.*, *9th*, 602 (1988).
15. As of the end of 1989, there were more than 50 patent references to modifications of the Rh system, primarily limited to the Japanese patent literature.
16. For a review of the synthesis of methyl formate and the rearrangement of methyl formate, see: Lee, J. S., Kim, J. C., and Kim, Y. G., *Appl. Catal.*, *57*, 1 (1990).
17. Dreyfus, H., U.S. Patent 1,697,109 (1929).
18. Roper, M., Elvevoll, E. O., and Leutgendorf, M., *Erdoel Kohle, Erdgas, Petrochem.*, *38*, 38 (1985).
19. Bryant, F. J., Johnson, W. R., and Singleton, T. C., *Prepr., Div. Pet. Chem., Am. Chem. Soc.*, *18*, 193 (1973).

20. Schreck, D. J., Busby, D. C., and Wegman, R. W., *J. Mol. Catal.*, *47*, 117 (1988).
21. Wegman, R. W., Busby, D. J., and Schreck, D. J., European Patent Appl. 146,823 (1985).
22. Wegman, R. W., PCT Int. Patent Appl. WO 86/889 (1986).
23. Hoeg, H. U., and Bub, G., Ger. Offen. 3,333,317 (1985).
24. Hoeg, H. U., and Bub, G., Ger. Offen. 3,236,351 (1984).
25. Drent, E., European Patent Appl. 118,151 (1984).
26. Wakamatsu, H., Sato, J., and Hamaoka, T., Ger. Offen. 2,109,025 (1971).
27. Drury, D. J., and Hamlin, J. E., European Patent Appl. 109,212 (1984).
28. Kojima, H., and Fujiwa, T., Ger. Offen. 3,512,246 (1985).
29. Mitsubishi Chemical Co. Ltd., JP 81/83439 (1981).
30. Pruett, R. L., and Kacmarcik, R. T., *Organometallics*, *1*, 1693 (1982).
31. Pruett, R. L., European Patent Appl. 45,637 (1982).
32. BP Chemicals, Ltd., JP 85/41,633 (1985).
33. Mitsubishi Chemical Industries Co. Ltd., JP 81/22745 (1981).
34. Leutgendorf, M., Elvevoll, E. O., and Roeper, M., *J. Organomet. Chem.*, *289*, 97 (1985).
35. Braca, G., Datillo, B., Guainai, G., Sbrana, G., and Valentini, G., *Actas Simp. Iberoam. Catal.*, *9th*, *1*, 416 (1984), (CA 102(4):26812r).
36. Rizkalla, N., Ger. Offen. 3,335,694 (1984).
37. Halcon SD Group, Inc., Neth. Appl. 82/2188 (1982).
38. Mitsubishi Chemical Industries Co. Ltd., JP 81/73040 (1981).
39. Cheong, M., Lee, S. H., Kim, J. C., Lee, J. S., and Kim, Y. G., *Chem. Commun.*, 661 (1990).
40. Kuraishi, M., Igarashi, T., Isogai, N., and Igasaki, T., JP 75/16773 (1975).
41. Isogai, N., JP 73/35053 (1973).
42. Teranishi, K., Shimizu, T., and Nakamata, T., JP 76/65703 (1976).
43. Wakamatsu, H., Sato, J., and Hamaoka, T., Ger. Offen. 2,240,778 (1973).
44. Drent, E., European Patent Appl. 134,601 (1985).

6

Production Economics

Peter N. Lodal
Eastman Chemical Company, Kingsport, Tennessee

I. INTRODUCTION

This chapter provides a brief discussion of acetic acid and acetic anhydride historical costs, as well as a discussion of process economics for acetic acid produced from partial oxidation of acetaldehyde (natural gas route) and via rhodium-catalyzed carbonylation of methanol (Monsanto process).

Additional discussion on the cost of separation of acetic acid and water is also provided.

II. ACETIC ACID PRICE AND VOLUME INFORMATION

Table 1 [1] shows volume estimates for production and consumption of acetic acid both worldwide and in the United States. Note the relative change away from oxidation-produced acetic acid toward carbonylated production since 1978.

Since its production by synthetic means early in this century, the price of acetic acid has tracked natural gas prices relatively closely, as shown in Figure 1 [2–4]. The advent of the Monsanto carbonylation process in the mid-1970s began to decouple this arrangement, and various periods of over- and undersupply have contributed to an increased variability in the pricing. However, since acetic acid is a commodity chemical (34th in U.S. output

Table 1 World and U.S. Supply and Demand for Acetic Acid in 1987, with Forecasts for 1992

World supply/demand for acetic acid

	United States		Western Europe		Japan		Other[a]		Total	
	1987	1992	1987	1992	1987	1992	1987	1992	1987	1992
(millions of pounds)										
Capacity	3391	3616	2584	3183	1219	1296	3097	3913	10,291	12,008
Production	3228	3582	2410	2590	809	783	2873	3419	9,319	10,375
Imports	35	20	Net 70	Net 154	110	110	Net 82	Net 165	974	1,093
Exports	102	100			86	44			999	1,168
Consumption	3266	3610	2354	2560	831	851	2939	3461	9,390	10,482
(thousands of metric tons)										
Capacity	1538	1640	1172	1444	553	588	1405	1775	4668	5447
Production	1464	1625	1093	1175	367	355	1303	1551	4227	4706
Imports	16	9	Net 32	Net 70	50	50	Net 37	Net 75	442	496
Exports	146	45			39	20			453	530
Consumption	1481	1637	1068	1161	377	386	1333	1571	4259	4755

U.S. capacity for acetic acid by major process

	Methanol carbonylation		Acetaldehyde[b]		n-Butane liquid phase oxidation		Other[c]		Total	
	Millions of pounds	Percent	Millions of Pounds	Percent	Millions of pounds	Percent	Millions of pounds	Percent	Millions of pounds	Percent[d]
1978	515	17	1210	41	1150	39	80	3	2955	100
1980	1700	42	1125	28	1150	28	80	2	4055	100
1982	1800	44	925	23	1250	31	60	2	4035	100
1985	1890	59	550	17	550	17	195	6	3185	100
1988	2390	81	350	12	0[e]	0	200	7	2940	100

[a]Other includes Andean countries, Argentina, Southeast Asia, Australia, Brazil, Bulgaria, Canada, Chile, People's Republic of China, Czechoslovakia, Egypt, German Democratic Republic, Hungary, India, Iran, Israel, Republic of Korea, Mexico, New Zealand, Poland, Rumania, Saudi Arabia, South Africa, Taiwan, USSR, and Yugoslavia.

[b]Acetic acid derived from TPA production prior to 1988 is included under the acetaldehyde process. In 1988, no acetic acid was derived from TPA production.

[c]Coproduct acetic acid from coal gas acetic anhydride operations is included under "Other" beginning in 1985. Also included in this category is production from peracetic acid manufacture.

[d]May not equal 100% because of rounding.

[e]Hoechst Celanese's 550 million pounds of acetic acid from n-butane liquid phase oxidation resumed production in January 1989.

Source: CEH estimates in conjunction with World Petrochemicals Program, SRI International. Reprinted by permission.

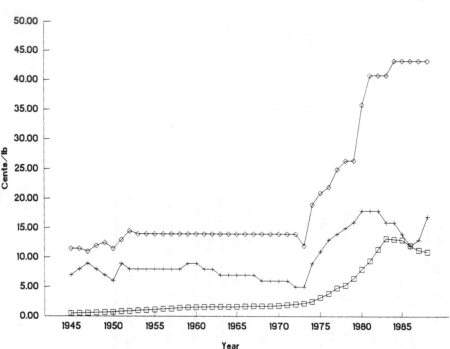

Figure 1   Historical price information 1945–1988. (□) Natural gas; (+) acetic acid; (◇) acetic anhydride.

in 1987, 36th in 1988, 34th in 1989, 33rd in 1990) [7,8], the price tends to travel in a narrowly defined range. Pricing has been further defined by the relatively large quantities of acid either coproduced in other operations (i.e., TPA manufacture) or produced as a byproduct of another acetyl-based operation (acetic anhydride manufacture, cellulose ester production).

III.   ACETIC ACID COST INFORMATION

Table 2 shows estimated cost figures for three acetic acid production methods. The oxidation processes are generally lower in direct capital cost, but have higher energy consumption and higher raw materials cost (due to higher value-added feedstocks) than carbonylation processes.

These cost figures do not reflect other charges, such as freight, sales, advertising, distribution, and administration costs (SADA), and other distributed charges. However, these charges should be independent of production method, and the differences in cost shown in Table 2 should remain the same.

Table 2 Major Cost Components for Synthetic Acid Production[a] [1,10]

	Bioderived process (optimistic)	Acetaldehyde oxidation (natural gas)	Methanol carbonylation, Rh catalyst (Monsanto)
Net variable costs	13.0	30.6	6.0
Raw materials	10.0	28.9	4.5
Utilities	3.0	2.6	1.5
By-products	0.0	(0.9)	0.0
Labor	0.3	0.3	0.5
Other costs (capital, catalyst)	10.0	4.3	4.6
Total production cost	23.0	35.2	11.1

[a]All costs in cents per pound of acid product. Based on 600 million lb/yr, plant at capacity.

Cost figures for butane and naphtha oxidation are not included because production economics for these processes are heavily dependent on valuation of other, nonacetyl by-products. If only acetic acid production is considered, these processes are not cost-competitive with carbonylation routes.

An excellent discussion of overall acetic acid production and consumption projections is given in the *Chemical Economics Handbook* [1].

IV. ACETIC ANHYDRIDE PRICE INFORMATION

Until the early 1980s, most of the world's acetic anhydride was produced via the ketene process. Although a majority of acetic anhydride is still produced by this method, commercialization of the Eastman-Halcon methyl acetate carbonylation process by Eastman Chemical Company (ECC) (United States) and BP Chemicals (United Kingdom) has made significant inroads into established ketene-based anhydride. Using a Texaco gasifier and coal instead of natural gas as a carbon source (in the case of ECC), the variable manufacturing costs for these processes has dropped significantly. Overall process economics appear to be slightly improved on a fully allocated basis and are much less subject to swings in the worldwide price of petroleum.

V. ACETIC ACID REFINING/WATER REMOVAL

A particular problem for acetic acid manufacturers and other acetyl (i.e., acetic anhydride) users is the relatively large quantity of acetic acid/water

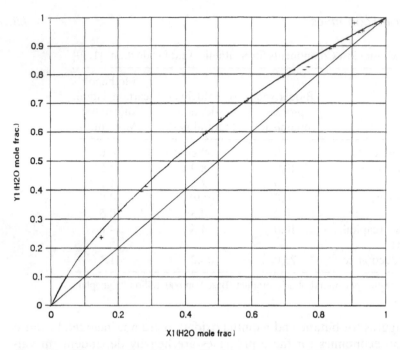

Figure 2 Water-acetic acid vapor/liquid composition (Wilson equation).

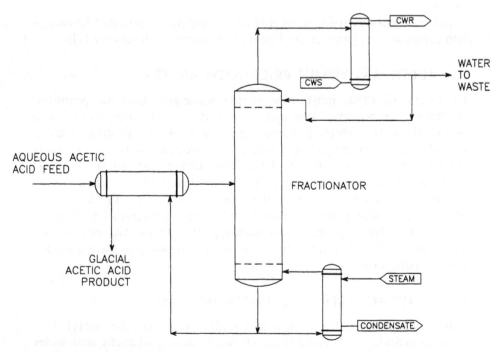

Figure 3 Flow diagram for acetic acid recovery by simple fractionation.

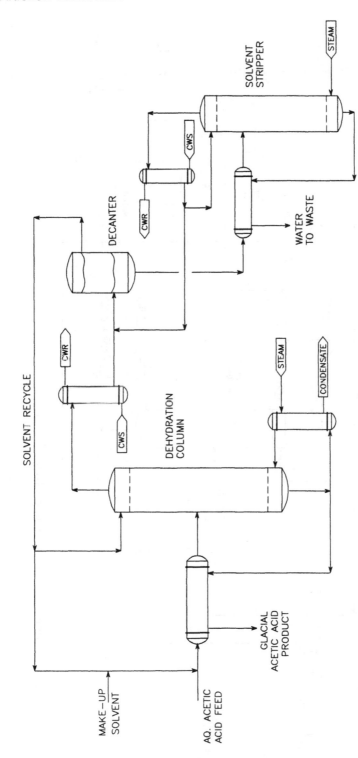

Figure 4 Acetic acid recovery by azeotropic distillator.

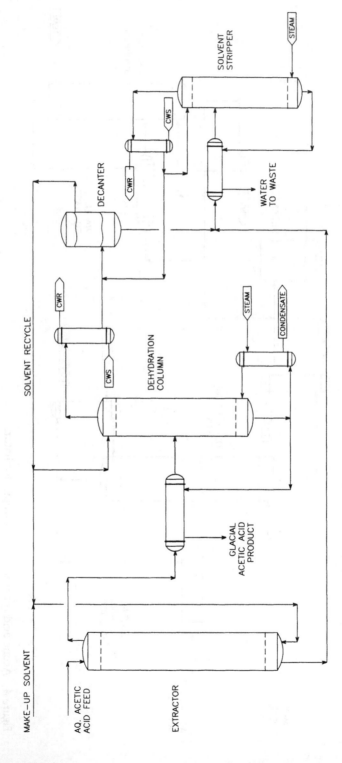

Figure 5   Solvent extraction with low boiling solvent.

mixture that must be rerefined to glacial acetic acid, either for sale or to be recycled back to other acetyl forms. Although acetic acid and water do not form a minimum boiling azetotrope, the acid rich end of the X-Y plot shows a marked pinch, as shown in Figure 2 [5,6], making conventional distillation extremely energy intensive to produce glacial (>99.7 wt% acid) acid, while at the same time removing pure or nearly pure water from the distillate.

Most acid recovery schemes, therefore, involve some type of third component, azeotropic distillation, extraction, or combination thereof, to minimize the energy consumed in refining the recycled acid. Many processes are also heat-integrated to minimize sensible heat loads by reusing either product heat or overhead condensing energy to minimize direct steam input. Typical extraction agents include ethyl acetate, *n*-propyl and isopropyl acetate, isoamyl ketone, methyl isobutyl ketone (MIBK), and other solvents, primarily ketones and esters. Azeotroping agents used, in addition to the above solvents, include ethyl ether, isopropanol, diethyl ketone (DEK), and benzene.

Effective use of these schemes can decrease the steam usage for generating a pound of acetic acid from as high as 5:1 for a straight, nonazeotropic distillation scheme down to 1.7:1 for a highly heat-integrated extraction/azeotropic distillation scheme. Shown in Figures 3, 4, and 5 are typical arrangements for simple distillation (Fig. 3), azeotropic distillation (dehydration) (Fig. 4), and extraction/distillation (Fig. 5) [9].

REFERENCES

1. CEH Marketing Research Report 602.5020, in *Chemical Economics Handbook*, SRI International, Menlo Park, CA. 1988, pp. 602.5020C, E.
2. *Natural Gas Annual*, Vol. II, Office of Oil and Gas, U.S. Department of Energy, Washington, D.C. 1988, pp. 188–191.
3. *Chemical Economics Handbook*, SRI International, Menlo Park, CA. 1988, p. 602.5021B.
4. *Chemical Economics Handbook*, SRI International, Menlo Park, CA. 1990, 603.50000I.
5. Bonauguri, E., Carpani, Z., and Dall Orto, D., *Chim. Ind.*, 28, 768 (1956).
6. Gmehling, J., Onken, U., and Arlt, W., *DECHEMA*, Vol. 1, Part 1a, 89–109 (1981).
7. *C&E News*, 11–12 (April 10, 1989).
8. *C&E News*, 13–14 (April 8, 1991).
9. Brown, W. V., *Chem. Eng. Prog.*, 59(10), 65–68 (1963).
10. Busche, R. M., in *Biotechnology Applications and Research* (P. N. Cheremisinoff and R. P. Ouellette, eds), Technomic Publishing, Lancaster, PA, 1985.

II

Thermophysical Properties of Acetic Acid and Its Aqueous Solutions

7
Acetic Acid

William H. Seaton*

Eastman Chemical Company, Kingsport, Tennessee

I. INTRODUCTION

A. The Need for Mathematical Models of Design Data

In 1992, Eastman Chemical Company produced about 11% of the 3.8 billion pounds of acetic acid that were produced in the United States and it contributed to the 3.9% annual growth rate of acetic acid production. Because of the great industrial importance of acetic acid, the literature contains a large amount of engineering design data related to acetic acid and its aqueous solutions; however, the publications are in several different languages, they often appear in journals that are not readily available, and they usually contain only a small number of measured data points for limited ranges of pressure, temperature, and concentration. To extend the usefulness of several important types of design data to engineers, the objectives of this project were to acquire the available data, to determine as well as possible those data sets which are most reliable in cases where there are conflicting data, and then to reduce the data to mathematical models to facilitate computerized design methods. The mathematical models chosen for this work were based on theoretically sound equations or well-established empirical equations so that reliable interpolation and extrap-

---

*Private consultant, retired from Eastman Chemical Company.

olation of the available data can be accomplished with a high level of confidence. In most cases, the ranges of validity and the accuracies of the equations have been defined.

B. On the Selection of Mathematical Models

In the selection of mathematical models for this chapter, preference was always given to those models which have a theoretical basis so that extrapolation beyond the range of available data can be done with a minimum risk of large error. When models based on theory were not available, preference was given to models that have been established as reliable by extensive use in the literature.

Preference was always given to published models when they were available, and the criterion of a fully successful model was that it could yield estimated data that were as accurate as the data on which its parameters were based. Although this degree of perfection was not achieved in all cases, all the equations in this chapter are believed to be sufficiently accurate for most engineering purposes. The ranges of validity and the accuracies of the equations are indicated either by statements in the text or by comparison of the best available data to estimated data, with the range of the estimated data indicating the recommended range of the equation.

C. On the Division of Properties in This Chapter

According to *Webster's Unabridged Dictionary*, any property that is not mental or spiritual is a physical property. By this definition, every property mentioned in this report is a physical property. Nevertheless, it is desirable to classify the properties somewhat more selectively so that users of this book will be able to locate the data they need with less effort. For this reason, the data have been classified into the following categories: equations of state, phase equilibrium data, transport properties, and physical properties. It is expected that the people who need the data in this report will recognize when the property they seek falls into any of the first three of these categories. Any other property, if available here, will therefore be in the fourth category. Because of its qualitative nature, the discussion about the molecular structures of acetic acid has been given a separate section.

II. EQUATIONS OF STATE FOR ACETIC ACID

A. Equations of State for Acetic Acid in the Gas State

Johnson and Nash [1] provide the equation of state for acetic acid in the gas state, which is presented here as Eq. (1)–(5). These equations were derived from the assumption that acetic acid exists in the gas state as a mixture of monomers, dimers, and trimers and that the components of the mixture are ideal gases.

$$z = P/P_I \tag{1}$$

$$K_2 = \exp(A/T + B) \tag{2}$$

$$K_3 = \exp(C/T + D) \tag{3}$$

$$\Phi P = \Phi p_{B1} + K_2(\Phi p_{B1})^2 + K_3(\Phi p_{B1})^3 \tag{4}$$

$$\Phi P_I = \Phi p_{B1} + 2K_2(\Phi p_{B1})^2 + 3K_3(\Phi p_{B1})^3 \tag{5}$$

where P = pressure, kPa

        P_I = ideal pressure of acetic acid, that is, the pressure that acetic acid would have if it were all in the form of monomers, kPa

        P_{B1} = partial pressure of acetic acid monomer, kPa

        T = temperature, K

        z = compressibility factor

        A = 6949.202

        B = −16.6412

        C = 11420.82

        D = −29.6306

        Φ = 9.8692 × 10⁻³

For 288 data points taken from the publications of Nicholls [2], MacDougall [3], Ritter and Simons [4], Johnson and Nash [1], and Ramsay and Young as presented by Timmermans [5], Eq. (1) had a mean error of 1.10% with a standard deviation of 2.43%. Typical experimental values for the compressibility factor of acetic acid vapor are compared in Table 1 with values estimated by means of Eq. (1).

Freeman and Wilson [6] developed an equation of state for water/acetic acid mixtures in the gas state. For their model, they assumed that acetic acid consists of a mixture of monomers and dimers at equilibrium in the gas state. They further assumed that the mixture of monomers and dimers deviates from ideal gas behavior because of physical interactions. Since their equation of state was developed for water/acetic acid mixtures, it is presented as Eq. (16) in Chapter 8. By specifying that the composition is 100% acetic acid, their equation of state can be compared to Eq. (1), as shown by the data in Table 1.

Lawrenson et al. [7] present experimentally measured data on the density of acetic acid in the gas state at temperatures between 544 and 603 K and at pressures between 2308 and 10603 kPa. However, none of their

Table 1 Compressibility Factor Data for Acetic Acid in the Gas State

Temperature (K)	Pressure (kPa)	Compressibility factor		
		Experimental	Eq. (1)	Eq. (16)[a]
298.20	0.400	0.604	0.633	0.587
298.20	2.000	0.541	0.554	0.538
303.20	0.533	0.616	0.639	0.596
303.20	2.666	0.546	0.557	0.542
313.20	0.533	0.661	0.695	0.649
313.20	3.066	0.574	0.579	0.563
357.10	5.932	0.710	0.719	0.712
373.15	6.140	0.718	0.797	0.795
373.15	27.300	0.664	0.655	0.659
373.15	53.710	0.591	0.603	0.609
378.50	5.932	0.814	0.826	0.825
413.15	11.790	0.874	0.899	0.901
413.15	95.350	0.706	0.696	0.702
413.15	182.430	0.614	0.635	0.636
424.30	12.305	0.918	0.925	0.926
457.90	153.626	0.821	0.812	0.805
463.15	118.590	0.900	0.855	0.847
463.15	299.920	0.757	0.756	0.743
463.15	588.120	0.636	0.682	0.650
466.90	100.071	0.877	0.881	0.873
503.15	157.200	0.909	0.925	0.907
503.15	296.470	0.859	0.877	0.851
503.15	683.960	0.756	0.791	0.747
503.15	1297.590	0.614	0.715	0.636

[a]Equation (16) of Chapter 8.
Source: Data from Refs. 1–6.

data are included in Table 1 because both Eq. (1) and Eq. (16) of Chapter 8 become unreliable at temperatures above about 500 K.

B. Equations of State for Acetic Acid in the Liquid State

Equation (6) was developed [8] as method for the estimation of the isothermal compressibility of acetic acid in the liquid state at atmospheric pressure.

$$\Phi k_T = A + B(T/100) + C(T/100)^2 \tag{6}$$

where k_T = isothermal compressibility, 1/Pa
         T = temperature, K

$$A = 157.3984 \qquad C = 28.3535$$
$$B = -105.4648 \qquad \Phi = 101.325 \times 10^9$$

Equation (6) is valid in the temperature range between 288 and 353 K. For eight data points over the specified temperature range, the equation had a standard error of 5.1×10^{-11} Pa^{-1}. This equation yields 87.75×10^{-11} Pa^{-1} at 288 K and 136.7×10^{-11} Pa^{-1} at 353 K. These values compare to 87.55×10^{-11} Pa^{-1} and 136.76×10^{-11} Pa^{-1}, respectively, reported by Tyrer [9].

Francis [10] published Eq. (7) to provide estimates of the orthobaric density of liquid acetic acid, and it is recommended for use in the temperature range between 289 and 573 K.

$$\Phi d = A - Bt - C/(D - t) \qquad (7)$$

where d = density, kg/m^3
$t = T - 273.15$
T = temperature, K
$A = 1.1275 \qquad C = 22.0$
$B = 9.1 \times 10^{-4} \qquad D = 385.0$
$\Phi = 0.001$

Francis [10] published Eq. (8) to provide estimates of the orthobaric density of liquid acetic acid at elevated temperatures, and it is recommended for use in the temperature range between 514 K and the critical temperature.

$$d = \Phi(A(T_c - T))^B + d_c \qquad (8)$$

where d = density, kg/m^3
$A = 1.18 \times 10^{-3} \qquad B = 0.3846$
$T_c = 594.45$ K $\qquad d_c = 350.6$ kg/m^3
$\Phi = 1000.0$

Some typical values estimated by means of Eq. (7) and (8) are compared in Table 2 with experimental data from Timmermans [5].

Lawrenson et al. [7] present experimentally measured data on the density of acetic acid in the liquid state at temperatures between 448 and 573 K and at pressures between 518 and 10,659 kPa. The data they measured at each temperature can be accurately correlated by means of the Tait equation, shown here as Eq. (9).

$$\frac{(V_0 - V)}{V_0} = \ln[(P + B)/(P_0 + B)]/C \qquad (9)$$

Table 2 Experimental and Estimated Orthobaric Density Data for Acetic Acid
in the Liquid State

		Density	
Temperature (K)	Experimental (kg/m³)	Estimated by Eq. (7) (kg/m³)	Estimated by Eq. (8) (kg/m³)
293.15	1049.1	1049.0	
333.15	1006.0	1005.2	
373.15	959.9	959.3	
413.15	909.1	910.3	
453.15	855.5	856.4	
493.15	794.1	794.0	
533.15	713.6	714.9	714.8
573.15	595.0	595.7	593.1
593.15	461.5		433.3
594.45	350.6		350.6

Source: Data from Ref. 5.

where V = volume $\qquad$ V_0 = volume at reference pressure
$\quad\quad\quad$ P = pressure $\qquad$ P_0 = reference pressure
$\quad\quad\quad$ B = constant, determined for each temperature
$\quad\quad\quad$ C = constant, determined for each temperature

Equation (9) is able to correlate pressure-volume data for liquids very
accurately if the constants B and C are determined from the data at each
temperature. However, the following equations were developed to extend
the data of Lawrenson et al. over the temperature range between 448 and
523 K and for pressures between the orthobaric pressure and 10,600 kPa.

$$B = a + bT + cT^2 \tag{10}$$

$$C = d \tag{11}$$

$$V_0 = e + fT + gT^2 \tag{12}$$

where T = temperature, K
$\quad\quad\quad$ $a = -228,963.0 \qquad e = 354.8062$
$\quad\quad\quad$ $b = 1322.966 \qquad\quad f = -1.316074$
$\quad\quad\quad$ $c = -1.639539 \qquad g = 0.151112 \times 10^{-2}$
$\quad\quad\quad$ $d = 7.74$

Equations (9)–(12) reproduced 45 molar volumes measured by Lawrenson et al., over the temperature and pressure ranges stated above, with
an average error of -0.02% and a standard deviation of 0.75%. For these

estimates, the reference pressures were obtained by means of Eq. (48). This suggests that Eq. (12) should yield saturated molar volumes for acetic acid. However, it was found by the author that Eq. (12) yields values for the molar volume that are consistently lower by about 2 1/2% than the values yielded by the more accurate Eq. (7) and (8). Therefore, Eq. (12) should be used only with Eq. (9).

Data from Lawrenson et al. [7] are compared in Table 3 with data estimated by means of Eq. (9)–(12).

By means of the Verschaffelt equation, Costello and Bowden [11] estimated the density of liquid acetic acid at absolute zero temperature to be 1291.0 kg/m³.

III. TRANSPORT PROPERTIES OF ACETIC ACID

A. Viscosity of Acetic Acid in the Liquid State

Thorpe and Rodger from Ref. 12 developed Eq. (13) to estimate the viscosity of acetic acid at normal pressures over the temperature range from its freezing point to its normal boiling point.

$$\mu = C/(1.0 + At + Bt^2) \tag{13}$$

Table 3 Molar Volumes of Acetic Acid in the Liquid State at Elevated Pressures and Temperatures

Temperature (K)	Pressure (kPa)	Molar volume Experimental (cm³/mol)	Eq. (9) (cm³/mol)
448.15	10659.9	65.79	66.24
448.15	3764.4	67.22	67.70
448.15	518.7	68.31	68.48
473.15	10656.2	68.13	67.86
473.15	3493.7	70.03	69.63
473.15	845.9	70.96	70.38
523.15	10641.6	74.68	75.49
523.15	3746.7	78.25	78.81
523.15	2045.6	79.54	79.82
548.15	10603.1	79.29	78.68
548.15	7155.5	81.75	81.82
548.15	3019.1	87.37	87.31

Source: Data from Ref. 7.

where μ = viscosity, mPa·s $A = 0.01826$
        $t = T - 273.15$ $B = 0.8537 \times 10^{-4}$
        T = temperature, K $C = 1.6867$

Experimental data presented by Timmermans [19] are compared in Table 4 with data estimated by means of Eq. (13).

The equation of Schonhorn [13] relating viscosity to surface tension was combined [8] with the equation relating surface tension to temperature of van der Waals [14] to derive Eq. (14). The constants for Eq. (14) were evaluated using the viscosity data for liquid acetic acid presented by Timmermans [5] and the viscosity of acetic acid at the critical point estimated by Abtiev [15].

$$\mu = \mu_c(T/T_c)^A + B/\ln(C(1.0 - T/T_c)^D) \tag{14}$$

where μ = viscosity, mPa·s
        T = temperature, K
        $\mu_c = 0.067$ mPa·s
        $T_c = 594.45$ K
        $A = 5.126$ $B = 1.4347$
        $C = 0.03941$ $D = -6.5222$

Experimental data presented by Timmermans [5] are compared in Table 4 with data estimated by means of Eq. (14).

Table 4   Experimental and Estimated Viscosity Data for Acetic Acid in the Liquid State at Orthobaric Pressures

	Viscosity		
Temperature (K)	Experimental (mPa·s)	Eq. (13) (mPa·s)	Eq. (14) (mPa·s)
288.15	1.314	1.304	1.317
304.01	1.025	1.026	1.000
321.62	0.806	0.809	0.780
330.63	0.721	0.724	0.698
357.68	0.534	0.535	0.523
376.04	0.445	0.446	0.442
385.72	0.406	0.407	0.407
523.15			0.170
548.15			0.151
594.15			0.067

Source: Data from Ref. 5.

B. Viscosity of Acetic Acid in the Gas State

Tsonopoulos and Prausnitz [16], in citing the work of several authors, showed that carboxylic acids exist in the gas state as mixtures of monomers, dimers, and trimers. They further noted that acetic acid, at pressures not significantly above atmospheric pressure, forms only tiny amounts of trimers. They also provided Eq. (15) to correlate the equilibrium constant for the dissociation of dimer to monomer.

$$\ln K = A - B/T \tag{15}$$

where K = equilibrium constant for the dissociation of dimer to monomer

   T = temperature, K

   $A = 28.149 \quad B = 6999.9$

Seaton [17] assumed that only monomers and dimers are present in acetic acid in the gas state at low pressures. He then used Eq. (15)–(18) to compute the concentrations of these two components of acetic acid in the gas state.

$$\alpha = 1/\sqrt{(1 + 4\,P/K)} \tag{16}$$

$$y_1 = 2\,\alpha/(1 + \alpha) \tag{17}$$

$$y_2 = (1 - \alpha)/(1 + \alpha) \tag{18}$$

where y_1 = concentration of monomer, mole fraction

   y_2 = concentration of dimer, mole fraction

   P = pressure, Pa

   K = equilibrium constant from Eq. (15)

With the composition of the gas state defined, Seaton then applied the equation of Chapman and Enskog, Eq. (19), to compute the viscosities of the separate components as a function of pressure and temperature. For this work, the collision integrals were computed by the equation of Neufeld, Eq. (20), and the viscosities of the mixtures were computed by the equation of Wilke, Eq. (21).

$$\mu_i = 2.669 \times 10^{-26}\, \sqrt{(M_i T)}/(\sigma_i^2 \Omega_i) \tag{19}$$

$$\Omega_i = 1.16145(T_i^*)^{0.14874} + 0.52487/\exp(0.77320\,T_i^*)$$
$$+ 2.16178/\exp(2.43787\,T_i^*) + \beta_i \tag{20}$$

$$\mu_m = \sum_{i=1}^{2} y_i \mu_i \bigg/ \left(\sum_{j=1}^{2} y_j \phi_{ij} \right) \tag{21}$$

where $\phi_{ij} = [1 + (\mu_i/\mu_j)^{1/2}(M_j/M_i)^{1/4}]^2/[8(1 + M_i/M_j)]^{1/2}$

   $\phi_{ji} = (\mu_j/\mu_i)(M_i/M_j)\phi_{ij}$

μ_i = viscosity of component i, Pa·s
μ_m = viscosity of mixture of monomer and dimer, Pa·s
T = temperature, K
$\sigma = GV_{ci}^{1/3}$ $G = 0.7671 \times 10^{-8}$
$T_i^* = T/(HT_{ci})$ $H = 0.9703$

To facilitate the calculations, Seaton used experimental data and established estimation methods to provide the following parameters.

Description	Symbol	Monomer	Dimer
Atm. boiling temperature, K	T_b	391.05	562.8
Critical temperature, K	T_c	592.71	772.9
Critical pressure, Pa	P_c	6997.6	4025.0
Critical volume, m³/mol	V_c	171.0×10^{-6}	342.0×10^{-6}
Molecular weight	M	60.05	120.1
Brokaw correction term	β	$7.3019/T$	0.0
Thermal conductivity parameter	k^*	0.013585	0.28710
Thermal conductivity parameter	n	1.2986	1.3791

Data estimated by means of Eq. (21) are compared in Table 5 to the experimental data of Timrot et al. [18].

C.   Thermal Conductivity of Acetic Acid in the Liquid State

Based on the "application of similarity methods," Usmanov and Mukhamedzyanov [19] developed Eq. (22) as a generalized correlation of the manner in which the thermal conductivities of liquids vary with their entropies.

$$k = k^*(1.0 + A(S^* - S)/R) \tag{22}$$

where k = thermal conductivity
        k^* = thermal conductivity at reference conditions
        S^* = entropy at reference conditions
        S = entropy
        R = universal gas constant
        A = constant determined by the nature of the liquid

Usmanov and Mukhamedzyanov recommend that $A = 0.028$ for nonassociated liquids and that $A = 0.015$ for associated liquids; however, their data for acetic acid are better fitted with $A = 0.027$.

Stull et al. [20] report that the entropy of acetic acid in the liquid state at 298 K is $S^* = 159.83$ J/(mol·K).

Table 5 Experimental and Estimated Viscosity Data for Acetic Acid in the Gas State

Temperature (K)	Pressure (kPa)	Viscosity Literature (μPa·s)	Eq. (21) (μPa·s)
307.1	1.82	7.50	7.52
307.1	2.53	7.44	7.46
307.1	2.84	7.43	7.45
314.1	1.72	7.77	7.81
314.1	2.53	7.72	7.72
314.1	3.34	7.63	7.67
324.6	2.53	8.13	8.15
324.6	4.15	8.01	8.02
324.6	5.07	7.97	7.98
324.6	7.50	7.91	7.90
344.1	1.82	9.18	9.20
344.1	1.93	9.16	9.17
344.1	3.34	8.90	8.95
344.1	6.89	8.69	8.70
344.1	13.17	8.51	8.51
344.1	16.41	8.47	8.45
363.1	1.82	10.07	10.14
363.1	3.75	9.81	9.85
363.1	6.08	9.59	9.64
363.1	11.86	9.34	9.36
363.1	18.24	9.15	9.20
363.1	35.06	8.94	9.00
383.3	1.82	10.90	11.04
383.3	5.27	10.65	10.70
383.3	29.08	9.94	9.94
383.3	55.73	9.64	9.68
383.3	69.00	9.57	9.60
401.0	2.33	11.48	11.67
401.0	4.86	11.38	11.51
401.0	19.35	10.96	10.97
401.0	26.95	10.79	10.81
401.0	80.05	10.26	10.30
401.0	124.22	10.07	10.12
421.8	2.23	12.23	12.41
421.8	6.28	12.15	12.27
421.8	35.67	11.71	11.68
421.8	96.26	11.17	11.18

Source: Data from Ref. 18.

Equation (44), for the heat capacity of acetic acid in the liquid state, was used [8] to derive Eq. (23), which yields the entropy of liquid acetic acid as a function of temperature at orthobaric pressures. Equation (23) is believed to be valid in the temperature range between 291 and 473 K.

$$\Phi S = A + B(T/100) + C(T/100)^2 + D(T/100)^3 \tag{23}$$

where S = entropy, J/(mol·K)
T = temperature, K
$A = -2.4442 \qquad B = 18.859$
$C = -2.2682 \qquad D = 0.17205$
$\Phi = 0.23901$

Frontas'ev and Gusakov [21] carefully determined the thermal conductivity of acetic acid in the liquid state at 298 K and found it to be $k^* = 0.159 \pm 0.00067$ W/(m·K).

The experimental data of Venart and Krishnamurthy [22] and of Jobst [23] are compared in Table 6 with data estimated by means of Eq. (22) and (23), using the indicated constants and $R = 8.3144$ J/(mol·K).

D.   Thermal Conductivity of Acetic Acid in the Gas State

Seaton [24] provided Eq. (24)–(33) to enable the estimation of thermal conductivity data for acetic acid in the gas state at pressures below 102

Table 6 Thermal Conductivity Data for Acetic Acid in the Liquid State

Temperature (K)	Thermal conductivity	
	Experimental (W/(m·K))	Eq. (22) (W/(m·K))
290.65	0.159	0.160
293.15	0.161	0.160
303.65	0.157	0.157
320.65	0.154	0.154
323.15	0.155	0.154
336.15	0.151	0.151
352.15	0.149	0.147
360.65	0.148	0.146

Source: Data from Refs. 22 and 23.

kPa and at temperatures in the range between 394 and 709 K. In these equations, subscript i represents the acetic acid monomer and subscript j represents the acetic acid dimer. The equations, parameters, and nomenclature defined in Section III.B are also used here.

$$\Gamma_i = T_{ci}^{1/6} M_i^{1/2} P_{ci}^{2/3} \tag{24}$$

$$p_i = y_i P \tag{25}$$

$$T_{ri} = T/T_{ci} \tag{26}$$

$$\theta_i = \exp(aT_{ri}) - \exp(bT_{ri}) \tag{27}$$

$$\tau_{ij} = (\Gamma_j/\Gamma_i)(\theta_i/\theta_j) \tag{28}$$

$$A_{ij} = [1 + (\tau_{ij})^{1/2} + (M_i/M_j)^{1/4}]^2 / [8(1 + M_i/M_j)]^{1/2} \tag{29}$$

$$k_i = k_i^*(T/273.2)^{ni} \tag{30}$$

$$k_F = \sum_{i=1}^{2} \left[y_i k_i \Big/ \sum_{j=1}^{2} y_j A_{ij} \right] \tag{31}$$

$$k_R = (C/T^{5/4})(p_1 p_2/(p_1 + 2p_2)^2) \tag{32}$$

$$k_M = k_F + k_R \tag{33}$$

where T = temperature, K
P = total system pressure, Pa
p_i = partial pressure of component i
k_F = nonreactive (frozen) component of the thermal conductivity, W/(m·K)
k_R = thermal conductivity due to monomer/dimer reaction, W/(m·K)
k_M = thermal conductivity of acetic acid in the gas state, W/(m·K)
$a = 0.0464$ $b = -0.2412$ $C = 580.44 \text{ wT}^{1/4}/\text{m}$

The data of Mashirov and Tarzimanov [25] and of Timrot and Makhrov [26] are compared in Table 7 with data estimated by means of Eq. [33].

E. Miscellaneous Transport Properties of Acetic Acid

Property	State	Value and units	Ref.
Viscosity at the critical point[a]	g/l	0.067 mPa·s	15

[a]Estimated from low-temperature viscosity data by means of a specially developed method.

Table 7   Thermal Conductivity of Acetic Acid in the Real Gas State

		Thermal conductivity	
Temperature (K)	Pressure (kPa)	Literature (mW/(m·K))	Eq. (33) (mW/(m·K))
298.5	0.47	69.34	69.98
328.1	0.47	70.65	75.62
351.6	0.47	38.98	50.51
328.1	1.01	83.74	78.51
337.5	0.96	80.10	75.96
366.3	0.96	44.82	49.02
396.8	3.87	43.40	47.90
337.5	10.13	55.01	65.54
351.8	9.93	70.87	73.08
366.3	9.87	79.70	76.38
405.5	10.40	59.12	58.39
351.8	19.07	59.08	67.73
366.3	18.80	72.23	73.96
383.8	22.26	78.30	75.99
405.1	20.53	71.58	69.00
366.3	30.00	65.04	70.74
366.3	32.40	63.99	70.16
383.8	31.33	78.48	75.50
383.8	40.26	76.43	74.55
405.1	42.26	76.96	75.06
394.1	56.26	75.10	75.21
424.5	100.53	76.69	75.42
468.5	99.99	57.06	58.57
506.5	99.99	43.28	44.52
546.8	99.99	39.75	39.13
709.5	100.25	41.97	47.20

Source: Data from Refs. 25 and 26.

IV.   PHYSICAL PROPERTIES OF ACETIC ACID

A.   Miscellaneous Physical Properties of Acetic Acid

Property	State	Value and units	Ref.
Heat of fusion	l/s	195.31 kJ/kg	30
Heat of combustion at 298 K	l	874.20 ± 0.21 kJ/mol	58
Cryoscopic constant		3.59 K	58
Melting temperature	l/s	289.76 K	3

Triple-point temperature	v/l/s	289.83 K	39
Atmospheric boiling temperature	l/v	391.10 K	48
Electrolyte conductivity at 298 K	l	112.0 pS/m	59
Dipole moment	g	5.804×10^{-30} C·m	60
Dielectric constant at 291 K	l	6.194	58
Ebullioscopic constant		2.530 K	58
Density in the solid state at the melting point	s	1265.85 kg/m³	5

B. Heat Capacity of Acetic Acid Monomer in the Ideal Gas State

Equation (34) was published by Kobe and Crawford [27] as a means for estimating the heat capacity of the acetic acid monomer in the ideal gas state.

$$\Phi C_{p1}^{0} = A + BT + CT^2 + DT^3 \tag{34}$$

where C_{p1}^{0} = heat capacity, J/(mol·K)
T = temperature, K
$A = 2.0142$ $\quad C = -34.0880 \times 10^{-6}$
$B = 56.0646 \times 10^{-3}$ $\quad D = 8.0208 \times 10^{-9}$
$\Phi = 0.23901$

In the temperature range between 300 and 1500 K, Eq. (34) is subject to a maximum error of 0.26% and an average error of 0.13%.

C. Heat Capacity of Acetic Acid Dimer in the Ideal Gas State

Equation (34) was developed [8] by regression analysis from the data of Weltner [28] as a means of estimating the heat capacity of the acetic acid dimer in the ideal gas state. The constants obtained were as follows:

$A = 6.5554$ $\quad C = -6.8210 \times 10^{-5}$
$B = 11.4216 \times 10^{-2}$ $\quad D = 1.5590 \times 10^{-8}$
$\Phi = 0.23901$

In the temperature range between 300 and 1500 K, Eq. (34), with the above constants, yielded, for eight data points, a mean error (with 95% confidence limits) of $-0.06 \pm 0.26\%$.

D. Heat Capacity of Acetic Acid Trimer in the Ideal Gas State

Weltner [28] provided Eq. (35) to enable the estimation of heat capacity data for the acetic acid trimer in the ideal gas state.

$$C_{p3}^{0} = 3(C_{p1} - 33.257) + 128.04 \tag{35}$$

where C_{p3}^0 = heat capacity of acetic acid trimer, J/(mol·K)
$\quad\quad\quad$ C_{p1}^0 = heat capacity of acetic acid monomer, J/(mol·K)

E. Heat Capacity of Acetic Acid in the Real Gas State

Tsonopoulos and Prausnitz [16] used gas density data to establish the fact that pure acetic acid exists in the real gas state as a mixture of monomers, dimers, and trimers, as indicated by Eq. (36) and (37). In these equations, the monomer, dimer, and trimer are assumed to be ideal gases.

$$2A_1 \rightleftharpoons A_2 \tag{36}$$

$$3A_1 \rightleftharpoons A_3 \tag{37}$$

Seaton [8] recognized that the real gas heat capacity of acetic acid at a specific pressure would be the sum of contributions of the monomer, dimer, and trimer as ideal gases plus the contributions of the changes of enthalpy-of-reaction with temperature for the two reversible reactions described by Eq. (36) and (37) plus a contribution from the change in volume of the system with the changing number of moles. This relationship is defined by Eq. (38).

$$C_p = n_1 C_{p1}^0 + n_2 C_{p2}^0 + n_3 C_{p3}^0 + H_2(dn_2/dT) + H_3(dn_3/dT)$$
$$+ RT(dn_1/dT + dn_2/dT + dn_3 dT) \tag{38}$$

where $\quad C_p$ = heat capacity of acetic acid in the real gas state, J/mol
$\quad\quad\quad$ C_{p1}^0 = heat capacity of acetic acid monomer, J/mol
$\quad\quad\quad$ C_{p2}^0 = heat capacity of acetic acid dimer, J/mol
$\quad\quad\quad$ C_{p3}^0 = heat capacity of acetic acid trimer, J/mol
$\quad\quad\quad$ H_2 = heat of reaction to form 1 mol of dimer from 2 mol of monomer, J/mol
$\quad\quad\quad$ H_3 = heat of reaction to form 1 mol of trimer from 3 mol of monomer, J/mol
$\quad\quad\quad$ n_1 = number of moles of monomer in the monomer, dimer, trimer mixture
$\quad\quad\quad$ n_2 = number of moles of dimer in the monomer, dimer, trimer mixture
$\quad\quad\quad$ n_3 = number of moles of trimer in the monomer, dimer, trimer mixture
$\quad\quad\quad$ R = gas constant, J/(mol·K)
$\quad\quad\quad$ T = temperature, K

Since this is a reactive system, the partial pressures of the dimer and of the trimer are related to the partial pressure of the monomer, as indicated in Eq. (39), and the total pressure of the system is related to the sum of the partial pressures, as indicated by Eq. (40).

$$p_i = K_i p_1^i \tag{39}$$

$$P = \Sigma p_i \tag{40}$$

The chemical equilibrium constants of Eq. (39) are related to temperature by means of Eq. (41).

$$K_i = \exp(A + B/T) \tag{41}$$

Substituting Eq. (39) and (41) into Eq. (40), an equation is obtained that relates the partial pressure of the monomer, p_1, to the total pressure and the temperature. This equation can be solved to obtain p_1. It can also be differentiated to obtain dp_1/dT.

Since the real gas is assumed to be a mixture of ideal gases, the number of moles of dimer and of trimer can be related to the number of moles of monomer, as indicated by Eq. (42).

$$n_i = (p_i/p_1)n_1 \tag{42}$$

If a total weight of M_1 is taken as the basis, then a material balance yields Eq. (43).

$$n_1 = 1/(1 + 2p_2/p_1 + 3p_3/p_1) \tag{43}$$

where P = total system pressure, Pa
  p_1 = partial pressure of component i, Pa
  T = temperature, K
  n_i = number of moles of component i in M_1 (0.06005 kg) of mass

This equation can then be differentiated at constant pressure to obtain dn_1/dT. Equation (43) can now be used with Eq. (42) to obtain the dn_2/dT and dn_3/dT, which are needed for Eq. (38).

To facilitate the use of Eq. (38), the following parameters are provided.

	A	B	Heat of reaction to form 1.0 mol, J/mol
Dimer	− 28.6212	7,040.956	− 58,540.6
Trimer	− 52.2823	11,342.62	− 94,305.9

The data of Weltner (30) are compared in Table 8 with data estimated by means of Eq. (38).

Table 8 Heat Capacity Data for Acetic Acid in the Real Gas State

| Temperature (K) | Pressure (kPa) | Heat capacity | |
		Experimental $(J/(mol \cdot K))$[a]	Eq. (38) $(J/(mol \cdot K))$
368.3	33.2	300.4	303.7
396.4	67.6	327.6	325.8
397.5	101.3	302.1	298.9
399.1	67.6	332.6	331.0
399.2	33.2	365.3	359.3
419.0	33.2	323.8	317.4
420.1	67.6	341.8	340.4
421.8	101.3	332.2	335.3
435.8	33.2	261.5	255.2
436.0	67.6	307.5	308.9
443.6	101.3	304.2	310.7
470.4	67.6	201.3	202.7
470.9	101.3	229.7	233.4
479.7	33.2	141.0	141.1
509.1	33.2	117.6	115.7

[a]Heat capacity per 0.06005 kg.
Source: Data from Ref. 28.

F. Heat Capacity of Acetic Acid in the Liquid State

Available heat capacity data for pure liquid acetic acid fall in the temperature range between 291 and 353 K. To estimate heat capacity data at temperatures beyond this range, use was made of the fact that the product of orthobaric liquid density and heat capacity varies only slightly with temperature for most liquids. Thus, the linear equation for the heat capacity of liquid acetic acid, which is presented in the *International Critical Tables* [29], was used [8] with Eq. (7) and (8) to develop Eq. (44). The upper temperature limit for Eq. (44) cannot be established at present; however, it is believed by the author to be useful in the temperature range from 289 to 493 K.

$$\Phi C_p = (A + BT)/(d/\beta) \tag{44}$$

where C_p = heat capacity, kJ/(kg·K)
T = temperature, K
d = density, kg/m^3
$A = 0.400705$ $\Phi = 0.23901$
$B = 0.000378$ $\beta = 1000.00$

Table 9 Experimental and Estimated Heat Capacity Data for Acetic Acid in the Liquid State

Temperature (K)	Heat capacity	
	Experimental (kJ/(kg·K))	Eq. (44) (kJ/(kg·K))
292.65	2.042	2.038
294.75	2.054	2.046
296.35	2.021	2.050
296.75	2.025	2.054
304.95	2.054	2.084
305.25	2.054	2.084
306.35	2.063	2.088
308.15	2.100	2.096
309.65	2.079	2.100
311.15	2.238	2.105
318.15	2.130	2.134
333.15	2.188	2.192
353.15	2.268	2.276
373.16		2.364
423.15		2.615
473.15		2.933

Source: Data from Refs. 9 and 29–31.

Experimental data from the *International Critical Tables* [58], from Parks and Kelley [30], from Pushin et al. [31], and from Tryer [9] are compared in Table 9 with data estimated by means of Eq. (44).

G. Heat Capacity of Acetic Acid in the Solid State

Equation (45) was developed [8] to enable the estimation of the heat capacity of acetic acid in the solid state at temperatures between 83.15 and 278.15 K.

$$\Phi C_p = A + Bt + C/(t_m - t) \tag{45}$$

where C_p = heat capacity, kJ/(kg·K)

$t = T - 273.15$

T = temperature, K

$A = 0.3057 \qquad C = 0.6446$

$B = 0.000640 \qquad t_m = 16.61$

$\Phi = 0.23901$

Table 10 Experimental and Estimated Heat Capacity Data for Acetic Acid in the Solid State

Temperature (K)	Heat capacity	
	Experimental (kJ/(kg·K))	Eq. (45) (kJ/(kg·K))
87.35	0.783	0.795
94.75	0.816	0.816
153.45	1.000	0.980
194.55	1.097	1.097
232.5	1.206	1.218
238.9	1.239	1.243
255.5	1.302	1.381
274.7	1.470	1.465

Source: Data from Ref. 30.

The terms in Eq. (45) were selected to account for the marked rise in the heat capacity that is observed in the case of acetic acid, as well as in the case of many other crystalline organic compounds, as the temperature approaches the melting point. Experimental data from Parks and Kelley [30] are compared in Table 10 with data estimated by means of Eq. (45).

H.   Heat of Vaporization of Acetic Acid

The Clapeyron equation, Eq. (46), is rigorous, and it can be applied to any vaporization equilibrium [32]. Its usefulness for calculating heats of vaporization is often restricted because it presupposes the availability of an accurate vapor pressure equation and accurate vapor and liquid molar volume data. In the present case, these values were obtained by means of Eq. (48), (1), (7), and (8).

$$H^v = (dP/dT)(T)(V - v) \tag{46}$$

where H^v = heat of vaporization, J/mol
   P = pressure, Pa
   T = temperature, K
   V = volume of vapor, m^3/mol
   v = volume of liquid, m^3/mol

Heat of vaporization data calculated by means of Eq. (46) are compared in Table 11 to calorimetric data reported by Weltner [28], Konicek and

Table 11 Heat of Vaporization Data for Acetic Acid

Temperature (K)	Heat of vaporization	
	Experimental (kJ/kg)	Eq. (46) (kJ/kg)
291.15	466.93	378.57
293.15	355.31	378.86
298.15	388.74	379.61
313.15	383.21	382.04
313.15	362.29	382.04
313.15	391.54	382.04
333.15	392.96	385.47
353.15	395.05	388.44
353.15	383.21	388.44
358.55	392.21	389.11
378.95	393.92	390.70
390.55	406.06	390.79
391.15	403.84	390.74
391.35	404.80	390.74
391.65	394.55	390.74
413.15	395.05	388.61
493.15	339.99	345.56

Source: Data from Refs. 28 and 33–37.

Wadso [33], Sabinin et al. [34], Brown [35], Kollar [36], and Lazeeva and Markuzin [37].

I. Heats of Dimerization and of Trimerization of Acetic Acid in the Gas State

The heats of dimerization and of trimerization of acetic acid in the gas state can be obtained from the equation of state, Eq. (1). As obtained by Johnson and Nash [1], the heat of dimerization is -57.86 kJ/mol and the heat of trimerization is -95.0 kJ/mol. Other investigators, who assumed that only dimerization occurs in the gas state, obtained the heats of dimerization presented in Table 12.

J. Surface Tension of Acetic Acid

Dass and Singh [43] published Eq. (47) as a general equation for the correlation of surface tension data. The presented constants are for acetic acid in the liquid state.

Table 12 Heat of Dimerization of Acetic Acid in
the Gas State

Median temperature of cited ref. (K)	Heat of dimerization (kJ/0.1201 kg)	Ref.
306.1	−68.62	3
360.1	−61.92	38
373.1	−64.02	39
378.1	−60.25	40
407.1	−60.67	4
413.1	−57.74	1
420.1	−57.74	41
456.1	−56.90	42

$$\ln \sigma = A + B(T - T_c)/(T - T_0) \tag{47}$$

where σ = surface tension, mN/m

T = temperature, K

$A = 1.3589 \qquad B = 4.5900$

$T_c = 594.80$ K (critical temperature) [44]

$T_0 = 999.89$ K

Jasper [45] recommended the data of Vogel [46] as the best available in the literature for the surface tension of acetic acid. The data of Vogel were used to determine the constants shown here for Eq. (47). Experimental data from Vogel [46] and from Bolz and Tuve [47] are compared in Table 13 with data estimated by means of Eq. (47).

K. Vapor Pressure of Acetic Acid

Values given in the literature for the normal boiling temperature of acetic acid range from about 391 K to about 392 K. Brown and Ewald [48] attributed this confusion to the difficulty of removing the last traces of water from acetic acid. Using extremely careful technique, they obtained the boiling point as a function of the number of times the acetic acid had been recrystallized and arrived at the conclusion that the normal boiling temperature of pure acetic acid is 391.10 K at 101.325 kPa. In a recent survey, Kudchadker et al. [44] concluded that the critical temperature of

Table 13 Experimental and Estimated
Surface Tension Data for Acetic Acid

| Temperature (K) | Surface tension | |
	Experimental (mN/m)	Eq. (47) (mN/m)
283.1	28.62	27.85
293.1	27.57	27.60
296.1	27.29	27.25
299.9	26.96	26.90
315.4	25.36	25.34
334.9	23.46	23.40
360.5	20.86	20.91
391.1	18.10	18.06

Source: Data from Refs. 45 and 47.

acetic acid is 594.45 K and that the critical pressure of acetic acid is 5785.7 kPa.

Miller [49] called attention to the fact that is it not possible for any equation with less than four constants to reproduce the correct curvature of a plot of ln P_r versus $1.0/T_r$ and to the fact that the reduced form of the Riedel equation fits experimental data well. The constants of Eq. (48) were obtained [8] using the vapor pressure data of Ramsay and Young [5] and of Potter and Ritter [50] with a regression technique that was constrained to pass through the normal boiling point and the critical point defined above.

$$P = P_c \exp(A + B/T_r + C \ln T_r + DT_r^6) \qquad (48)$$

where P = vapor pressure, kPa
T_r = reduced temperature, T/T_c
T = temperature, K
T_c = 594.45 K
P_c = 5785.7 kPa
A = 10.08590 C = -3.87306
B = -10.37932 D = 0.29342

Experimental data from Taylor [39], Ramsay and Young [5], Potter and Ritter [50], and Brown and Ewald [48] are compared in Table 14 with data estimated by means of Eq. (48).

Table 14 Experimental and Estimated Vapor Pressure
Data for Acetic Acid

Temperature (K)	Vapor pressure	
	Experimental (kPa)	Eq. (48) (kPa)
273.15 supercooled liquid	0.467	0.467
289.83 triple point	1.27	1.28
293.15	1.57	1.56
353.15	26.94	27.30
391.10 normal boiling point	101.325	101.325
423.15	249.98	250.29
463.15	631.01	638.77
503.15	1386.82	1394.15
543.15	2753.24	2723.52
583.15	4925.33	4936.59
594.45 critical point	5785.66	5785.66

Source: Data from Refs. 5, 39, 48, and 50.

V. COMPOSITION OF ACETIC ACID

A. Composition of Acetic Acid in the Gas State

Weltner [28] examined the infrared spectrum of acetic acid in the gas state
in an effort to identify absorption peaks that could be attributed to possible
trimers and/or tetramers. He concluded that only monomers and dimers
exist in the vapor phase in the temperature range between 298 and 448 K
and at pressures below 2.7 kPa. Derissen [51] used gas electron diffraction,
spectral, and dipole moment data to deduce that acetic acid exists in the
gas state as an equilibrium between the monomer and the planar centro-
symmetric cyclic dimer, as indicated by the following equation.

Vapor density data measured for acetic acid at pressures below 1 atm
have been successfully correlated by molecular models that are based on
the assumption that the gas is composed only of monomers and dimers
[3,38,39]. However, it has been found that in order to correlate vapor
density data measured at pressures above 1 atm, it is necessary to assume

the existence of either trimers or tetramers in addition to monomers and dimers [1,4].

B. Composition of Acetic Acid in the Liquid State

Vilcu and Lucinescu [52] used spectral data to deduce that liquid acetic acid exists as the following system of equilibria. They show that the system of equilibria shifts to the left as temperature is increased. Suzuki et al. [53] have shown that the equilibria are shifted to the right by pressure up to about 251 MPa. At higher pressures, the equilibria are shifted back to the left by increasing pressure. It is expected that for those conditions in which an appreciable amount of the open-chain dimer exist, there will also be appreciable amounts of trimers, tetramers, and higher polymers. Vilcu and Lucinescu [52] report the use of spectral data to determine that

$$2CH_3\overset{\overset{\displaystyle O}{\|}}{C}-OH \rightleftarrows CH_3-\overset{\overset{\displaystyle O}{\|}}{C}\underset{O-H-\!\!\!-O=\overset{\overset{\displaystyle H-O}{|}}{C}-CH_3}{}$$

$$\rightleftarrows CH_3-C\overset{O-\!\!\!-H-O}{\underset{O-H-\!\!\!-O}{<}}C-CH_3$$

the degree of association at 273 K is 1.88 and at 373 K it is 1.45. These values lead to a heat of dimerization of 46.86 kJ/mol of dimer.

Corsaro and Atkinson [54] have shown that the energy of ultra-high-frequency sound is able to shift the equilibrium between the open-chain dimer and the cyclic dimer. Using relaxation methods, they determined that, at 298 K, the equilibrium constant for the formation of cyclic dimer from open-chain dimer is 9.70 and the heat of this reaction is −16.48 kJ/mol. These results are consistent with the results of Traynard [55], who used spectrographic data to show that a small amount of the monomer is present even in 100% acetic acid.

C. Composition of Acetic Acid in the Solid State

Jones and Templeton [56] used X-ray diffraction data to show that crystalline acetic acid is made up of long chains of molecules linked by hydrogen bonds. In a more recent X-ray study, Nahringbauer [57] found that no phase transition occurs in crystalline acetic acid in the temperature range between 289 K and 83 K. Nahringbauer [57] reported carefully determined bond lengths for acetic acid monomer and acetic acid dimer in both the gas and solid states.

REFERENCES

1. Johnson, E. W., and Nash, L. K., *J. Am. Chem. Soc.*, 72, 547–556 (1950)
2. Nicholls, J. J., M.Sc. thesis, Queen's University, Kingston, Ontario, Canada, 1962.
3. MacDougall, F. H., *J. Am. Chem. Soc.*, 58, 2585 (1936).
4. Ritter, H. L., and Simons, J. H., *J. Am. Chem. Soc.*, 67, 757–762 (1945).
5. Timmermans, J., *Physico-chemical Constants of Pure Organic Compounds*, Elsevier, New York, 1950.
6. Freeman, J. R., and Wilson, G. M., *High Temperature PVT Properties of Acetic Acid/Water Mixtures* and *High Temperature Vapor-Liquid Equilibrium Measurements on Acetic Acid/Water Mixtures*, AIChE Symposium Series 244, Vol. 81 (M. S. Benson and D. Zudkevitch, eds.), American Institute of Chemical Engineers, 345 East 47 St., New York, NY, 10017, 1985, pp. 1–25.
7. Lawrenson, I. J., Lee, D. A., and Lewis, G. B., "The PVT properties of acetic acid," Report no. NPL-Report-Chem-76, Division of Chemical Standards, National Physical Laboratory, Teddington, England, October 1977.
8. Seaton, W. H., unpublished work for the Design Data Research Laboratory, Tennessee Eastman Co., Kingsport, TN, 1970–1974.
9. Tyrer, D., *J. Chem. Soc.*, 105, 2534 (1914).
10. Francis, A. W., *Chem. Eng. Sci.*, 10 37 (1959).
11. Costello, J. M., and Bowden, S. T., *Rec. Trav. Chim.*, 78, 391–403 (1959).
12. Bretsznajder, S., *Prediction of Transport and Other Physical Properties of Fluids*, International Series of Monographs in Chemical Engineering (P. V. Danckwerts, ed.), Pergamon Press, New York, 1971, p. 189.
13. Schonhorn, H., *J. Chem. Eng. Data*, 12(4), 524 (1967).
14. van der Waals, J. D., *Z. Physik. Chem.*, 13, 716 (1894).
15. Abtiev, K. T., *Vestnik Adad. Nauk Kazakh. S.S.R.*, 15(12), 72–77 (1959).
16. Tsonopoulos, C., and Prausnitz, J. M., *Chem. Eng. J.*, 1, 273–278 (1970).
17. Seaton, W. H., *Can. J. Chem. Engr.*, 57, 523–526 (1979).
18. Timrot, D. L., Serednitskaya, M. A., and Bespalov, M. S., *Teploovizika Vysokikh Temperatur* (Institute of High Temperatures, USSR Academy of Sciences), 14(6), 1192–1196 (Nov.–Dec. 1976).
19. Usmanov, A. G., and Mukhamedzyanov, G. Kh., *International Chem. Eng.*, 3(3), 369–374 1963).
20. Stull, D. R., Westrum, E. F., and Sinke, G. C., *The Chemical Thermodynamics of Organic Compounds*, Wiley, New York, 1969.
21. Frontas'ev, V. P., and Gusakov, M. Ya., *Uch. Zap. Saratovsh. Gos. Univ.*, 69, 237–238 (1960).
22. Venart, J. E. S., and Krishnamurthy, C., in (D. R. Flynn and B. A. Peavy, Jr., eds.), National Bureau of Standards Special Publication 302, Washington, D.C., 20402, 1968, pp. 659–669.
23. Jobst, W., *Int. J. Heat Mass Transfer*, 7, 725–732 (1964).
24. Seaton, W. H., *Can. J. Chem. Eng.*, 58, 416–418 (1980).
25. Mashirov, V. E., and Tarzimanov, A. A., *Tr. Kazan Khim. Tekhnol. Inst.*, 37, 245–251 (1968).

26. Timrot, D. L., and Makhrov, V. V., *Inzh-Fiz Zh*, *31*(6), 965–972 (1976); English translation: *J. Eng. Phys.*, *31*(6), 1383–1388 (1976).
27. Kobe, K. A., and Crawford, H. R., *Pet. Refiner*, *37*(7), 125–130 (1958).
28. Weltner, W. J., *J. Am. Chem. Soc.*, *77*, 3941–3950 (1955).
29. *International Critical Tables*, McGraw-Hill, New York, 1926.
30. Parks, G. S., and Kelley, K. K., *J. Am. Chem. Soc.*, *47*, 2089 (1925).
31. Pushin, N. A., Fedjushkin, A. V., and Krgovic, B., *Bull. Soc. Chim. Belgrade*, *11*(1/2), 12–24 (1940–1946).
32. Hougen, O. A., Watson, K. M., and Ragatz, R. A., *Chemical Process Principles*, 2nd ed., Part 1, Wiley, New York, 1966, Chapter 13.
33. Konicek, J., and Wadso, I., *Acta Chem. Scand.*, *24*, 2612–2616 (1970).
34. Sabinin, V. E., Belousov, V. P., and Morachevskii, A. G., *Izv. Vysshikh Ucheb. Zaved.*, *Khim. I Khim. Tekhnol.*, *9*(6), 889–891 (1966).
35. Brown, J. C., *J. Chem. Soc.*, *83*, 987 (1903).
36. Kollar, G., *Magyar Kemiai Folyoirat*, *72*(7), 314–316 (1966).
37. Lazeeva, M. S., and Markuzin, N. P., *J. Appl. Chem.*, *U.S.S.R.*, *46*, 579–582 (1973).
38. Barton, J. R., and Hsu, C. C., *J. Chem. Eng. Data*, *14*, 184–187 (1969).
39. Taylor, M. D., *J. Am. Chem. Soc.*, *73*, 315 (1951).
40. Levy, B., and Davis, T. W., *J. Am. Chem. Soc.*, *76*, 3268–3269 (1954).
41. Fenton, T. M., and Garner, W. E., *J. Chem. Soc.*, *1930*, 694–700 (1930).
42. Glien, W., and Opel, G., *Wiss. Z. Iniv. Rostock, Math.-Naturwiss. Reihe*, *18*, 877–880 (1969).
43. Dass, N., and Singh, O., *J. Phys. Soc. Jpn.*, *28*, 806 (1970).
44. Kudchadker, A. P., Alani, G. H., and Zwolinski, B. J., *Chem. Rev.*, *68*, 659–735 (1968).
45. Jasper, J. J., *J. Phys. Chem. Ref. Data*, *1*(4), 841–1009 (1972).
46. Vogel, A. I., *J. Chem. Soc.*, *1948*, 1809 (1948).
47. Bolz, R. E., and Tuve, G. L., eds., *Handbook of Tables for Applied Engineering Science*, Chemical Rubber Co., 18901 Cranwood Pkwy., Cleveland, OH 44128, 1970.
48. Brown, I., and Ewald, A. H., *Aust. J. Sci. Res.—Series A—Phys. Sci.*, *3*, 306–323 (1950).
49. Miller, D. G., *Ind. Eng. Chem.*, *56*(3), 46–57 (1964).
50. Potter, A. E., and Ritter, H. L., *J. Phys. Chem.*, *58*, 1040–1042 (1954).
51. Derissen, J. L., *J. Mol. Struct.*, *7*, 67–80 (1971).
52. Vilcu, R., and Lucinescu, E., *Rev. Roum. Chim.*, *14*, 283–290 (1969).
53. Suzuki, K., Taniguchi, Y., and Watanabe, T., *J. Phys. Chem.*, *77*(15), 1918–1922 (1973).
54. Corsaro, R. D., and Atkinson, G., *J. Chem. Phys.*, *55*(1), 1971–1978 (1971).
55. Traynard, Ph., *Bull. Soc. Chim. France*, Series 5, Vol. 14, pp. 316–321 (1947).
56. Jones, R. E., and Templeton, D. H., *Acta Cryst.*, *11*, 484 (1944).
57. Nahringbauer, I., *Acta Chem. Scand.*, *24*, 453–462 (1970).
58. Lagowski, J. J., *The Chemistry of Nonaqueous Solvents*, Vol. 3, Academic Press, New York, 1970, p. 244.

59. Dean, J. A., ed., *Lange's Handbook of Chemistry*, 12th ed., McGraw-Hill, New York, 1979, pp. 6–36.
60. Nelson, R. D., Lide, D. R., and Maryott, A. A., "Selected values of electric dipole moments in the gas phase," National Reference Data Series—National Bureau of Standards, Bulletin no. 10, Sept. 1, 1967.

8

Aqueous Solutions

William H. Seaton*
Eastman Chemical Company, Kingsport, Tennessee

I. INTRODUCTION

A. Importance of Aqueous Solutions of Acetic Acid

The manufacturing processes of Eastman Chemical Company generate several billion pounds of water/acetic acid mixtures with an average concentration of 30% acetic acid and another several more billion pounds of water/acetic acid mixtures in the concentration range between 93 and 96% acetic acid, from which the acid was recovered and recycled. Although these data are for only one company, they call attention to the fact that nearly all industrial processes that use acetic acid must deal with the problem of recovering the acid from its aqueous solutions. Clearly, the recovery of acetic acid from its aqueous solutions is a problem of enduring importance in the chemical process industries.

B. On the Selection of Mathematical Models

In the selection of mathematical models for this section, preference was always given to those models which have a theoretical basis so that extrapolation beyond the range of available data can be done with a minimum risk of large error. When models based on theory were not available,

*Private consultant, retired from Eastman Chemical Company.

preference was given to models that have been established as reliable by extensive use in the literature. In a few cases, it was necessary to devise models for the properties of mixtures. In these cases, the properties were assumed to be the fractional sum of the properties of the pure components, at the same conditions of temperature and pressure, plus the contribution of a deviation function. Models of this type are seldom subject to large error if the properties of the pure components are available. Mathematical models of the needed properties of water have been included at the end of this chapter to facilitate the use of the mixture models of this type. Equation (i) was found to be a useful deviation function because of its simplicity and because of its ability to show a maximum (or minimum), which can be a function of temperature, at any composition between the pure components.

$$y = Kx^n(1 - x) \tag{i}$$

where y = property
          x = fractional composition
          K, n = constants that may be functions of temperature

C. On the Division of Properties in This Section

As with the chapter on the thermophysical properties of acetic acid, the properties in this chapter have been classified into categories to make it easier to locate specific data. The categories of this section are as follows: Equations of State, Phase Equilibrium Data, Transport Properties, and Physical Properties. It is expected that the people who need the data in this section will recognize when the property they seek falls into any of the first three of these categories. Any other property, if available here, will therefore be in the fourth category. Because of their qualitative nature, the discussions of the molecular structures of the components and their mixtures are given separate sections.

II. EQUATIONS OF STATE FOR WATER AND ACETIC ACID MIXTURES

A. Equation of State for Water/Acetic Acid Mixtures in the Gas State

Equation (1) provides a good approximation for gas density data of the water/acetic acid system.

$$z = y_A z_A + y_B z_B \tag{1}$$

where z = compressibility factor of the water/acetic acid mixture at a specified pressure and temperature

y_A = concentration of water in the mixture, mole fraction
y_B = concentration of acetic acid in the mixture, mole fraction
z_A = compressibility factor of water at the same pressure and temperature as the water/acetic acid system
z_B = compressibility factor of acetic acid at the same pressure and temperature as the water/acetic acid system

The experimental data of Freeman and Wilson [1] are compared in Table 1 with data estimated by means of Eq. (1).

B. DIPPR Equation of State for Water/Acetic Acid Mixtures in the Gas State

With funding from the Design Institute for Physical Property Data (DIPPR), Freeman and Wilson [1] measured new gas density data for the water/acetic acid system and then used Eq. (2)–(16) to correlate their data.

In the following system of equations, the subscripts represent the three components (1) water, (2) acetic acid monomer, and (3) acetic acid dimer.

$$s_j = b_j \exp(C_j T_{cj}/T) \tag{2}$$

where T = temperature, K
        T_{cj} = critical temperature of component j, K

	b	C	T_c
Water	0.93025×10^{-5}	1.05	647.27
Acetic acid monomer	0.38521×10^{-4}	0.72	482.17
Acetic acid dimer	0.79290×10^{-4}	0.474	648.78

$$A_{jj} = RT(\alpha_j + \beta_j T_{cj}/T + \Gamma_j(T_{cj}/T)^2 + \delta(T_{cj}/T)^2)/(s_j) \tag{3}$$

where T = temperature, K T_{cj} = critical temperature of component j, K
        R = 0.0083144

	α	β	Γ	δ
Water	-0.208704	1.343966	2.793454	-0.4372572
Acetic acid monomer	-2.952821	7.005197	-1.066291	0.4048984
Acetic acid dimer	-3.397608	7.467332	-1.005363	0.3372111

$$A_{ij} = (A_{ii} + A_{jj})/2 \tag{4}$$

$$K_D = \Theta \exp(\tau_1 + \tau_2/T + \tau_3/T^2) \tag{5}$$

where $\tau_1 = -16.637$
        $\tau_2 = 1457.7$
        $\tau_3 = 1.132 \times 10^6$
        $\Theta = 7.50062$

$$K_P = K_D \phi_2^2/\phi_3 \tag{6}$$

Table 1 Experimental and Estimated Compressibility Data for Mixtures of Water and Acetic Acid in the Gas State

| Temp. (K) | Conc. of acetic acid (wt%) | Pressure (kPa) | Compressibility factor | | |
			Expt'l	Eq. (1)	Eq. (16)
373.15	10.00	11.721	0.997	0.990	0.997
373.15	10.00	25.028	1.003	0.986	0.995
373.15	10.00	36.887	0.996	0.983	0.992
373.15	10.00	48.746	0.990	0.981	0.990
373.15	10.00	60.053	0.978	0.978	0.988
373.15	90.00	12.204	0.797	0.801	0.828
373.15	90.00	24.338	0.796	0.754	0.782
373.15	90.00	40.265	0.754	0.724	0.750
373.15	90.00	50.469	0.734	0.712	0.737
373.15	90.00	60.536	0.721	0.702	0.726
413.15	30.00	29.716	1.017	0.977	0.993
413.15	30.00	63.707	0.992	0.965	0.986
413.15	30.00	103.352	0.983	0.956	0.979
413.15	30.00	180.504	0.966	0.944	0.966
413.15	30.00	254.760	0.948	0.934	0.955
413.15	70.00	19.029	0.979	0.941	0.969
413.15	70.00	55.089	0.944	0.895	0.935
413.15	70.00	89.218	0.917	0.873	0.914
413.15	70.00	186.227	0.870	0.839	0.875
413.15	70.00	246.418	0.844	0.826	0.858
463.15	50.00	165.474	0.955	0.952	0.975
463.15	50.00	324.742	0.937	0.928	0.955
463.15	50.00	479.184	0.921	0.910	0.937
463.15	50.00	771.521	0.889	0.885	0.906
463.15	50.00	905.968	0.870	0.874	0.893
503.15	9.12	381.279	0.980	0.981	0.983
503.15	9.12	1005.941	0.950	0.952	0.954
503.15	9.12	1618.883	0.920	0.922	0.925
503.15	9.12	2650.335	0.833	0.868	0.873
503.15	90.00	206.153	0.956	0.930	0.930
503.15	90.00	388.863	0.901	0.887	0.887
503.15	90.00	563.989	0.871	0.857	0.852
503.15	90.00	723.258	0.838	0.834	0.826
503.15	90.00	877.010	0.813	0.816	0.802
503.15	90.00	1023.868	0.791	0.801	0.781
503.15	90.00	1151.420	0.762	0.789	0.763
503.15	90.00	1291.384	0.748	0.777	0.745
503.15	90.00	1413.420	0.728	0.768	0.730
503.15	90.00	1526.494	0.708	0.760	0.715
503.15	90.00	1634.052	0.689	0.753	0.702
503.15	90.00	1693.346	0.654	0.749	0.694

Source: Data from Ref. 1.

$$p_2 = 2Py_B/(1 + (1 + 4K_P(2 - y_B)Py_B)^{1/2}) \tag{7}$$

$$p_3 = K_P p_2^2 \tag{8}$$

$$p_1 = P - p_2 - p_3 \tag{9}$$

where P = system pressure, kPa
  p_i = partial pressure of component i, kPa
  y_B = observed mole fraction of acetic acid in the mixture (counting all acetic acid as monomer)

$$y_i = p_i/P \tag{10}$$

where y_i = mole fraction of component i in the gas phase mixture

$$B = \sum_{i=1}^{3} y_i b_i \tag{11}$$

$$S = \sum_{i=1}^{3} y_i s_i \tag{12}$$

$$Z^* = V/(V - B) - V \left(\sum_{j=1}^{3} \sum_{k=1}^{3} y_j s_j y_k s_k A_{jk} \right) \Big/ (RT(V + S - B)^2) \tag{13}$$

$$\phi_k = \exp\left\{ -\ln Z^* + b_k \sum_{i=1}^{3} y_i \Big/ \left(V - \sum_{j=1}^{3} b_j y_j \right) \right.$$

$$- 2s_k \sum_{i=1}^{3} (y_i s_i A_{i,k}/RT) \Big/ \left[V + \sum_{j=1}^{3} (y_j(s_j - b_j)) \right]$$

$$+ \ln\left(V \Big/ \left(V - \sum_{j=1}^{3} y_j b_j \right) \right) + \left[(s_k - b_k) \sum_{i=1}^{3} \sum_{j=1}^{3} y_i s_i y_j s_j A_{ij}/RT \right]$$

$$\div \left[V + \sum_{j=1}^{3} (y_j(s_j - b_j)) \right]^2 \right\} \tag{14}$$

$$V = Z^*RT/P \tag{15}$$

$$Z = PZ^*/(P + p_3) \tag{16}$$

To solve for the compressibility factor, Z, of Eq. (16), the fugacity coefficients, ϕ_1, ϕ_2, and ϕ_3 are initially set equal to 1.0 and the volume is estimated. Equations (6)–(15) are then iterated until the change in volume, V, becomes acceptably small. The compressibility factor of the mixture is then obtained from Eq. (16).

The data measured by Freeman and Wilson [1] are compared in Table 1 with data estimated by Eq. (16).

C.   Density of Water/Acetic Acid Mixtures in the Liquid State

Equations (17)–(19) were developed to provide a means for estimating the orthobaric densities of liquid state mixtures of water and acetic acid as functions of composition and temperature [2].

$$v_M = x_1 v_1 + x_2 v_2 + v_e \qquad (17)$$

$$v_e = (A + BT)x_1 x_2^{(C+DT)} \qquad (18)$$

$$\Phi d_M = (18.02 x_1 + 60.05 x_2)/v_M \qquad (19)$$

where x_1 = concentration of water, mole fraction
   x_2 = concentration of acetic acid, mole fraction
   T = temperature, K
   v_1 = molal volume of water at the specified temperature, m³/mol
   v_2 = molar volume of acetic acid at the specified temperature, m³/mol
   d_M = density of the mixture, kg/m³
   $A = -4.34565 \times 10^{-6}$
   $B = -2.68152 \times 10^{-10}$
   $C = 0.69329$
   $D = 7.68148 \times 10^{-4}$
   $\Phi = 1000.0$

Density data estimated by means of Eq. (17)–(19) are compared in Table 2 with experimental data obtained from the *International Critical Tables* [3] and from the publications of Ragni et al. [4] and of Campbell and Kartzmark [5]. The values used in Eq. (17) for the molal volumes of pure water, v_1, and pure acetic acid, v_2, were estimated by dividing the respective molecular weights by density values that were estimated by means of Eq. (34) and (35) and of Eq. (7) and (8) of Chapter 7.

Korpella [78] measured the compression of water/acetic acid mixtures at three temperatures. He then correlated his data using the Tait equation. The Tait equation is presented as Eq. (9) of Chapter 7. The molar volume data reported by Korpella for the pressure of 101.325 MPa are presented in Table 3 along with molar volumes computed by means of Eq. (9) of Chapter 7, using the parameters presented by Korpella. The molar volumes at other pressures, and at the specified temperatures, can be estimated by means of these parameters.

Gibson [6] provides data on the compressibility of water/acetic acid mixtures and he describes an equation that is able to correlate the specific

Table 2 Orthobaric Liquid Phase Density
Data for the System: Water/Acetic Acid

Temp. (K)	Conc. of water (wt%)	Density Experimental (kg/m³)	Density Eq. (19) (kg/m³)
283.1	98.0	1002.9	1003.1
283.1	63.0	1052.4	1050.5
283.1	42.0	1071.5	1070.9
283.1	7.0	1073.9	1071.1
298.1	89.7	1011.0	1011.4
298.1	79.8	1024.0	1023.6
298.1	69.8	1036.0	1035.0
298.1	59.7	1045.2	1045.1
298.1	42.2	1058.4	1059.2
298.1	33.0	1063.0	1064.0
298.1	12.8	1062.0	1061.4
298.1	0.093	1044.0	1043.8
328.1	59.7	1022.5	1023.3
328.1	42.2	1031.6	1033.4
328.1	0.0	1009.5	1010.7
388.1	59.7	968.4	967.7
388.1	42.2	974.1	971.8

Source: Data from Refs. 3–5.

volume data as a function of pressure at constant temperature. The equation is limited to low concentrations of acetic acid.

III. PHASE EQUILIBRIUM DATA FOR THE WATER/ACETIC ACID SYSTEM

A. Vapor/Liquid Equilibrium Data for the Water/Acetic Acid System

More than 30 publications in the literature report experimentally measured vapor/liquid equilibrium data at constant pressure for the water/acetic acid system. Among them are publications that provide data at the following pressures, kPa (ref.): 2.667 [7], 9.333 [8], 13.332 [9], 26.664 [8], 46.662 [9], 53.328 [10], 79.993 [7], 99.765 [11], 101.325 [12], 273.711 [13], 718.741 [13], 1013.25 [14], 4053.0 [14], 5066.2 [14], 6079.5 [14], 7092.7 [14].

Table 3 Molar Volumes of Water/Acetic Acid Mixtures in the Liquid State at Elevated Pressures

Temp. (K)	Conc. of water (wt. %)	B[a] (kPa)	C[a]	Molar vol. at 101.33 kPa (cm³/mol)	Molar vol. at 101.33 MPa Experimental (cm³/mol)	Eq. (9)[a] (cm³/mol)
298.2	100.0	293,842.5	7.5125	18.07[b]	17.36	17.36
298.2	90.0	289,586.8	7.8856	19.18	18.45	18.45
298.2	80.0	273,577.5	8.4530	20.48	19.71	19.71
298.2	70.0	251,893.9	9.0050	22.04	21.21	21.21
298.2	60.0	228,487.9	9.5503	23.95	23.03	23.03
298.2	50.0	201,028.8	10.2292	26.33	25.28	25.28
298.2	40.0	177,521.4	10.7397	29.32	28.09	28.09
298.2	30.0	158,067.0	11.0914	33.23	31.74	31.74
298.2	20.0	138,308.6	11.3484	38.47	36.61	36.61
298.2	10.0	118,550.3	11.5014	45.93	43.47	43.46
298.2	0.0	101,325.0	10.7447	57.55	53.85	53.84
313.1	100.0	306,812.1	7.4086	18.16	17.46	17.46
313.1	90.0	309,345.2	7.4638	19.30	18.57	18.57
313.1	80.8	296,983.6	7.7346	20.50	19.72	19.72
313.1	70.0	275,908.0	8.0906	22.25	21.39	21.39
313.1	60.0	245,307.8	8.6304	24.21	23.24	23.24
313.1	50.0	216,734.2	9.1300	26.63	25.51	25.51
313.1	40.0	193,936.0	9.4679	29.70	28.38	28.38
313.1	30.0	169,922.0	9.8359	33.68	32.08	32.08
313.1	20.0	146,617.3	10.1615	39.03	37.02	37.02
313.1	10.0	123,515.2	10.3814	46.62	43.94	43.94
313.1	0.0	92,003.1	10.7247	58.50	54.46	54.46
328.1	100.0	298,503.4	7.5868	18.28	17.57	17.57
328.1	90.0	299,820.7	7.5643	19.45	18.71	18.70
328.1	80.0	282,899.4	7.8373	20.83	20.02	20.01
328.1	70.0	255,440.3	8.2976	22.47	21.57	21.57
328.1	60.0	225,549.4	8.8357	24.48	23.45	23.45
328.1	50.0	204,169.9	9.1228	26.96	25.77	25.77
328.1	40.0	184,918.1	9.3147	30.36	28.94	28.94
328.1	30.0	167,794.2	9.3753	34.14	32.42	32.43
328.1	20.0	137,092.7	10.0200	39.60	37.40	37.42
328.1	10.0	99,602.5	11.2651	47.39	44.42	44.44
328.1	0.0	80,654.7	10.9076	59.48	55.04	55.05

[a]Refer to Eq. (9) of Chapter 7.
[b]V_0 of Eq. (9) of Chapter 7 for $P_0 = 101.325$ kPa.
Source: Data from Ref. 78.

Vapor/liquid equilibrium data at constant temperature have been reported for the following temperatures, K (ref.): 293.2 [15], 298.2 [16], 315.2 [17], 342.9 [18], 353.1 [18], 363.1 [18], 372.8 [1], 412.6 [1], 462.1 [1], 502.9 [1].

No low-boiling azeotrope has been observed for this system; however, Chalov and Aleksandrova [7] analyzed the available data and concluded that a high-boiling azeotrope probably exists for the system in the temperature range between 283 and 293 K and that it should contain about 90 wt% acetic acid.

Tsonopoulos and Prausnitz [19] provide a generalized molecular model for the correlation of vapor/liquid equilibrium data for water/carboxylic acid systems. Their model is based on the assumption that the vapor is composed of water monomers and dimers plus acid monomers, dimers, trimers, and a cross-dimer of water and acid. They apply their model to correlate vapor/liquid equilibrium data for the water/acetic acid system.

Freeman and Wilson [1] use vapor phase fugacities estimated by means of Eq. (2)–(16) and liquid phase activity coefficients estimated by means of a modified Wilson equation to correlate the vapor/liquid equilibrium data they measured at 372.8, 412.6, 462.1, and 502.9 K.

The vapor/liquid equilibrium data measured at 101.325 kPa by Brown and Ewald are presented in Table 4. In spite of the complexity of the system, these data are rather well represented by the simple relationship shown here as Eq. (20).

Table 4 Experimental and Estimated Vapor/Liquid Equilibrium Data for the System Water/Acetic Acid at 101.325 kPa

Temp. (K)	Conc. of water in the liquid phase (mole fraction)	Conc. of water in the vapor phase	
		Experimental (mole fraction)	Eq. (20) (mole fraction)
391.11	0.000210	0.000226	0.000348
391.07	0.000213	0.000347	0.000353
390.79	0.003400	0.006890	0.005624
390.66	0.00545	0.01122	0.00900
388.18	0.04741	0.09790	0.07623
384.66	0.1497	0.2382	0.2259
382.99	0.2198	0.3273	0.3183
381.31	0.2917	0.4071	0.4057
374.39	0.8251	0.8783	0.8866
373.22	0.9891	0.9921	0.9933

Source: Data from Ref. 12.

$$y = \alpha x/[1 + (\alpha - 1)x] \tag{20}$$

where x = concentration of water in the liquid phase, mole fraction
        y = concentration of water in the vapor phase, mole fraction
        α = relative volatility = 1.658

Data estimated by means of Eq. (20) are compared in Table 4 with the data of Brown and Ewald [12].

B. Liquid/Solid Equilibrium Data for the Water/Acetic Acid System

Hirado et al. [20] published the liquid/solid equilibrium data shown in Table 5 for the water/acetic acid system. They observed that for a broad range of initial compositions, the eutectic mixture is composed of 59.0 wt% acid and it melts at 246.6 K. This composition and temperature are in good agreement with the earlier work of Dahms [as reported in Ref. 3] and of Pickering [21]; however, Hirado et al. observed that the melting temper-

Table 5 Liquid/Solid Equilibrium Data for the System: Water/Acetic Acid

Temp. (K)	Conc. of water in the liquid phase	
	Experimental (wt%)	Eq. (21) (wt%)
273.15	100.0	100.0
271.45	95.0	94.2
269.85	90.0	88.7
266.45	80.0	79.8
262.65	70.0	70.6
257.95	60.0	59.8
252.95	50.0	50.7
246.55	41.0[a]	41.0
247.15	40.0	40.0
255.75	30.0	29.9
265.05	20.0	19.1
275.15	10.0	9.5
281.35	5.0	5.0
289.75	0.0	0.0

[a]Eutectic point.
Source: Data from Ref. 20.

ature of the eutectic decreases as the initial composition approaches high concentrations of either component.

When mixtures of water and acetic acid, which are on the water-rich side of the eutectic composition, are cooled sufficiently, crystals of essentially pure ice will separate from the liquid phase. When the composition of the mixture is on the acetic acid-rich side of the eutectic composition, crystals of essentially pure acetic acid will separate from the liquid. Based on these facts, Null [22] derived Eq. (21), which relates the activity of either component in the liquid phase to the temperature of the liquid/solid system at equilibrium.

$$a_i = \Gamma_{Li} x_{Li} = (T/T_i^*)^{(C_{Li} - C_{Si})/R}$$
$$\times \exp(((T - T_i^*)/RT)(H_{Fi}/T_i^* - C_{Li} + C_{Si})) \quad (21)$$

The activity coefficient of component i in the liquid phase can be related to its mole fraction in the liquid phase by means of the van Laar equation, shown here as Eq. (22).

$$\ln \Gamma_{Li} = A_{ij}((A_{ji}x_{Lj})/(A_{ij}x_{Li} + A_{ji}x_{Lj}))^2 \quad (22)$$

Substituting Γ from Eq. (22) into Eq. (21), an equation is obtained that relates the liquid phase composition to the equilibrium temperature of the two-phase system. In the resulting equation, the two van Laar constants are the only adjustable parameters. These parameters were determined by applying the equation to the two components at the eutectic composition specified above. The resulting values are presented below with the physical property data provided by Null [22] for Eq. (21).

Gas constant
   $R = 8.31444$ J/(mol·K)
van Laar constants
   $A_{12} = 0.797299$ $A_{21} = 0.992570$
Heat of fusion, H_{Fi}
   water, 6009.48 J/mol   acetic acid, 11,715.2 J/mol
Triple-point temperature, T^*
   water, 273.16 K   acetic acid, 289.76 K
Liquid phase heat capacity, C_L
   water, 76.00 J/(mol·K)   acetic acid, 121.353 J/(mol·K)
Solid phase heat capacity, C_S
   water, 37.095 J/(mol·K)   acetic acid, 86.178 J/(mol·K)
Other terms
   x_{Li} concentration of component i in the liquid phase, mole fraction
   Γ_{Li} activity coefficient of component i in the liquid phase
   T temperature, K

The above values were used in Eq. (21) and (22) to estimate the liquid phase compositions that correspond to the equilibrium temperatures of Hirado et al. [20]. The results are presented in column 3 of Table 5.

IV. TRANSPORT PROPERTIES OF MIXTURES OF WATER AND ACETIC ACID

A. Diffusivity of Acetic Acid in Water in the Liquid State

Bidstrup and Geankoplis [23] used the diaphragm cell technique with a sintered glass diaphragm to measure the diffusivity of acetic acid in water at low concentrations and at temperatures in the range between 283 and 298 K. Thorat and Nageshwar [24] used a similar technique to measure the diffusivity of acetic acid in water over a wide range of concentrations and at temperatures in the range between 303 and 318 K. Equation (23) was developed to correlate and extend these measured values.

$$(D_{21}T/\mu) \times 10^9 = A + Bw + Cw^2 + D\mu \tag{23}$$

where D_{21} = diffusivity of acetic acid in water, m^2/s
         μ = viscosity of the mixture, mPa·s
         w = concentration of water, weight fraction
         T = temperature, K
         $A = 1.21184 \times 10^{-2}$
         $B = -1.77810 \times 10^{-2}$
         $C = 9.20684 \times 10^{-3}$
         $D = 2.45288 \times 10^{-4}$

Experimental data presented by Bidstrup and Geankoplis [23] and by Thorat and Nageshwar [24] are compared in Table 6 with data estimated by means of Eq. (23).

B. Electrical Conductivity of Water/Acetic Acid Mixtures in the Liquid State

Perrault [25] reported measured data on the electrical conductivity of the water/acetic acid system as a function of composition and of temperature. His data are presented in graph form. The curves are interpreted in terms of the ionic species that are believed to be present.

Tabular data on the specific electrical conductance of water/acetic acid mixtures at 297.15 K are presented by Wolf et al. [26].

Table 6 Diffusion Coefficient Data for the System:
Water/Acetic Acid

Temp. (K)	Conc. of water (wt. fract.)	Diffusion coefficient	
		Experimental (m²/s) × 10⁹	Eq. (23) (m²/s) × 10⁹
282.85	0.997	0.77	0.83
285.65	0.999	0.91	0.90
287.15	0.970	0.90	0.89
293.15	0.999	1.01	1.11
298.15	0.997	1.27	1.25
303.15	0.493	1.07	1.05
303.15	0.751	1.13	1.05
303.15	0.984	1.45	1.38
308.15	0.652	1.21	1.16
308.15	0.850	1.30	1.28
308.15	0.942	1.43	1.44
308.15	0.986	1.61	1.55
313.15	0.526	1.29	1.33
313.15	0.745	1.36	1.34
313.15	0.813	1.39	1.39
313.15	0.912	1.48	1.54
313.15	0.991	1.83	1.75
318.15	0.647	1.40	1.48
318.15	0.735	1.47	1.50
318.15	0.807	1.50	1.56
318.15	0.876	1.60	1.65
318.15	0.955	1.75	1.83
318.15	0.976	1.91	1.89
318.15	0.994	2.10	1.96

Source: Data from Refs. 23 and 24.

C. Thermal Conductivity of Water/Acetic Acid Mixtures in the Gas State

A careful search of the literature failed to locate any experimentally measured data on the thermal conductivity of water/acetic acid mixtures in the gas state [2]. It can be anticipated that the thermal conductivity of water/acetic acid mixtures in the gas state will be abnormally high because thermal energy will be transported not only because of thermal gradients, but also because of concentration gradients that will result from the shifting of the equilibria between the various species as a consequence of temperature differences. The latter effect will result in latent enthalpies of reaction being transported from points of higher temperature to points of lower temperature by means of concentration gradients.

D. Thermal Conductivity of Water/Acetic Acid Mixtures in the Liquid State

Usmanov [27] measured the thermal conductivity of various liquid phase mixtures of water and acetic acid and then used the equation of Filippov [28] to correlate his data. The resultant equation is presented here as Eq. (24).

$$k_M = w_1 k_1 + w_2 k_2 - A w_1 w_2 (k_1 - k_2) \tag{24}$$

where $A = 0.3$
k_M = thermal conductivity of the mixture, $W/(m \cdot K)$.
k_1 = thermal conductivity of water, $W/(m \cdot K)$.
k_2 = thermal conductivity of acetic acid, $W/(m \cdot K)$.
w_1 = concentration of water, weight fraction
w_2 = concentration of acetic acid, weight fraction

The experimental data of Usmanov [27] are compared in Table 7 with data estimated by means of Eq. (24). For these calculations, the thermal

Table 7 Atmospheric Pressure Thermal Conductivity Data for the Liquid Phase System: Water/Acetic Acid

Temp. (K)	Conc. of water (wt%)	Thermal conductivity	
		Experimental ($W/(m \cdot K)$)	Eq. (24) ($W/(m \cdot K)$)
293.1	0.0	0.1675	0.1595
293.1	40.0	0.3081	0.3052
293.1	80.0	0.4886	0.4936
303.1	0.0	0.1641	0.1578
303.1	40.0	0.3115	0.3086
303.1	80.0	0.5058	0.5041
313.1	0.0	0.1616	0.1557
313.1	40.0	0.3199	0.3115
313.1	80.0	0.5104	0.5137
333.1	0.0	0.1570	0.1516
333.1	40.0	0.3270	0.3161
333.1	80.0	0.5292	0.5292
333.1	90.0	0.5757	0.5899
333.1	100.0	0.6490	0.6536

Source: Data from Ref. 27.

conductivity of pure water was estimated by means of Eq. (38) and the thermal conductivity of pure acetic acid was estimated by means of Eq. (22) of Chapter 7.

E. Viscosity of Water/Acetic Acid Mixtures in the Gas State

A careful search of the literature failed to locate any experimentally measured data on the viscosity of water/acetic acid mixtures in the gas state [2]. Flynn et al. [as reported in Ref. 29] have shown that the viscosity of a polar gas at low pressure is a linear function of temperature. To the extent that this system can be represented by an equation of state such as that of Tsonopoulos and Prausnitz [19], the viscosity of mixtures of water and acetic acid in the gas state at low pressures can be estimated to a first approximation by Eq. (190) in the book by Golubev [30].

F. Viscosity of Water/Acetic Acid Mixtures in the Liquid State

Equation (25) was developed as a means for estimating the viscosity of water/acetic mixtures in the liquid state at orthobaric pressures [2].

$$\mu_M = x_1\mu_1 + x_2\mu_2 + (A + BT + CT^2)x_2x_1^{(D + T/E)} \tag{25}$$

where μ_M = viscosity of mixture, mPa·s
μ_1 = viscosity of water, mPa·s
μ_2 = viscosity of acetic acid, mPa·s
T = temperature, K
x_1 = concentration of water, mole fraction
x_2 = concentration of acetic acid, mole fraction
A = 164.9425
B = −0.9173
C = 0.001285
D = 1.03
E = 10,000.0

Viscosity data estimated by means of Eq. (25) are compared in Table 8 with experimental data from the *International Critical Tables* [3] and from the paper of Campbell and Kartzmark [5]. In these calculations, the viscosity of pure water was estimated by means of Eq. (39) and the viscosity of pure acetic acid was estimated by means of Eq. (13) and (14) of Chapter 7.

V. PHYSICAL PROPERTIES OF MIXTURES OF WATER AND ACETIC ACID

A. Miscellaneous Physical Property Data

Lemlich et al. [31] published an enthalpy/composition chart for the water/acetic acid system, which covers the solid, liquid, and vapor phases.

Table 8 Orthobaric Liquid Phase Viscosity
Data for the System: Water/Acetic Acid

Temp. (K)	Conc. of water (wt%)	Viscosity	
		Experimental (mPa·s)	Eq. (25) (mPa·s)
288.2	90.0	1.359	1.342
288.2	10.0	2.680	2.605
298.2	89.70	1.197	1.215
298.2	79.85	1.407	1.437
298.2	69.83	1.640	1.678
298.2	61.32	1.833	1.894
298.2	51.98	2.038	2.132
298.2	42.81	2.259	2.360
298.2	33.00	2.473	2.566
298.2	23.36	2.635	2.669
298.2	12.76	2.449	2.496
298.2	0.093	1.268	1.268
303.2	10.0	1.839	1.892
318.2	10.0	1.345	1.367
333.2	10.0	1.029	1.002
368.2	0.0	0.490	0.475

Source: Data from Refs. 3 and 5.

Konicek and Wadso [32] report calorimetrically measured data on the heat of solution at infinite dilution and on the partial molar heat capacity at infinite dilution of acetic acid in water.

Gray [33] published a detailed description of the manner in which the heat capacity of saturated vapor and the heat capacity of saturated liquid can be derived from constant pressure heat capacity data plus vapor pressure, volumetric, and dimerization equilibrium data.

B. Heat Capacity of Water/Acetic Acid Mixtures in the Gas State

A careful search of the literature failed to locate any experimentally measured data on the heat capacity of water/acetic acid mixtures in the gas state [2]. It can be anticipated that the heat capacities of water/acetic acid mixtures in the gas state will be abnormally high because of the shifting equilibria between the various associated species in the mixture. To the extent that the system is represented by equilibria, such as that proposed by Tsonopoulos and Prausnitz [19], the heat capacity of the mixture can be estimated by summing the heat capacities of the species in the mixture and the enthalpies of reaction as the various equilibria shift with temperature. The gas state heat capacities of the acetic acid monomer, dimer, and trimer are presented in Chapter 7.

C. Heat Capacity of Water/Acetic Acid Mixtures in the Liquid State

Equations (26) and (27) were developed to provide a means for estimating the heat capacity of liquid phase mixtures of water and acetic acid at orthobaric pressures as a function of composition and temperature [2].

$$C_{pM} = w_1C_{p1} + w_2C_{p2} + C_e \tag{26}$$

$$\Phi C_e = (A + BT)w_2w_1^{(C+DT)} \tag{27}$$

where C_{pM} = heat capacity of the mixture, kJ/(kg·K)
C_{p1} = heat capacity of pure water, kJ/(kg·K)
C_{p2} = heat capacity of pure acetic acid, kJ/(kg·K)
w_1 = concentration of water, wt fraction
w_2 = concentration of acetic acid, wt fraction
$A = 4.8493 \times 10^{-2}$
$B = 2.9507 \times 10^{-4}$
$C = 1.5002$
$D = 5.0808 \times 10^{-4}$
$\Phi = 0.23901$

Heat capacity data estimated by means of Eq. (26) and (27) are compared in Table 9 with experimental data obtained from Richards and

Table 9 Liquid Phase Heat Capacity Data for the System Water/Acetic Acid

Temp. (K)	Conc. of water (wt%)	Heat capacity	
		Experimental [kJ/(kg·K)]	Eq. (26) [kJ/(kg·K)]
287.1	88.23	3.991	3.985
288.1	94.79	4.101	4.099
288.1	70.33	3.634	3.636
298.1	99.83	4.174	4.175
298.1	23.08	2.577	2.586
311.1	40.16	3.054	3.014

Source: Data from Refs. 3, 34, 35, and 75.

Gucker [34], Bury and Davies [35], the *International Critical Tables* [3], and Parker [36]. The heat capacity of water was obtained by means of Eq. (47) and the heat capacity of acetic acid was obtained by means of Eq. (44) of Chapter 7 to generate the estimated data of Table 9.

Casanova et al. [37] developed Eq. (28) to provide an accurate means for estimating the heat capacity of liquid phase mixtures of water and acetic acid at orthobaric pressures and at 298.15 K. Equation (28) is used with Eq. (26) and with $C_{p1} = 4.17825$ kJ/(kg·K) and with $C_{p2} = 2.03347$ kJ/(kg·K). Data estimated by means of Eq. (26) and (28) are presented in Table 10.

$$\Phi C_e = x_1 x_2 \sum_{i=0}^{4} A_i (1.0 - 2.0 x_2)^i \tag{28}$$

where
$$x_1 = (w_1/18.02)/(w_1/18.02 + w_2/60.05)$$
$$x_2 = (w_2/60.05)/(w_1/18.02 + w_2/60.05)$$
$$\Phi = 18.02 x_1 + 60.05 x_2$$
$$A_0 = 9.083$$
$$A_1 = -2.466$$
$$A_2 = 5.549$$
$$A_3 = 17.526$$
$$A_4 = 14.715$$

Table 10 Liquid Phase Heat Capacity Data for the System Water/Acetic Acid at 298.15 K

Conc. of acetic acid (wt%)	Heat capacity	
	Experimental [kJ/(kg·K)]	Eq. (26) [kJ/(kg·K)]
93.754	4.086	4.085
82.409	3.896	3.889
69.246	3.620	3.624
42.770	3.029	3.025
22.926	2.582	2.583
15.049	2.406	2.408
5.794	2.187	2.187
0.751	2.054	2.055

Source: Data from Ref. 37.

D. Heat of Mixing of Water and Acetic Acid in the Liquid State

Vilcu and Lucinescu [38] provide Eq. (29) as a means for estimating the heat of mixing of water and acetic acid at 298.6 K.

$$\Phi H_M = x_2(1 - x_2)(A + B(2x_2 - 1) + C(2x_2 - 1)^2 + D(2x_2 - 1)^3)$$

$$(29)$$

where H_M = heat of mixing, J/mol of mixture
x_2 = concentration of acetic acid, mole fraction
$A = 315.740$
$B = 152.870$
$C = -28.417$
$D = 434.109$
$\Phi = 0.23901$

If the heat of mixing at a temperature other than 298.6 K is desired, it can be obtained by the imagined process of cooling (or heating) the pure components from the desired temperature to 298.6 K, mixing at 298.6 K, and then heating (or cooling) the mixture to the desired temperature. The calculations for this process can be accomplished by means of Eq. (44) of Chapter 7 plus Eq. (26) and (47) of this chapter. Experimental heat of mixing data from Vilcu and Lucinescu [38] and from Sabinin et al. [39] are compared in Table 11 with data estimated by this procedure.

Table 11 Heat of Mixing of Water and Acetic Acid in the Liquid State

Temp. (K)	Conc. of water (wt fraction)	Heat of mixing per mole of mixture	
		Experimental (J/mol)	Estimated (J/mol)
290.5	0.872	-52.72	-39.71
290.5	0.097	312.04	343.13
298.5	0.309	270.16	282.38
307.4	0.421	215.56	208.07
313.1	0.411	221.75	230.41
333.1	0.320	359.82	346.44
353.1	0.170	502.08	411.16
353.1	0.025	242.67	211.96

Source: Data from Refs. 38 and 39.

E.   Heat of Vaporization of Water/Acetic Acid Mixtures

Van Sickle and Seaton [40] developed Eq. (30) to estimate the enthalpy
of vaporization of water/acetic acid mixtures at temperatures between 313
and 704 K. The mixture data of Maklov et al. [41] are compared in Table
12 with data estimated by means of Eq. (30).

$$H^v = x_1 H_1^v + x_2 H_2^v + (a + bT)x_1 x_2^{(cT)} \tag{30}$$

Table 12   Experimental and Estimated Enthalpy
of Vaporization Data for Mixtures of Water and
Acetic Acid

Temp. (K)	Conc. of water (mole fraction)	Enthalpy of vaporization	
		Expt'l (kJ/mol)	Eq. (30) (kJ/mol)
382.25	0.3665	30.40	30.61
382.35	0.3553	30.21	30.42
382.55	0.3473	30.08	30.28
377.15	0.6723	35.37	35.54
377.15	0.6662	35.27	35.45
377.25	0.6631	35.22	35.40
375.25	0.8428	38.14	38.15
375.25	0.8395	38.09	38.10
375.25	0.8395	38.09	38.10
375.35	0.8374	38.05	38.06
375.45	0.8288	37.91	37.93
375.45	0.8299	37.93	37.95
375.55	0.8360	38.03	38.04
375.65	0.8168	37.72	37.75
374.45	0.9121	39.27	39.22
376.45	0.9109	39.25	39.12
376.45	0.9101	39.23	39.10
374.45	0.9069	39.18	39.14
378.65	0.9051	39.15	38.93
373.95	0.9749	40.29	40.21
373.95	0.9736	40.27	40.19
373.95	0.9753	40.29	40.22
373.95	0.9743	40.28	40.21
374.05	0.9717	40.24	40.16

Source: Data from Ref. 41.

where H^v = enthalpy of vaporization of the water/acetic acid mixture, kJ/mol

x_1 = concentration of water, mole fraction

x_2 = concentration of acetic acid, mole fraction

H_1^v = enthalpy of vaporization of pure water, kJ/mol

H_2^v = enthalpy of vaporization of pure acetic acid, kJ/mol

T = temperature, K

$a = 14.9231 \quad b = -0.0280834 \quad c = 0.0039797$

F. Dielectric Constant and Dissociation Constant of Acetic Acid in Water/Acetic Acid Mixtures

Kilpi and Lindell [42] measured the dielectric constant and the dissociation constant of acetic acid in water/acetic acid mixtures. The results are shown in Table 13. It should be noted that in the concentration range used, the

Table 13 The Dielectric Constant and the Dissociation Constant of Acetic Acid in Water/Acetic Acid Mixtures at 298.15 K

Conc. of water (wt%)	Dielectric constant	pK$_d$
100.0	78.5	4.754[a]
99.4	82.0	4.678
98.5	83.0	4.680
97.1	83.5	4.669
95.0	83.5	4.646
93.2	83.5	4.660
90.0	83.0	4.67
80.0	78.5	4.75
70.0	72.0	4.85
60.0	63.3	4.97
50.0	56.0	5.18
40.0	45.5	5.44
30.0	38.5	5.89
20.0	27.0	6.65
10.0	17.0	8.10
5.0	13.32	9.64
2.5	9.76	10.95
1.0	7.62	12.13
0.0	6.21	13.92

[a]pK$_d$ = $-\log(K)$, where K is the dissociation constant.
Source: Data from Ref. 42.

value of the dielectric constant could not be determined directly and had to be deduced from the electrolyte effect on the acid dissociation constant, as calculated by the Debye-Huckel equation.

Lown and Thirsk [43] found that the dissociation constant of acetic acid in water is between 4 and 26% greater at the solution vapor pressure than at 1 atm, in the temperature range between 523 and 647 K.

G. Refractive Index of Water/Acetic Acid Mixtures in the Liquid State

Arich and Tagliavini [44] developed Eq. (31) to correlate the refractive index of water/acetic acid mixtures as a function of concentration.

$$n_D{}^{20} = A - Bw_1 + s(1 - s)(C + D(2s - 1)^2) \tag{31}$$

where $s = w_1/(2 - w_1)$
        w_1 = concentration of water, weight fraction
        $A = 1.37161$
        $B = 0.03861$
        $C = 0.0563$
        $D = 0.02$

With the temperature fixed at 293.15 K and using the sodium-D line as the light source, they obtained the listed constants for Eq. (31). According to Arich and Tagliavini, Eq. (31) with these constants yields the refractive index of water/acetic acid mixtures at 293.15 K to within ±0.00005.

Garwin and Haddad [45] and Campbell and Gieskes [46] reported the values tabulated in Table 14 for the refractive indices of various water/acetic acid mixtures, using the sodium-D line as the light source.

Levitman and Ermolenko [47] measured the refractive index at 287.15, 293.15, and 303.15 K at 39 different concentrations between pure water and pure acetic acid. They noted that at all temperatures the curves show maxima at approximately 67 mole % acetic acid.

H. Surface Tension of Mixtures of Water and Acetic Acid

An equation first derived by Butler [48] was transformed by Eriksson [49] into Eq. (32) to provide a means for estimating the surface tension of a mixture of water and acetic acid as a function of its composition and temperature.

$$x_1 \exp(A(\sigma_M - \sigma_1)/T) + x_2 \exp(B(\sigma_M - \sigma_2)/T) = 1.0 \tag{32}$$

Table 14 Refractive Index Data for the System: Water/Acetic Acid

Conc. of water (wt. fraction)	Refractive index[a] at 298.15 K using sodium-D light	Conc. of water (wt. fraction)	Refractive index[b] at 298.15 K using sodium-D light
1.000	1.3325	1.0000	1.3328
0.900	1.3395	0.9055	1.3395
0.800	1.3462	0.8140	1.3458
0.700	1.3528	0.7835	1.3480
0.600	1.3588	0.7293	1.3514
0.500	1.3640	0.6211	1.3585
0.400	1.3686	0.5411	1.3623
0.300	1.3724	0.4509	1.3668
0.200	1.3750	0.3747	1.3699
0.150	1.3755	0.3318	1.3718
0.100	1.3756	0.2979	1.3729
0.050	1.3745	0.1759	1.3759
0.000	1.3691	0.1531	1.3759
		0.1530	1.3761
		0.1294	1.3761
		0.1135	1.3757
		0.0910	1.3756
		0.0506	1.3743
		0.0306	1.3732
		0.0000	1.3702

[a]Data from Ref. 45.
[b]Data from Ref. 46.

where
x_1 = concentration of water, mole fraction
x_2 = concentration of acetic acid, mole fraction
σ_1 = surface tension of water, mN/m
σ_2 = surface tension of acetic acid, mN/m
σ_M = surface tension of the mixture, mN/m
T = temperature, K
A = 12.31987
B = 31.16203

Inherent in the constants A and B are the assumptions by Eriksson that the molar areas of water and acetic acid do not vary with concentration and that they are 0.17 nm^2 for water and 0.43 nm^2 for acetic acid.

Some experimental data from the *International Critical Tables* [3] and from Catchpole and Ellis [50] are compared in Table 15 with surface tension data estimated by means of Eq. (32). The corresponding surface tension

Table 15 Surface Tension Data for the System:
Water/Acetic Acid

| Temp. (K) | Conc. of water (mole fraction) | Surface tension | |
		Experimental (mN/m)	Eq. (32) (mN/m)
298.1	0.9933	65.32	63.42
298.1	0.9642	52.79	53.09
298.1	0.7558	37.81	38.24
298.1	0.4042	31.11	30.97
303.1	0.9969	67.98	66.35
303.1	0.9844	60.11	58.25
303.1	0.9497	50.53	50.21
303.1	0.8856	43.60	43.88
303.1	0.7694	38.38	38.23
308.1	0.9933	64.16	62.55
308.1	0.9642	52.12	52.33
308.1	0.8245	39.23	40.01
308.1	0.6396	34.16	34.06
308.1	0.2393	28.03	27.85
373.1	0.9750	39.50	49.13
373.3	0.9500	35.50	44.70
373.5	0.9000	32.50	39.50
374.3	0.7999	30.50	33.72
376.3	0.5999	27.00	27.50
383.0	0.2000	20.00	20.76

Source: Data from Refs. 3 and 50.

values for pure acetic acid, σ_2, were estimated by means of Eq. (47) of
Chapter 7 and the values for pure water, σ_1, were estimated by means of
Eq. (48).

VI. COMPOSITION OF MIXTURES OF WATER AND ACETIC ACID

A. Composition of Water/Acetic Acid Mixtures in the Gas State

Levy and Davis [51] used vapor density data for the water/acetic acid
system to show that acetic acid dimer dissociates into two molecules of
acetic acid monomer rather than into a molecule of water plus a molecule
of acetic anhydride. Arich and Tagliavini [44] found that the Duhem-
Margules equations are unable to correlate observed activity coefficient
data for the water/acetic acid system, in the temperature range between

343 and 363 K, even if vapor phase polymerization of acetic acid is taken into account. They concluded that a significant amount of hydration of acetic acid occurs in the vapor phase. Christian et al. [52] used vapor density and infrared data to conclude that trifluoroacetic acid forms a cyclic dihydrate in the vapor phase. The analogous structure, if formed from water and acetic acid, is shown below.

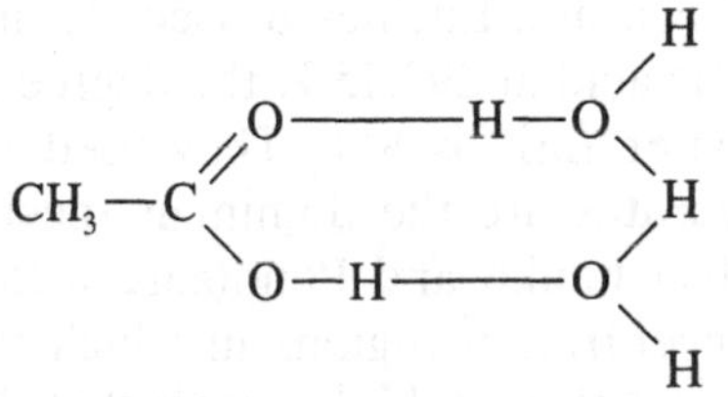

B. Composition of Water/Acetic Acid Mixtures in the Liquid State

Vilcu and Lucinescu [53] used spectral, cryoscopic, refractive index, density, and viscosity data to show that the following system of equilibria exists for mixtures of water and acetic acid in the liquid state.

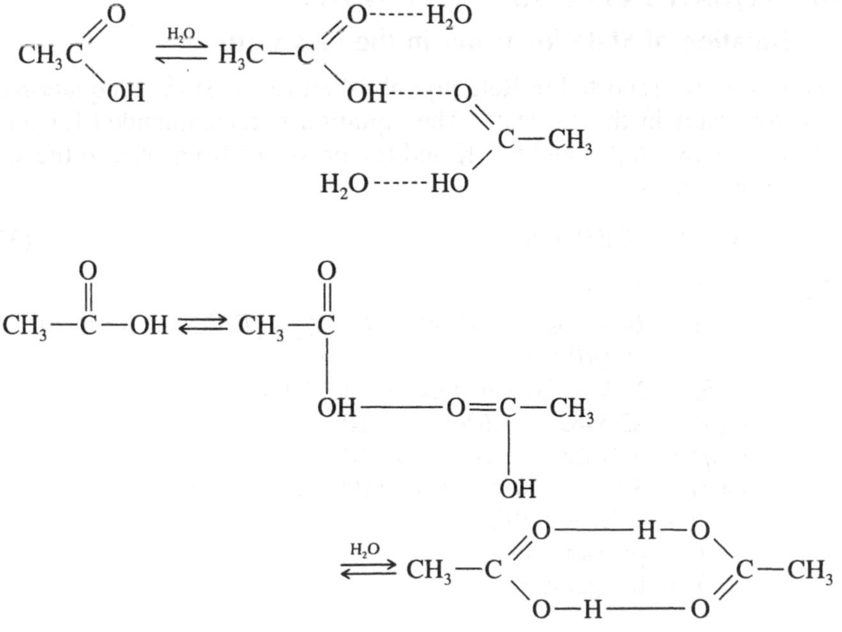

The system of equilibria shifts to the right as the concentration of acetic acid is increased and it shifts to the left as the temperature is increased. It is expected that for those concentration regions in which appreciable amounts of the open-chain dimer exist, there will also be trimers, tetramers, and higher polymers. Vilcu and Lucinescu [53] use cryoscopic data to show that for acid concentrations in excess of 95 wt%, the acid exists almost entirely in the form of the cyclic dimer and that the water that is present is largely not associated with the acid. Vilcu and Lucinescu used Raman spectral data to show that in 98 wt% acetic acid at 293.15 K the degree of association is 1.88 and the dimer/monomer ratio is 8:1. They used refractive index data to show that the hydrates are the dominant species between 31 and 50 wt% acetic acid. Mishchenko and Ponomareva [54] used entropy data to determine the concentration regions in which the various species are dominant. Corsaro and Atkinson [55] suggest that the surprising equilibrium at the left of the above sequence is made possible by the ability of water molecules to form cages around the acetic acid monomers.

Kilpi and Lindell [56] present data on the acid dissociation constant for acetic acid as a function of concentration in water. Appreciable ionization occurs only in very dilute solutions.

Lown and Thirsk [57] and North [58] present data on the influence of pressure on the dissociation constant of acetic acid in water.

VII. EQUATIONS OF STATE FOR WATER

A. Equation of State for Water in the Gas State

Keyes et al. [as reported in Ref. 59] published Eq. (33) as an equation of state for water in the gas state. The equation is recommended for temperatures between 273 and 617 K and for pressures from zero to the saturation pressure.

$$z = (AT/P + B)(PM/RT) \tag{33}$$

where
$$A = 461.5394$$
$$B = B_0 + B_0^2 g_1(q)qP/\Phi + B_0^4 g_2(q)(qP/\Phi)^3 - B_0^{13} g_3(q)$$
$$\times (qP/\Phi)^{12}$$
$$B_0 = 1.89 - 2641.62q \exp((431.5206q)^2)$$
$$g_1(q) = 82.546q - 1.6246q^2 \times 10^5$$
$$g_2(q) = 0.21828 - 1.2697q^2 \times 10^5$$
$$g_3(q) = 3.635 \times 10^{-4} - 6.768(100.0q)^{24} \times 10^{16}$$
$$q = 1/(T + 0.01)$$
$$P = \text{pressure, kPa}$$
$$T = \text{temperature, K}$$

Table 16 Compressibility Factor Data for Water in the Gas State

Temp. (K)	Pressure (kPa)	Compressibility factor	
		Literature	Eq. (33)
273.34	0.607	0.999	0.999
311.11	6.545	0.997	0.997
366.66	79.479	0.986	0.987
422.22	461.504	0.957	0.957
533.33	4688.419	0.804	0.804
477.77	689.465	0.963	0.964
516.66	3447.360	0.843	0.843
588.88	3585.253	0.913	0.914

Source: Data from Ref. 60.

$$M = 18.02$$
$$R = 8314.44$$
$$\Phi = 101.325$$
$$z = \text{compressibility factor}$$

Data from the *Steam Tables* by Keenan et al. [60] are compared in Table 16 with data estimated by means of Eq. (33).

B. Equations of State for Water in the Liquid State

Kravchenko [61] published Eq. (34) to provide estimates of the density of liquid water at orthobaric pressures, and it is recommended for use in the temperature range between 263 and 373 K.

$$\Phi d = A - (t - B)^2/(C + Dt - Bt^2) \tag{34}$$

where d = density, kg/m^3
  $t = T - 273.15$
  T = temperature, K
  $A = 1.0$
  $B = 4.0$
  $C = 1.190 \times 10/s5$
  $D = 1.365 \times 10^3$
  $\Phi = 0.001$

Keenan and Keyes [62] published Eq. (35) to provide estimates of the orthobaric density of liquid water at all temperatures between 273 and 634

K. However, Eq. (35) is somewhat less accurate than Eq. (34) below 373 K, so it is recommended for use in the temperature range between 373 and 634 K.

$$\Phi d = [1.0 + A(t_c - t)^{1/3} + B(t_c - t)]$$
$$\div [v_c + C(t_c - t)^{1/3} + D(t_c - t) + E(t_c - t)^4] \quad (35)$$

where d = density, kg/m^3
         $t = T - 273.15$
         T = temperature, K
         $A = 0.1342489$
         $B = -3.9462630 \times 10^{-3}$
         $C = -0.3151548$
         $D = -1.2033740 \times 10^{-3}$
         $E = 7.489080 \times 10^{-13}$
         $t_c = 374.11$
         $v_c = 3.1975$
         $\Phi = 0.001$

Some typical values estimated by means of Eq. (34) and (35) are compared in Table 17 with data from Perry [63] and from Keenan et al. [64].

According to Harned and Owen [66], the Tait equation [Eq. (9) of Chapter 7] does an excellent job of correlating the molar volume of water as a function of pressure at constant temperature.

By means of the Verschaffelt equation, Costello and Bowden [65] estimated the density of liquid water at absolute zero temperature to be 1150.0 kg/m^3.

Table 17 Experimental and Estimated Orthobaric Density Data for Water

		Density	
Temp. (K)	Experimental (kg/m^3)	Estimated by Eq. (34) (kg/m^3)	Estimated by Eq. (35) (kg/m^3)
263.2	998.10	998.10	
273.2	999.90	999.90	
293.2	998.20	998.20	
373.2	958.30	957.20	957.80
477.6	859.50		859.10
533.2	784.10		783.80
616.5	599.30		597.90

Source: Data from Refs. 63 and 64.

VIII. TRANSPORT PROPERTIES OF WATER

A. Thermal Conductivity of Water in the Gas State

Vargaftik et al. [67] developed Eq. (36) to correlate the thermal conductivity of steam at a pressure of 100 kPa for the temperature range between 373 and 973 K.

$$\Phi k^* = A + Bt + Bt^2 + Ct^3 \tag{36}$$

where k^* = thermal conductivity at 100 kPa, W/(m·K)
$t = T - 273.15$
T = temperature, K
$A = 4.2037 \times 10^{-5}$
$B = 1.4020 \times 10^{-7}$
$C = 2.4840 \times 10^{-10}$
$D = -1.0772 \times 10^{-3}$
$\Phi = 2.3884 \times 10^{-3}$

Kestin et al. [68] developed Eq. (37) to correlate the difference between the thermal conductivity of superheated steam at elevated pressures and the thermal conductivity at a pressure of 100 kPa. Equation (37) is valid in the pressure range from 100 to 50,000 kPa and in the temperature range from 373 to 973 K.

$$\Phi(k - k^*) = (A + Bt + Ct^2)(d/\beta) + (D/t^n)(d/\beta)^2 \tag{37}$$

where k = thermal conductivity, $W/(m·K)$
$A = 2.4723 \times 10^{-4}$
$B = 1.0027 \times 10^{-6}$
$C = -6.6184 \times 10^{-11}$
$D = 5.1309 \times 10^8$
$n = 4.20$
d = density, kg/m³
$\beta = 1000.0$

Thermal conductivity data estimated by means of Eq. (36) and (37) are compared in Table 18 to data from the skeleton tables adopted by the Sixth International Conference on the Properties of Steam [69]. For the calculations, the density term of Eq. (37) was obtained from the equation of state for steam, Eq. (33).

B. Thermal Conductivity of Water in the Liquid State

Bruges [70] developed Eq. (38) as a correlation of the thermal conductivity of water in the liquid state in the pressure range from the saturation pressure to 50 MPa and in the temperature range from 273 to 623 K.

Table 18 Adopted Experimental and Estimated Thermal Conductivity Data for Water in the Superheated Vapor State at Elevated Pressures and Temperatures

		Thermal conductivity	
Pressure (kPa)	Temp. (K)	Experimental (W/(m·K))	Eq. (36) and (37) (W/(m·K))
100.0	373.15	0.02479	0.02483
100.0	473.15	0.03316	0.03320
100.0	573.15	0.04338	0.04342
100.0	773.15	0.06737	0.06737
100.0	973.15	0.09429	0.09424
500.0	473.15	0.03379	0.03383
500.0	573.15	0.04379	0.04379
500.0	773.15	0.06766	0.06774
500.0	973.15	0.09458	0.09458

Source: Data from Ref. 69.

$$
k = A_0 + A_1(T/T_0) + A_2(T/T_0)^2 + A_3(T/T_0)^3 + A_4(T/T_0)^4
$$
$$
+ ((P - P_0)/E)[B_0 + B_1(T/T_0) + B_2(T/T_0)^2 + B_3(T/T_0)^3]
$$
$$
+ ((P - P_0)/E)^2[C_0 + C_1(T/T_0) + C_2(T/T_0)^2 + C_3(T/T_0)^3] \quad (38)
$$

where k = thermal conductivity, $W/(m·K)$
T = temperature, K
P = pressure, kPa
P_0 = saturation pressure, kPa
$A_0 = -9.2247 \times 10^{-1}$
$A_1 = 2.8395$
$A_2 = -1.8007$
$A_3 = 5.2577 \times 10^{-1}$
$A_4 = -7.3440 \times 10^{-2}$
$B_0 = -9.4730 \times 10^{-4}$
$B_1 = 2.5186 \times 10^{-3}$
$B_2 = -2.0012 \times 10^{-3}$
$B_3 = 5.1536 \times 10^{-4}$
$C_0 = 1.6563 \times 10^{-6}$
$C_1 = -3.8929 \times 10^{-6}$
$C_2 = 2.9323 \times 10^{-6}$
$C_3 = -7.1693 \times 10^{-7}$
$E = 100.00$
$T_0 = 273.15$

Thermal conductivity data adopted by the Sixth International Conference on the Properties of Steam [69] are compared in Table 19 with data estimated by means of Eq. (38). For the calculations, the saturation pressure of water was estimated by means of Eq. (44).

C. Viscosity of Water in the Liquid State

The Sixth International Conference on the Properties of Steam—Transport Properties of Water Substance, as reported by Kestin and Whitelaw [69], adopted Eq. (40) as a means for estimating the viscosity of water in the liquid state at pressures in the range from the saturation pressure to 80,000 kPa and in the temperature range from 273 to 573 K.

$$\mu = A \exp(B/(T - C)) \times (1.0 + D\Phi(P - P_s) \times (T - E)) \qquad (39)$$

where μ = viscosity, mPa·s
T = temperature, K
P = pressure, kPa
P_s = saturation pressure, kPa

Table 19 Experimental and Estimated Thermal Conductivity Data for Water in the Liquid State

Temp. (K)	Pressure (kPa)	Thermal conductivity	
		Experimental (W/(m·K))	Eq. (38) (W/(m·K))
273.15	100.0	0.5690	0.5687
273.15	10,000.0	0.5769	0.5770
273.15	20,000.0	0.5849	0.5849
273.15	30,000.0	0.5916	0.5924
273.15	50,000.0	0.6058	0.6061
373.15	10,000.0	0.6879	0.6877
373.15	20,000.0	0.6946	0.6945
373.15	30,000.0	0.7009	0.7009
373.15	50,000.0	0.7126	0.7126
473.15	10,000.0	0.6720	0.6720
473.15	20,000.0	0.6808	0.6806
473.15	30,000.0	0.6887	0.6888
473.15	50,000.0	0.7038	0.7044
623.15	20,000.0	0.4538	0.4539
623.15	30,000.0	0.4957	0.4960
623.15	50,000.0	0.5518	0.5518

Source: Data from Ref. 69.

$$A = 2.414 \times 10^{-2}$$
$$B = 570.58$$
$$C = 140.0$$
$$D = 1.39553 \times 10^{-9}$$
$$E = 305.0$$
$$\Phi = 7.5006$$

Equation (39) reproduces the best available data to within ±2.5% in the pressure range between the saturation pressure and 35,000 kPa (350 bar) and to within ±4.0% in the pressure range between 35,000 kPa and 80,000 kPa (800 bar).

D. Viscosity of Water in the Gas State

The Sixth International Conference on the Properties of Steam—Transport Properties of Water Substance, as reported by Kestin and Whitelaw [69], adopted Eq. (40) and (41) for estimating the viscosity of steam at the specified conditions.

Equation (40) was developed to estimate the viscosity of steam at 100 kPa (1.0 bar) in the temperature range between 373 and 973 K.

$$\mu^1 = A + Bt \tag{40}$$

where μ^1 = viscosity at 1 bar, mPa·s
$t = T - 273.15$
T = temperature, K
$A = 0.00804$
$B = 0.0000407$

Equation (41) was developed to estimate the viscosity of superheated steam in the pressure range between 1 bar and the saturation pressure in the temperature range between 373 and 573 K.

$$\mu = \mu^1 - \Phi d(A - Bt) \tag{41}$$

where μ = viscosity, mPa·s
μ^1 = low-pressure viscosity from Eq. (40), mPa·s
$t = T - 273.15$
T = temperature, K
d = density, kg/m^3
$A = 0.1858$
$B = 0.00059$
$\Phi = 0.001$

Equation (42) was developed to estimate the viscosity of superheated steam in the pressure range between 1 bar and 800 bar (789.5 atm) and in the temperature range between 653 and 973 K.

$$\mu = \mu^l + A(\Phi d) + B(\Phi d)^2 + C(\Phi d)^3 \tag{42}$$

where μ = viscosity, mPa·s
μ^l = low-pressure viscosity from Eq. (40), mPa·s
d = density, kg/m³
$A = 0.0353$
$B = 0.06765$
$C = 0.01021$
$\Phi = 0.001$

For the specified conditions, Eq. (41) reproduces the best available data to within ±1.0%. For the specified conditions, Eq. (42) reproduces the best available data to within ±4.0%.

Theiss and Thodos [77] critically evaluated most of the available data and arrived at Eq. (43) as a means for estimating the viscosity of water in the gas state at pressures in the range between 20 and 505 kPa and at temperatures below 973 K.

$$\mu = AT_r^B \tag{43}$$

where μ = viscosity, mPa·s
T_r = reduced temperature, T/T_c
T = temperature, K
T_c = critical temperature, 647.2 K
$A = 2.25 \times 10^{-2}$
$B = 1.07$

By using reduced state correlations of the viscosity of water as a function of pressure and temperature, Theiss and Thodos deduced that the viscosity of water at the critical point is $\mu_c = 43.0 \times 10^{-3}$ mPa·s.

IX. PHYSICAL PROPERTIES OF WATER

A. Vapor Pressure of Water

Keenan et al. [60] published Eq. (44) as a means for estimating the vapor pressure of pure water, and the equation is recommended for use in the temperature range between 273.15 and 647.29 K.

$$P = P_c \exp\Bigg((t_c - t)$$
$$\times \left(\sum_{i=1}^{8} F_i(0.65 - 0.01t)^{(i-1)} \right) \Bigg/ (100.0(t + 273.15)) \Bigg) \quad (44)$$

where P = vapor pressure, kPa
$t = T - 273.15$
T = temperature, K
$F_1 = 741.9242$
$F_2 = -29.72100$
$F_3 = -11.55286$
$F_4 = -0.8685635$
$F_5 = 0.1094098$
$F_6 = 0.439993$
$F_7 = 0.2520658$
$F_8 = 0.05218684$
$P_c = 1635.0673$ kPa
$t_c = 374.136$

Some data from the *Steam Tables*, of Keenan et al. [60], are compared in Table 20 with data estimated by means of Eq. (44).

Table 20 Experimental and Estimated Vapor Pressure Data for Water

Temp. (K)	Vapor pressure	
	Experimental (kPa)	Eq. (44) (kPa)
273.15	0.611	0.611
310.85	6.551	6.551
344.25	32.716	32.716
373.15	101.325	101.325
455.35	1,054.339	1,054.339
588.65	10,624.850	10,625.020
644.25	21,304.873	21,306.662
647.29	22,087.943	22,087.943

Source: Data from Ref. 64.

B. Heat Capacity of Water in the Ideal Gas State

Kobe et al., as reported by Hougen et al. [71], published Eq. (45) as a means for estimating the ideal gas heat capacity of water monomer in the temperature range between 273 and 3800 K.

$$\Phi C_P^0 = A + BT + CT^2 + DT^3 \tag{45}$$

where C_P^0 = heat capacity, J/(mol·K)
   T = temperature, K
   $A = 7.700$
   $B = 4.594 \times 10^{-4}$
   $C = 2.521 \times 10^{-6}$
   $D = -8.587 \times 10^{-10}$
   $\Phi = 0.23901$

In the specified temperature range, Eq. (45) is subject to a maximum error of 0.53% and an average error of 0.24%.

C. Heat Capacity of Water Dimer in the Ideal Gas State

Chao et al. [72] present data on the thermodynamic properties of water dimer in the ideal gas state. Their heat capacity data were correlated by means of Eq. (46).

$$\Phi C_P^0 = A + B(T/100.0) + C(T/100.0)^2 + D(T/100.0)^3 \tag{46}$$

where C_P^0 = heat capacity, J/(mol·K)
   T = temperature, K
   $A = 11.44238$
   $B = 0.91346$
   $C = 0.022082$
   $D = -0.001653$
   $\Phi = 0.23901$

Equation (46) is recommended for use in the temperature range between 200 K and 1500 K. Data presented by Chao et al. are compared in Table 21 with data estimated by means of Eq. (46).

D. Heat Capacity of Water in the Liquid State

After an extensive evaluation of published heat capacity data for water in the liquid state and at orthobaric pressures, Touloukian and Makita [73] developed the following two sets of constants for Eq. (47) to represent the data that in their judgment are the best available. These equations were

Table 21 Experimental and Estimated
Heat Capacity Data for Water Dimer in
the Ideal Gas State

| Temp. (K) | Heat capacity data | |
	Experimental (J/(mol·K))	Estimated by Eq. (46) (J/(mol·K))
200.00	55.789	55.789
298.15	59.986	59.907
300.00	60.224	59.982
500.00	68.187	68.425
700.00	76.864	˙76.781
900.00	84.843	84.709
1100.00	91.872	91.885
1500.00	102.784	102.646

Source: Data from Ref. 72.

found to reproduce the selected data with mean deviations of 0.14 and 0.59% and maximum deviations of 1.83 and 2.42%, respectively. These equations yield 4.209 kJ/(kg·K) at 366.48 K and 5.632 kJ/(kg·K) at 566.48 K.

$$\Phi C_P = A + BT + CT^2 + DT^3 \tag{47}$$

where C_P = heat capacity, kJ/(kg·K)
        T = temperature, K

For the temperature range between 273 K and 410 K,

$$A = 2.13974 \qquad C = 2.68536 \times 10^{-5}$$
$$B = -9.68137 \times 10^3 \qquad D = -2.42139 \times 10^8$$
$$\Phi = 0.23901$$

For the temperature range between 410 K and 590 K,

$$A = -11.1558 \qquad C = -1.74799 \times 10^{-4}$$
$$B = 7.96443 \times 10^{-2} \qquad F = 1.29156 \times 10^{-7}$$

E. Surface Tension of Water

Weissberger [74] published Eq. (48) to provide estimates of the surface tension of water, and it is recommended that it be used in the temperature range between 263 and 373 K.

Table 22 Experimental and Estimated Surface Tension Data for
Water

| Temp. (K) | Surface Tension | |
	Experimental (mN/m)	Eq. (48) (mN/m)
268.15	76.40	76.36
273.15	75.60	75.68
278.15	74.90	74.98
283.15	74.20	74.26
288.15	73.50	73.53
291.15	73.10	73.08
293.15	72.80	72.78
298.15	72.00	72.01
303.15	71.20	71.23
313.15	69.60	69.62
323.15	67.90	67.94
333.15	66.20	66.21
343.15	64.40	64.43
353.15	62.60	62.60
373.15	58.90	58.79

Source: Data from Ref. 75.

$$\sigma = A + Bt + Ct^2 + Dt^3 \tag{48}$$

where σ = surface tension, mN/m
$t = T - 273.15$
T = temperature, K
$A = 75.680$
$B = -0.138$
$C = -3.56 \times 10^{-4}$
$D = 4.70 \times 10^{-7}$

Some typical values estimated by means of Eq. (48) are compared in Table 22 with experimental data from Weast [75].

Dass and Singh [76] provided Eq. (49) as a means of estimating the surface tension of water in the temperature range between 273 and 373 K; however, Eq. (49) was found by the author to be less accurate than Eq. (48).

$$\ln \sigma = A + B(T - T_c)/(T - T_0) \tag{49}$$

where σ = surface tension, mN/m
T = temperature, K
A = 3.6385172 T_c = 647.2
B = -1.2181227 T_0 = 390.0

REFERENCES

1. Freeman, J. R., and Wilson, G. M., *High Temperature PVT Properties of Acetic Acid/Water Mixtures* and *High Temperature Vapor-Liquid Equilibrium Measurements on Acetic Acid/Water Mixtures*, AIChE Symposium Series 244, Vol. 81 (M. S. Benson and D. Zudkevitch, ed.), American Institute of Chemical Engineers, 345 East 47 Street, New York, NY, 10017, 1985, pp. 1–25.
2. Seaton, W. H., unpublished work for the Design Data Research Laboratory, Tennessee Eastman Co., Kingsport, TN, 1970–1974.
3. *International Critical Tables*, McGraw-Hill, New York, 1926.
4. Ragni, A., Ferrari, G., and Papoff, P., *Ann. Chim. (Rome)*, *45*, 960–969 (1955).
5. Campbell, A. N., and Kartzmark, E. M., *Can. J. Res.*, *28B*, 161–169 (1950).
6. Gibson, R. E., *J. Am. Chem. Soc.*, *57*, 284–293 (1935).
7. Chalov, N. V., and Aleksandrova, O. A., *Gidroliz. Lesokhim. Prom.*, *10*(6), 10–12 (1957).
8. Ito, T., and Yoshida, F., *J. Chem. Eng. Data*, *8*, 315–320 (1963).
9. Keyes, D. B., *Ind. Eng. Chem.*, *25*, 569 (1933).
10. Marek, J., *Collect. Czech. Chem. Commun.*, *21*, 269–280 (1956).
11. Marek, J., *Chem. Listy*, *49*, 957–969 (1955).
12. Brown, I., and Ewald, A. H., *Aust. J. Sci. Res.—Series A—Phys. Sci.*, *3*, 306–323 (1950).
13. Othmer, D. F., Silvis, S. J., and Spiel, A., *Ind. Eng. Chem.*, *44*, 1864–1872 (1952).
14. Ermolaev, M. I., Kapitanov, V. F., and Nesterova, A. K., *Zh. Fiz. Khim.*, *46*(3), 808 (1972).
15. Christian, S. D., *J. Phys. Chem.*, *61*, 1441 (1957).
16. Hansen, R. S., Miller, F. A., and Christian, S. D., *J. Phys. Chem.*, *59*, 391–395 (1955).
17. Wrewsky, M. S., Mischenko, K. P., and Muromzew, B. A., *Z. Phys. Chem.*, *13*, 362–369 (1928).
18. Arich, G., and Tagliavini G., *Ric. Sci.*, *28*, 2493–2500 (1958).
19. Tsonopoulos, C., and Prausnitz, J. M., *Chem. Eng. J.*, *1*, 273–278 (1970).
20. Hirado, M., Hirose, Y., Ohmi, A., and Kobayashi, J., *Kagaku Kogaku*, *23*(6), 403–405 (1959).
21. Pickering, S. U., *J. Chem. Soc.*, *63*, 998–1027 (1893).
22. Null, H. R., *Phase Equilibrium in Process Design*, Wiley-Interscience, New York, 1970.
23. Bidstrup, D. E., and Geankoplis, C. J., *J. Chem. Eng. Data*, *8*(2), 170–173 (1963).

24. Thorat, R. R., and Nageshwar, G. D., *Chem. Petro-Chem. J.*, *8*(10), 23 (1977).
25. Perrault, G. G., *Compt. Rend.* *260*(11)(Group 7), 3049–3052 (1965).
26. Wolf, A. V., Brown, M. G., and Prentiss, P. G., in *Handbook of Chemistry and Physics*, 57th ed. (R. C. Weast, ed.), Chemical Rubber Publishing Co., Cleveland, OH, 1976, p. D-220.
27. Usmanov, I. U., *Izv. Akad. Nauk Uzb. SSR, Ser. Tekhn. Nauk*, *12*(6), 61–64 (1968).
28. Filippov, L. P., *Int. J. Heat Mass Transfer*, *11*, 331–345 (1968).
29. Bretsznajder, S., *Prediction of Transport and Other Physical Properties of Fluids*, Pergamon Press, New York, 1971, p. 159.
30. Golubev, I. F., *Viscosity of Gases and Gas Mixtures*, translated from the Russian language by the Israel Program for Scientific Translations, available from the U.S. Department of Commerce as TT 70-50022, 1970.
31. Lemlich, R., Gottschlich, C., and Hoke, R., *J. Chem. Eng. Data*, *2*(1), 32–35 (1958).
32. Konicek, J., and Wadso, I., *Acta Chem. Scand.*, *25*, 1541–1551 (1971).
33. Gray, P., in *Progress in International Research on Thermodynamic and Transport Properties* (J. F. Masi and D. H. Tsai, eds.), Academic Press, New York, 1962, pp. 100–106.
34. Richards, T. W., and Gucker, F. T., *J. Am. Chem. Soc.*, *47*, 1876–1893 (1925).
35. Bury, C. R., and Davies, D. G., *J. Chem. Soc.*, *1932*, 2413–2417 (1932).
36. Parker, V. B., in *Handbook of Chemistry and Physics*, 57th ed. (R. C. Weast, ed.), Chemical Rubber Publishing Co., Cleveland, OH, 1976, p. D-103.
37. Casanova, C., Wilhelm, E., Grolier, J-P. E., and Kehiaian, H. V., *J. Chem. Thermodyn.*, *13*, 241–248 (1981).
38. Vilcu, R., and Lucinescu, E., *Bull. Insti. Politeh. Din Isai*, *16*(1–2), 29–39 (1970).
39. Sabinin, V. E., Belousov, V. P., and Morachevskii, A. G., *Izv. Vysshikh Ucheb. Zaved., Khim. I Khim. Tekhnol.*, *9*(6), 889–891 (1966).
40. Van Sickle, D. E., and Seaton, W. H., unpublished work of the Design Data Research Laboratory, Tennessee Eastman Co., Kingsport, TN, (1974).
41. Maklov, E. A., Shaburov, M. A., and Poplavskii, Yu. V., *Tsentr. Nauchno-Issled. Lesokhim. Inst., Khimki, (USSR). Sb. Tr. Tsentr. Nauchno-Issled. Proektn. Inst. Lesokhim. Prom-sti.*, *25*, 94–99 (1976).
42. Kilpi, S., and Lindell, E., *Ann. Acad. Sci. Fennicae, Ser. A. II*, *136*, 10 pp. (1967); *Chem. Abstr.*, *67*, 26381 (1967).
43. Lown, D. A., and Thirsk, H. R., *J. Chem. Soc., Faraday Trans.*, *I*, *68*, 1982–1986 (1972).
44. Arich, G., and Tagliavini, G., *Ric. Sci.*, *28*, 2493 (1958).
45. Garwin, L., and Haddad, P. O., *Anal. Chem.*, *25*(3), 435–437 (1953).
46. Campbell, A. N., and Gieskes, J. M. T. M., *Can. J. Chem.*, *43*, 1004–1011 (1965).
47. Levitman, S. Y., and Ermolenko, N. F., *Zhur. Obshchel Khim. (J. Gen. Chem.)*, *18*, 1567–1572 (1948).
48. Butler, J. A. V., *Proc. Roy. Soc. Lond.*, *A135*, 348 (1932).

49. Eriksson, J. C., *Adv. Chem. Phys.*, 6, 145–174 (1964).
50. Catchpole, J. P., and Ellis, S. R. M., *J. Chem. Eng. Data*, 8, 418–419 (1963).
51. Levy, B., and Davis, T. W., *J. Am. Chem. Soc.*, 76, 3268–3269 (1954).
52. Christian, S. D., Affsprung, H. E., and Ling, C., *J. Chem. Soc.*, 1965, 2378–2381 (1965).
53. Vilcu, R., and Lucinescu, E., *Rev. Roum. Chim.*, 14, 283–290 (1969).
54. Mischenko, K. P., and Ponomareva, A. M., *Thermodinam. i Stoenie Rasvorov, Akad. Nauk S.S.S.R., Otdel. Khim. Nauk, Khim. Fak. Mosk. Gosudarst. Univ., Trudy Soveshcaniya. Mosc.*, 245–251 (1958); *Chem. Abstr.*, 54, 21966 (1960).
55. Corsaro, R. D., and Atkinson, G., *J. Chem. Phys.*, 55(1), 1971–1978 (1971).
56. Kilpi, S., and Lindell, E., *Ann. Acad. Sci. Fenn., Ser. A.II, 136*, 10 pp. (1967).
57. Lown, D. A., and Thirsk, H. R., *J. Chem. Soc., Faraday Trans., 1*, 68, 1982–1986 (1972).
58. North, N. A., *J. Phys. Chem.*, 77(7), 931–934 (1973).
59. Keenan, J. H., and Keyes, F. G., *Thermodynamic Properties of Steam*, Wiley, New York, 1936, p. 15.
60. Keenan, J. H., Keyes, F. G., Hill, P. G., and Moore, J. G., *Steam Tables*, Wiley, New York, 1969.
61. Kravchenko, V. S., *At. Energ. (USSR)*, 20(2), 168 (1966); *Chem. Abstr.*, 64, 18895 (1966).
62. Keenan, J. H., and Keyes, F. G., *Thermodynamic Properties of Steam*, Wiley, New York, 1936.
63. Perry, J. H., ed., *Chemical Engineers' Handbook*, 4th ed., McGraw-Hill, New York, 1963.
64. Keenan, J. H., Keyes, F. G., Hill, P. G., and Moore, J. G., *Steam Tables*, Wiley, New York, 1969.
65. Costello, J. M., and Bowden, S. T., *Rec. Trav. Chim.*, 78, 391–403 (1959).
66. Harned, H. S., and Owen, B. B., *The Physical Chemistry of Electrolytic Solutions*, 3rd ed., Reinhold, New York, 1958, Chapter 8.
67. Vargaftik, N. V., Tarzimanov, A. A., and Oleschuk, O. N., *Draft Skeleton Tables on Steam—Thermal Conductivity*, report prepared for the Meeting of the International Thermal Conductivity Panel of the 6th International Conference on the Properties of Steam, Paris, France, 1964. See Ref. 69.
68. Kestin, J., Whitelaw, J. H., and Zien, T. F., *Thermal Conductivity of Superheated Steam*, report prepared on behalf of the U.S. Delegation for the Third Formal Conference of the International Commission on the Properties of Steam, Providence, RI, 1963. See Ref. 69.
69. Kestin, J., and Whitelaw, J. H., *J. Eng. Power*, 88(1), 82–104 (1966).
70. Bruges, E. A., Report prepared for the Meeting of the International Thermal Conductivity Panel of the 6th International Conference on the Properties of Steam, Paris, France, 1964. See Ref. 69.
71. Hougen, O. A., Watson, K. M., and Ragatz, R. A., *Chemical Process Principles*, 2nd ed., Part 2, Wiley, New York, 1966, Appendix.
72. Chao, J., Wilhoit, R. C., and Zwolinski, B. J., *J. Chem. Thermo.*, 1971(3), 195–201 (1971).

73. Touloukian, Y. S., and Makita, T., *Thermophysical Properties of Matter*, Vol. 6, IFI/Plenum, New York, 1970, p. 102.
74. Weissberger, A., *Physical Methods for Organic Chemistry*, Vol. 1, Interscience, New York, 1945, p. 163.
75. Weast, R. C., ed., *Handbook of Chemistry and Physics*, 52nd ed, Chemical Rubber Publishing Co., Cleveland, OH, 1976, p. F-30.
76. Dass, N., and Singh, O., *J. Phys. Soc. Jpn.*, *28*, 806 (1970).
77. Theiss, R. V., and Thodos, G., *J. Chem. Eng. Data*, *8*, 390–395 (1963).
78. Korpella, J., *Acta Chem. Scand.*, *25*(8), 2852–2864 (1971).

III

Selected Acetic Acid Derivatives

9

Acetic Anhydride

Steven L. Cook
Eastman Chemical Company, Kingsport, Tennessee

I. INTRODUCTION

Acetic anhydride is a colorless liquid with a pungent odor similar to acetic acid. Acetic anhydride has been manufactured and used for many years. The first synthesis was reported in 1852 and was accomplished by the reaction of potassium acetate with benzoyl chloride [1]. This method has been replaced by numerous industrial processes for the large-scale production of acetic anhydride. Nearly a million metric tons are produced annually, and it is used primarily for acetylation of the hydroxyl group of various substrates. The purpose of this chapter is to provide chemical and physical information on acetic anhydride. Commercial production of anhydride is also covered, and emphasis is placed on emerging technology related to this important industrial chemical.

II. PHYSICAL PROPERTIES

Acetic anhydride has the empirical formula $C_4H_6O_3$, M.W. = 102.09, and the structural formula $(CH_3CO)_2O$. It is frequently abbreviated Ac_2O and has been referred to by the synonyms "acetyl oxide," "ethanoic anhydride," "acetic oxide," "acetyl ether," "acetyl acetate," and "acetyl anhydride." Table 1 lists physical properties compiled from various sources.

Table 1 Key Physical Properties of Acetic Anhydride

Property	Value	Ref.
Boiling point (at 1 atm)	138.6°C	2
Melting point	−71.3°C	2
Specific gravity, liquid	1.08 (15.6°C)	3
Vapor density (air = 1)	3.52	3
Solubility in water	Decomposes	—
Flash point (tagged closed cup)	52°C	3
Autoignition temp., ASTM D 2155	332°C	3
Flammability limits		
$\quad$ LEL	2.8% at 81°C	3
$\quad$ UEL	12.4% at 129°C	3
Viscosity, cP	0.90 at 20°C	4
Specific heat	1817 J/kg at 20°C	5
Refractive index, n_D^{20}	1.3892	6
Heat of vaporization	406.6 J/g	6
Surface tension, dyn/cm	32.65 at 20°C	7
Dipole moment, μ	9.3×10^{-30} C·m	8
Dielectric constant, ϵ^{20}	22.1	9

From a reactivity standpoint, acetic anhydride is considered stable. It is incompatible with strong oxidizing materials and has the flammability rating of a combustible liquid. It is slightly soluble in water and may react violently after an induction period to produce acetic acid as heat is evolved. This induction effect is probably related to the formation of the acetic acid. As the hydrolysis takes place, the acetic acid tends to act as a cosolvent, which allows more anhydride to dissolve. The effect of increasing solubility, rising temperature, and perhaps the catalytic effect of the forming acid combine to cause an exponential increase in the reaction rate. The heat of reaction of acetic anhydride with water is −13.96 kcal/g-mol at 25°C [10].

Commercially produced acetic anhydride is available in purities up to 99.7%. The primary contaminant is acetic acid. Several schemes have been devised to remove this residual acid [11,12], but the commercial grade is adequate for most uses. Traces of substances that reduce potassium permanganate may also be present, such as mesityl oxide and 2,4-pentanedione (~100–200 ppm), and may be determined by titration with dilute permanganate solution. Methods exist for removal of these impurities [13]. Traces of diketene may also be present in the range of 5–100 ppm, and this impurity is claimed to be removable by photolysis [14]. Assay can be performed by various techniques [15], the most convenient of which is gas chromatography.

III. CHEMICAL PROPERTIES AND MAJOR USES

The principal reaction in which acetic anhydride employed as the key reagent is the acetylation of the hydroxyl group to form the acetate:

$$(CH_3CO-)_2O + R-OH \longrightarrow CH_3CO_2R + CH_3CO_2H \tag{1}$$

Many of these esterifications proceed without the requirement of a catalyst, thereby eliminating the potential need for a catalyst separation step. For the commercial production of large quantities of acetyl esters, however, the continuous, acid-catalyzed reaction of acetic acid with the alcohol is more economical. Acetic anhydride reacts in a manner analogous to acetyl chloride, but at a slower rate, and acetic acid is formed rather than highly corrosive HCl. (Acylating agents, RCOX, generally follow the reactivity series X = OH < NR_2 < OR < OCOR < halogen [16].) When mixed with strong acids, acetic anhydride ionizes to the acetylium ion, which participates in typical electrophilic aromatic substitution reactions [17]. Acetic anhydride reacts with hydrogen peroxide to form peracetic acid [18]. Care must be taken, however, to avoid conditions that result in formation of the highly explosive diacetyl peroxide.

The chief use for acetic anhydride is in the conversion of wood cellulose to cellulose acetate. This reaction generates one mole of acetic acid per mole of converted hydroxyl group. For this reason, much of the anhydride produced in commercial plants is captive, and the acetic acid is recovered and reconverted to anhydride. (See the chapter on cellulose esters for more information on this topic.) Somewhat related to this use is the treatment of wood and wood products with anhydride to impart improved chemical, weathering, and biological decay resistance [19–23]. This type of treatment also improves the mechanical properties [24,25] and the dimensional stability of the wood substrate, especially with regard to the effects of water in liquid or vapor form [26]. Wood particles treated with anhydride have even been used as fillers for plastics [27]. Acetylated wood is also more resistant to attack by termites [28]. Coals have also been acetylated with anhydride for modification [29] and analytical reasons [30].

In the area of drug and food applications, a well-known and beneficial use for high-purity acetic anhydride is in the manufacture of aspirin via the acetylation and salicylic acid [31–33]. Anhydride/propylene carbonate mixtures have been used for drug analysis [34]. Structural studies of nucleoprotein particles are accomplished with the aid of anhydride [35]. Citrate esters, which serve as plasticizers for medical articles, have been prepared via acetylation with acetic anhydride [36]. Acetylated plant proteins for the food industry have been reported [37,38]. Acetoxy derivatives of α-lipoic acid can be prepared by the mild acetylation of β-lipoic acid with acetic anhydride [39].

In chemical synthesis, cyclic polyimides are prepared using acetic anhydride [40]. Cyanoacrylic acid amides are prepared with the aid of anhydride [41]. Cyclic ethers can be cleaved with a mixture of magnesium bromide and acetic anhydride [42]. Acetic anhydride is also useful for the synthesis of enol acetates. Potassium acetate has been used as a catalyst [43], but an improvement utilizes N,N-dimethylaminopyridine [44]:

$$(CH_3CO-)_2O + CH_3(CH_2)_2CHO \xrightarrow[Et_3N]{DMAP} CH_3CH_2CH{=}CHOAc \qquad (2)$$

Although applicable to aldehydes only, this reaction proceeds under mild conditions and is thus useful for thermally sensitive compounds.

Another key use for acetic anhydride is in the synthesis of higher anhydrides. When acetic anhydride is mixed with a fatty acid, disproportionation occurs, and an equilibrium mixture of both acids, the mixed anhydride and the symmetrical anhydrides, form. The most volatile component, acetic acid, is removed from the refluxing mixture by use of an efficient fractionating column, and the reaction is thus driven to completion.

A breakdown of consumption of acetic anhydride in the United States is provided in Table 2 [45].

Table 2   U.S. Consumption of Acetic Anhydride, 1988

Use	Consumption (millions of pounds)	Percent
Cellulose acetate		
Cigarette filter tow	810	48
Filament yarn	330	19
Flake export	270	16
Other		
Cellulose triacetate	50	2.9
Cellulose mixed ester	50	2.9
Coatings	35	2.0
Aspirin	16	0.9
Acetaminophen	12	0.7
Miscellaneous	130–135	7.8
Total	1700	100

Source: Used with the permission of SRI International.

IV. COMMERCIAL ROUTES TO ACETIC ANHYDRIDE

The first commercial route to acetic anhydride was via the reaction of phosphorous oxychloride with potassium acetate. Variations on this general procedure were used through the early part of the 1900s. The most recent version of this route utilizes a solid-phase copolymer of 4-vinylpyridine to catalyze the formation of a variety of anhydrides from a mixture of thionyl chloride and the carboxylic acid [46]. Work toward more practical commercial routes to anhydride eventually led to four distinct processes. The starting materials for these processes consist of acetone, acetaldehyde, acetic acid, and methyl acetate. The first three routes have been discussed extensively in the literature and will therefore be described only briefly.

A. Acetone Process

The acetone process utilizes the thermal decomposition of acetone to ketene and methane as the key initial step:

$$CH_3COCH_3 \xrightarrow{\Delta} CH_2{=}C{=}O + CH_4 \tag{3}$$

This is an endothermic reaction and is typically carried out in heated metal tubes at a temperature of 700–800°C and atmospheric pressure. The ketene is then passed through a packed absorber fed with acetic acid to give acetic anhydride:

$$CH_2{=}C{=}O + CH_3CO_2H \longrightarrow (CH_3CO)_2O \tag{4}$$

Details of the design and construction of an acetone-based anhydride process have been published [47]. A major drawback to the acetone process is related to the formation of coke in the reactor tubes. A variety of metals and feed additives have been tested in an attempt to reduce this problem. Optimized conditions require that the conversion be kept low (<25%), so a considerable amount of acetone recycle is required.

B. Acetic Acid Process

This process has a major similarity to the acetone process in that ketene is generated and reacted with acetic acid in an absorber to produce acetic anhydride. The process begins with a step in which glacial acetic acid is vaporized and mixed with about 0.25% triethylphosphate (TEP). The mixture is then preheated and fed into a tubular reactor where the acid is cracked to ketene and water:

$$CH_3CO_2H \xrightarrow{\Delta} CH_2{=}C{=}O + H_2O \tag{5}$$

The TEP acts to increase the selectivity to ketene. The metal tubes are usually constructed of a chrome-iron alloy and can be externally heated by burning natural gas. The temperature is controlled at about 750°C, and the pressure maintained at 200 torr. To deter reversal of the reaction to reform acetic acid, ammonia—about 0.05–0.06 wt%—is injected into the hot gas as it exits the reactor. The conversion of acetic acid at this stage is about 90%. The reactor gas is then cooled, aqueous acetic acid is condensed from the stream, and the gas is then sent to a series of four absorbers. The first absorber is fed with a 15:85 acetic acid/acetic anhydride mixture. The effluent from this scrubber is passed through a second unit that is fed with glacial acetic acid. A third unit is fed with crude anhydride to scrub out trace acetic acid, and the fourth tower is used to condense trace anhydride from the off-gas. The pressure drop through this equipment is such that an 80-torr pressure at the vacuum source is required to maintain the 200-torr pressure at the vaporizer. The various streams from the scrubber system are combined and sent to the refining section of the plant. The key goals at this stage are to recover acetic acid for recycle and purify the anhydride. The required stills can be operated at atmospheric pressure, but operation at reduced pressure results in less decomposition of the anhydride to tars. Trace diketene, presumably resulting from slight decomposition of the anhydride to acetic acid and ketene followed by dimerization of the ketene, is also reduced when the stills are operated at reduced pressure. Anhydride is preferentially taken as a sidedraw from the final purification column. An article by Hunter (48) is a good initial reference to process details.

C. Acetaldehyde Process

A route to anhydride that does not depend on the generation of ketene utilizes acetaldehyde as the raw material. The conversion is accomplished by contacting acetaldehyde with air in a liquid-phase reaction [49] as outlined below:

$$CH_3CHO + O_2 \longrightarrow CH_3CO_3H$$

$$CH_3CO_3H + CH_3CHO \longrightarrow \qquad\qquad (6)$$

$$(AcO)_2O + H_2O \qquad 2CH_3CO_2H$$

The acetaldehyde monoperacetate appears to quantitatively decompose to anhydride, acetic acid, and water. The commercial success of this route is surprising at the outset, given that an equimolar amount of water is generated with the anhydride. To minimize product hydrolysis, a low reaction temperature ($\sim 50°C$) is utilized and the water is driven from the reactor by entrainment with the air used for the oxidation. The reaction can be catalyzed by a mixture of cobalt and copper acetates.

D. Methyl Acetate Carbonylation Process

The newest addition to commercially viable acetic anhydride production plants depends on a homogeneous catalysis system to convert methyl acetate to acetic anhydride:

$$CH_3CO_2CH_3 + CO \xrightarrow{\text{catalyst}} (CH_3CO)_2O \qquad (7)$$

This reaction seems to represent a carbonyl insertion in a formal sense; that is, a carbonyl group appears to have been inserted between the CH_3—O bond of methyl acetate. The mechanism by which this transformation occurs is actually more complicated and relies on a catalyst system that can include a group VIII noble metal, methyl halide/ionic halide mixture, and an acetate salt.

The first literature that described the carbonylation of methyl acetate appeared in a 1951 BASF patent [50] and has spawned numerous modified and improved versions of the general concept [51–59]. The key feature of a practical carbonylation process would include a metal complex that is capable of undergoing facile oxidative addition with a methyl halide, CO insertion between the methyl-metal bond, and reductive elimination of the acetyl group as the acetyl halide. This general requirement is illustrated below:

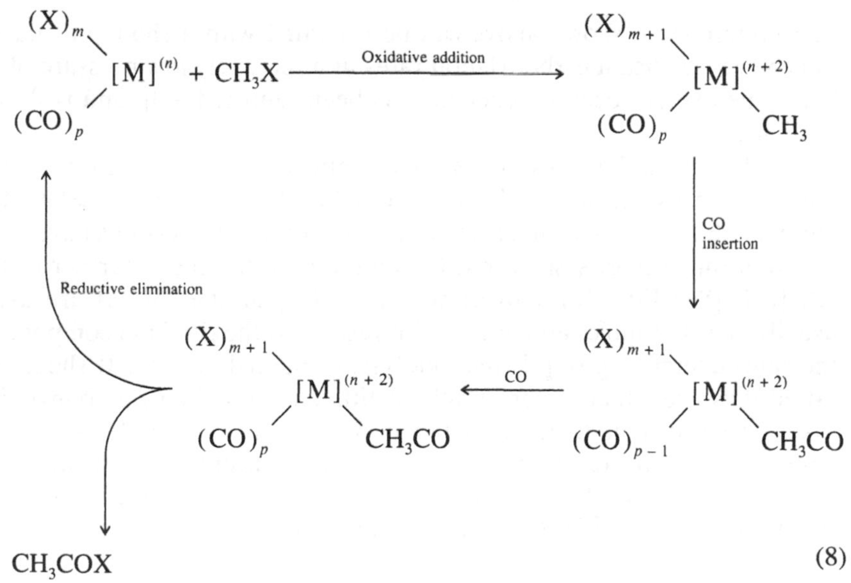

$$
\begin{array}{c}
(X)_m \\
\diagdown \\
[M]^{(n)} + CH_3X \quad \xrightarrow{\text{Oxidative addition}} \\
\diagup \\
(CO)_p
\end{array}
\qquad
\begin{array}{c}
(X)_{m+1} \\
\diagdown \\
[M]^{(n+2)} \\
\diagup \quad \diagdown \\
(CO)_p \qquad CH_3
\end{array}
$$

CO insertion

Reductive elimination

$$
\begin{array}{c}
(X)_{m+1} \\
\diagdown \\
[M]^{(n+2)} \\
\diagup \quad \diagdown \\
(CO)_p \qquad CH_3CO
\end{array}
\quad \xleftarrow{\text{CO}} \quad
\begin{array}{c}
(X)_{m+1} \\
\diagdown \\
[M]^{(n+2)} \\
\diagup \quad \diagdown \\
(CO)_{p-1} \qquad CH_3CO
\end{array}
$$

$$CH_3COX \qquad (8)$$

Thus, the group VIII metal complex bearing halide $(X)_m$ and $(CO)_p$ ligands reacts with the methyl halide to give an intermediate in which the metal, M. is higher in oxidation state $(n + 2)$. A CO insertion occurs (simultaneous with or followed by CO addition to the metal complex), and reductive elimination follows. The acetyl halide is liberated, and the original catalyst complex is regenerated. This can be referred to as the "organometallic" cycle in the process.

Another requirement of the overall process involves a series of reactions of the organic halide/halide salt mixture with methyl acetate and can be referred to as the "organic" cycle in the process. In this series of reactions, the acetyl halide generated in the organometallic sequence reacts with an acetate-bearing moiety to generate acetic anhydride. The process is illustrated below for a methyl iodide/lithium iodide/lithium acetate system:

$$CH_3COI + CH_3CO_2Li \longrightarrow (CH_3CO)_2O + LiI \tag{9}$$

$$LiI + CH_3CO_2CH_3 \longrightarrow CH_3I + CH_3CO_2Li \tag{10}$$

Thus, acetyl iodide reacts with lithium acetate to generate anhydride and lithium iodide. The lithium iodide reacts with methyl acetate to regenerate lithium iodide and furnish methyl iodide, which enters the organometallic cycle of reactions to form acetyl iodide. In the absence of lithium, the transformation can be achieved at a slower rate with proton as the counterion:

$$CH_3COI + CH_3CO_2H \longrightarrow (CH_3CO)_2O + HI \tag{11}$$

$$HI + CH_3CO_2CH_3 \longrightarrow CH_3I + CH_3CO_2H \tag{12}$$

The lithium cycle shown above can be operated with a rhodium catalyst to achieve significant carbonylation rates at a relatively low pressure of 50 bars. The overall reaction sequence has been reported [60], and is shown in Figure 1.

The lithium cation thus exerts a large effect on the overall rate of the reaction because of the iodolysis of methyl acetate to lithium acetate and methyl iodide, especially in an acetic acid solvent. Other cations have been tested in this type of system, but the equilibrium lies much further to the left [61]. [See Eq. (10) above.] When small quantities of LiI are used, usually in a 10:1 mole ratio or less with respect to the rhodium component, the rate-determining step is the iodolysis of methyl acetate. It should be noted, however, that at zero levels of lithium (or sodium) carbonylation takes place at a finite rate, so the non–alkali-assisted cycle is a lesser contributor to the reaction rate [62]. As the concentration of lithium is increased, the rate-determining step becomes the oxidative addition of methyl iodide to $Li[Rh(CO)_2I_2]$. A small amount of hydrogen is mixed

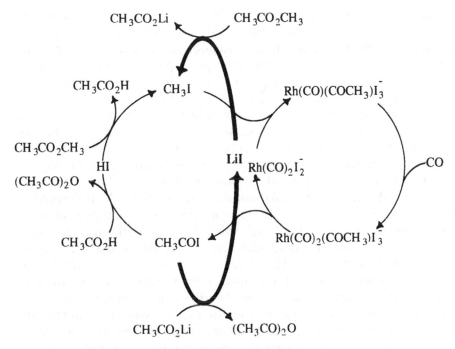

Figure 1 Mechanism for the carbonylation of methyl acetate to acetic anhydride.

with the carbon monoxide feed to induce and maintain this Rh(I) component and avoid formation of the inactive Rh(III) complex Li[Rh(CO)$_2$I$_4$]. A drawback to this essential additive is that some of the acetic anhydride is hydrogenated to ethylidene diacetate (EDA). The process can be carefully adjusted, however, so that EDA formation is minimized without sacrificing good catalyst activity.

Eastman Chemical Company, an Eastman Kodak Company based in Kingsport, Tennessee, practices the carbonylation of methyl acetate on an industrial scale. In this liquid phase process, the initial carbonylation mixture can consist of almost any rhodium salt (RhCl$_3$, RhI$_3$, etc.), lithium iodide, methyl iodide, methyl acetate, and acetic acid solvent. Following passage of the reactants through the reactor system at 50 bars of CO/H$_2$ pressure and 190°C, the liquid mixture is depressurized. A mixture of acetic anhydride, acetic acid, methyl acetate, and methyl iodide is stripped from the carbonylation fluid by flash distillation, and the nonvolatile carbonylation heel is combined with methyl acetate and various recycle streams and returned to the pressurized reactor system. A portion of the volatile components, chiefly methyl iodide and methyl acetate, is separated from the crude distillate. Acetic anhydride is isolated via a series of distillations, including steps to separate the remaining volatiles, as well as acetic acid and EDA. The refined anhydride is taken as a sidedraw from the final column. All other streams with the exception of EDA are returned to the process. If methanol is added at an appropriate point in the process, acetic acid and methyl acetate are produced. The additional methyl acetate is carbonylated to more acetic anhydride, and the acetic acid is removed as a coproduct and purified via the main anhydride-refining portion of the process. Even though the acetic anhydride is produced in an iodine-rich environment, the refined product contains <1 ppm total iodine. The iodine level can be reduced to <10 ppb by the use of proprietary technology.

Other by-products of the carbonylation process include "tars" or "heavies," formed partly by the breakdown of hot acetic anhydride. If the product is obtained by distillation from reaction mixtures, as is typical of many homogeneous reactions, the "tars" would build up in the reactor system as the catalyst containing distillation residues are returned and eventually lead to process shutdown. Therefore, a provision must be made for the removal of these "tars" and separation of the valuable occluded rhodium from these residues. The process used by Eastman Chemical Company in handling these "tars" is proprietary. The existing system is complex and is the result of over a decade of diligent research and development to generate and perfect the technologies and art for the recovery of the catalyst components which now permits the recovery of >99.9% of the catalyst components.

Although the exact process used by Eastman is not accessible, there are numerous patent references which describe several useful techniques including extraction [63,68–70], precipitation [64–67], and even an electrochemical method [71]. A particularly interesting approach was to partition a sample of the tar between an organic halide, preferably methyl iodide, and an HI solution. An aqueous layer containing rhodium and organic layer containing "tar" which is significantly reduced in rhodium content are obtained. A flow diagram of this extraction process has been published [61].

The carbonylation plant described above operates in combination with a variety of other processes such that the feedstock for production of acetyl is oxygen and the carbon and hydrogen derived from coal gasification. Key features of this industrial complex have been described [72]. The integrated anhydride production facility consists of a syngas complex to supply carbon monoxide and hydrogen to the chemical plants. The syngas is produced by two high-temperature, high-pressure Texaco gasifiers. These units operate at over 1000°C and are designed to use high-sulfur coal. This coal, at the rate of 900 tons/day, is ground and fed as a water slurry with pure oxygen to the gasifiers. The raw gas is scrubbed with water to lower the temperature and remove ash. Some of the gas is passed through a water-gas shift reactor to enrich the stream in hydrogen so that the stoichiometry required for methanol synthesis is matched. The gas is then scrubbed free of hydrogen sulfide via a Rectisol process developed by Linde AG. In this step, cold methanol absorbs H_2S and CO_2. The hydrogen sulfide gas is then passed through a Claus unit where it is converted to elemental sulfur, which is sold as a pure by-product. The off-gas from this operation is treated in a Shell-Claus absorber such that 99.7% of the total sulfur output is recovered. The cleansed gas from the Rectisol process is then processed at cryogenic temperatures in a Linde-developed "cold box" to further purify the CO and H_2, and heat integration is utilized extensively. The heat removed throughout the gas purification operations is used to generate process steam such that the thermal efficiency of the overall process is optimized. The refined CO/H_2 mixture is passed through a Lurgi AG low-pressure methanol process, which efficiently produces a pure stream of methanol. The methanol is reacted with acetic acid in a unique reactor-distillation column [73] in which the reactants are fed countercurrently. The sulfuric acid–catalyzed esterification proceeds through a series of countercurrent flashing stages. Above the reaction section, acetic acid flowing down the column extracts water and methanol, thus eliminating an otherwise troublesome azeotrope with methyl acetate. Above this section, acetic acid is separated from the methyl acetate. Below the reactor section, methanol is forced up the column by the water of reaction, which exits the column base. Methyl acetate is removed from the top of the column and is adequate for carbonylation with no additional purification steps. The design of this reactor column, coupled with careful control of the feed ratios, is such that the reactants are captured in the core of the unit until they react in the desired fashion. (For more details on this process, see Chapter 14.) A simplified flow diagram of the integrated plants is presented in Figure 2. The acetic anhydride generated by this complex is primarily used to convert wood cellulose to cellulose acetate. This plastic is thus produced entirely from raw materials that are in abundant supply and available near the plant site.

Several other group VIII metals, especially cobalt [74,75] and nickel [76–80], have been studied for methyl acetate carbonylation purposes. Of these two, the most promising metal is nickel, often in combination with another transition metal such as chromium [81–83]. The catalyst system has been reported to operate at conditions comparable to the rhodium version, although details on the recyclability, recoverability, and stability of the nickel component have not been published. It should also be noted

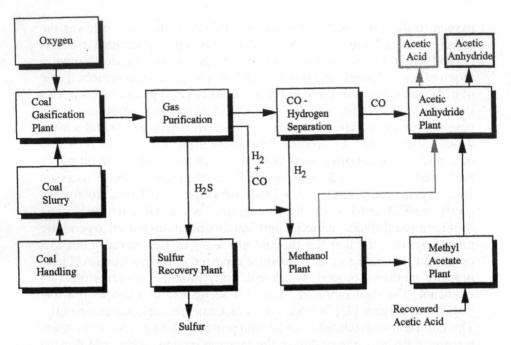

Figure 2   Block diagram of main steps required to produce acetic anhydride from coal.

Table 3   World Producers of Acetic Anhydride

Country	Annual capacity (millions of pounds)	Remarks
Australia	29	
Brazil	57	
Canada	64	Acetic acid/anhydride
France	287	
Germany	252	
India	4	
Italy	55	
Japan	338	
Mexico	173	
Rumania	33	Methanol carbonylation
Sweden	—	plants on standby
Switzerland	44	
United Kingdom	406	
United States	2620	
USSR	270	
Total	>4632	

Source: Used with permission of SRI International.

Table 4 U.S. Producers of Acetic Anhydride

Company	Annual capacity (millions of pounds)	Remarks
Tennessee Eastman	1820	1120 Mlb via methyl acetate carbonylation
Hoechst Celanese	800	300 Mlb via butane oxidation
Union Carbide	(250)	Butane oxidation, currently shut down

Source: Used with the permission of SRI International.

that most carbonylation systems will work with a dimethyl ether feed rather than methyl acetate, but the reaction rates are slower, the operating pressures are higher, and most users of anhydride on a large-scale, captive basis generate an acetic acid stream that can readily be esterified with methanol and carbonylated as methyl acetate.

V. PRODUCTION VOLUME

The estimated production volume of acetic anhydride on a worldwide basis is shown in Table 3. Over half the world anhydride capacity is in the United States [45]. Most of the U.S. anhydride production capacity is at Tennessee Eastman Division, the largest plant site of Eastman Chemical Company, a division of the Eastman Kodak Company. The estimated U.S. production capacity is summarized in Table 4 [45].

REFERENCES

1. Gerhardt, C., *Ann. Chim. Phys.*, *37*(3), 285 (1852).
2. McDonald, R. A., Shrader, S. A., and Stull, D. R., *J. Chem. Eng. Data*, *4*, 311 (1959).
3. *Acetic Anhydride Material Safety Data Sheet*, Eastman Chemical Co., Kingsport, TN, 1990.
4. Wright, F. J., *J. Chem. Eng. Data*, *6*, 454 (1961).
5. *Acids and Anhydrides*, Union Carbide Chemical Corp., New York, 1960.
6. Wright, F. J., *Rec. Trav. Chim. Pays Bas*, *79*, 784 (1960).
7. Timmermans, J., *Physico-Chemical Constants of Pure Organic Compounds*, Elsevier, New York, 1950.
8. Lippert E., and Moll, F., *Z. Electrochem.*, *58*, 718 (1954).
9. Jander, G., and Surawski, H., *Z. Electrochem.*, *65*, 3 (1961).
10. Pedley, P. B., Naylor, R. D., and Kirby, S. P., *Thermo-chemical Data of Organic Copounds*, 2nd ed., Chapman and Hall, New York, 1986, pp. 405.

11. Wagner, F. S., Jr., *Encyclopedia of Chemical Technology*, 3rd ed., Vol. 1, Wiley, New York, 1978, pp. 151–160.
12. Salton, J. H., and Withrow, L. L., *J. Am. Chem. Soc.*, *45*, 2689 (1923).
13. Rastogi M. K., and Bisarya, S. C., *Res. Ind.*, *33*, 25 (1988).
14. Miura, H., Jap. Patent 62,246537 (1987).
15. Joy, E. F., and Barnard, A. J., Jr., *Encylopedia of Industrial Chemical Analysis*, Vol. 4, Wiley, New York, 1967, pp. 102–107.
16. Sutherland, I. O., ed., *Comprehensive Organic Chemistry*, Vol. 2, Pergamon Press, Oxford, 1979, p. 875.
17. Schmidt, H., Wittkopf, I., and Jander, G., *Z. Anorg. Chem.*, *256*, 113 (1948).
18. Swern, D., ed., *Organic Peroxides*, Vol. 1, Interscience, New York, 1970, Chapter 6.
19. Sadate, M., Kobayashi, H., and Shigeru, H., Jap. Patent 63199604 (1988) (CA 110:9962n).
20. Rowell, R. M., Esenther, Gr. R., Darrel, D. D., and Nilsson, T., *J. Wood Chem. Technol.*, *7*, 427 (1987).
21. Rowell, R. M., Simonson, R., and Tillman, A. M., *Pap. Puu*, *68*, 740 (1986).
22. Shiraishi, N., Onodera, S., Otani, K., Masumoto, K., Tsubochi, K., and Mayahara, T., Jap. Patent 61171701 (1986) (CA 105:174671w).
23. Rowell, R. M., Tillman, A. M., and Simonson, R., *J. Wood Chem. Technol.*, *6*, 93 (1986).
24. Rowell, R. M., Imaura, Y., Kawa, S., and Norimoto, M., *Wood Fiber Sci.*, *21*, 67 (1989).
25. Rowell, R. M., and Banks, W. B., *Br. Polym. J.*, *19*, 479 (1987).
26. Imagawa, T., and Tsubochi, K., Jap. Patent 62236702 (1987) (CA 108:152417c).
27. Sato, T., Jap. Patent 62225303 (1987) (CA 108:58199m).
28. Imamura, Y., and Nishimoto, K., *Wood Res.*, *72*, 37 (1986).
29. Bailey, T. N., Lawson, G. J., and Monsef-Mirzai, P., *Fuel*, *69*, 533 (1990).
30. Baltisberger, J. R., Patel, K. M., Woolsey, N. F., and Stenberg, V. I., *Fuel*, *61*, 848 (1982).
31. Candoros, F., Rom. Patent 85726 (1984) (CA 103:104715y).
32. Drevina, V. M., Nesterov, V. M., and Markitanova, L. I., *Khim. Farm. Zh.*, *10*, 120 (1976).
33. Vodnar, I., Costin, D., and Joza, P., *Rev. Chim. (Bucharest)*, *20*, 401 (1969).
34. Evstratova, K. I., and Bakholdina, L. A., *Farmatsiya (Moscow)*, *39*, 62 (1990).
35. Nieto, M. A., and Palacian, E., *Biochem. J.*, *241*, 621 (1987).
36. Hull, E. H., Ger. Patent 3520750 (1986) (CA 105:60480c).
37. Schwenke, K. D., Kim, Y. H., and Krajewski, A., E. Ger. Patent 240125 (1986) (CA 106:212803w).
38. Schwenke, K. D., Nitecka, E., Kujawa, M., and Macholz, R., E. Ger. Patent 236448 (1986) (CA 106:31602j).
39. Saito, I., and Fukui, S., *J. Vitaminol.*, *12*, 244 (1966).
40. Maruyama, Y., Komoriya, H., and Tsutsumi, K., Jap. Patent 2053825 (1990) (CA 113:79257x).

41. Stockinger, F., Lohse, F., and Moser, R., European Patent 88047 (1983) (CA 100:7875f).
42. Campbell, G. R., *Diss. Abstr. Int. B 1975*, *35*, 5323 (1974).
43. Bedoukian, P. Z., *Organic Synthesis*, Collect. Vol, III, Wiley, New York, 1955, p. 127.
44. Cousineau, T. J., Cook, S. L., and Secrist, J. A., III, *Synth. Comm.*, *9*, 157 (1979).
45. *Chemical Economics Handbook*, *SRI International* (T. C. Gunn, director), Chemical Economics, Menlo Park, 1990.
46. Fife, W. K., and Zhang, Z-d., *Tet. Lett.*, *27*, 4933 (1986).
47. Jeffreys, G. V., *The Manufacture of Acetic Anhydride*, 2nd ed., Institution of Chemical Engineers, London, 1964.
48. Hunter, W., published by U.S. Dept. of Commerce as PB 69,123, from BIOS Field Report 1050, Feb. 2, 1947.
49. Allen, G. C., and Aguilo, A., *Adv. Chem. Ser.*, *76*, 363 (1968).
50. BASF, Fr. Patent 1073437 (1951).
51. Ajinomoto KK, Jap. Patent 50030820 (1973).
52. Showa Denko KK, Jap. Patent 50047922 (1973).
53. Halcon Int. Inc., Fr. Patent 2242362 (1973).
54. Hoechst AG, Fr. Patent 2289480 (1974).
55. Mitsubishi Gas Chem. Ind., Fr. Patent 2408571 (1977).
56. Rhone-Polenc, European Patent 48210 (1980).
57. Eastman Kodak Co., European Patent 64986 (1980).
58. Eastman Kodak Co., European Patent 64989 (1980).
59. Larkins, T. H., Polichnowski, S. W., Tustin, G. C., and Young, D. A., U.S. Patent 4,374,070 (1983).
60. Polichnowski, S. W., *J. Chem. Ed.*, *63*, 206 (1986).
61. Gauthier-Lafaye, J., and Perron, R., *Methanol and Carbonylation* (Eng. Trans.), Editions Technip, Paris, and Rhone-Polenc Recherches, Courbevoie, France, 1987, Chapter 7.
62. Zoeller, J. R., Agreda, V. H., Cook, S. L., Lafferty, N. L., Polichnowski, S. W., and Pond, D. M., *Catalysis Today*, *13*, 73 (1992).
63. Hembre, R. T., and Cook, S. L., U.S. Patent 4388217 (1983).
64. Erpenbach, H., Gehrmann, K., and Prinz, P., U.S. Patent 4746640 (1988).
65. Larkins, T. H., U.S. Patent 4434241 (1984).
66. Zoeller, J. R., U.S. Patent 4650649 (1987).
67. Pugach, J., European Patent 210,017 (1987).
68. Barnes, R. L., U.S. Patent 4364904 (1982).
69. Poslusky, J. V., and Palmer, B. J., U.S. Patent 4476238 (1984).
70. Gulliver, D. J., European Patent 255,389 (1988).
71. Pardy, R. B. A., U.S. Patent 4871432 (1989).
72. Agreda, V. H., *Chemtech*, *18*, 250 (1988).
73. Agreda, V. H., U.S. Patent 4435595 (1984).
74. BASF, U.S. Patent 2730546 (1956).
75. BASF, U.S. Patent 2789137 (1957).
76. BASF, U.S. Patent 2729651 (1956).

77. Rhone-Poulenc, U.S. Patent 4511517 (1985).
78. Rhone-Poulenc, U.S. Patent 4618460 (1986).
79. Mitsubishi, U.S. Patent 4544511 (1985).
80. Shell, U.S. Patent 4536354 (1985).
81. Halcon, U.S. Patent 4115444 (1978).
82. Halcon, U.S. Patent 4002678 (1977).
83. Halcon, U.S. Patent 4002677 (1977).

10

Ketene

Peter W. Raynolds
Eastman Chemical Company, Kingsport, Tennessee

I. PREPARATION OF KETENE AND OVERVIEW

Ketene is a thermodynamically stable, highly reactive, colorless, gaseous (b.p. $-48°C$) compound with a sharp, overpowering odor. It has been prepared in the laboratory by the pyrolysis of acetone [1] and, more conveniently in recent years, by the pyrolysis of diketene at ca. 400–500°C [2]. Ketene is made commercially by cracking acetic acid, as described in a previous chapter.

   Practical use of ketene as a reactant on a commercial scale has been
limited to a few specific processes for two reasons. First, the rate of self-
condensation of ketene is similar to that of many desired reactions, and
any process is complicated by the formation of diketene and tarry oligomers. Second, many products that could, in principle, be derived from
ketene are more easily made from acetic acid or anhydride. Nevertheless,
it is possible to design robust processes, and ketene has found commercial
utility in the preparation of acetic anhydride and diketene; these reactions
are described in other chapters. Here, we concern ourselves with the reactivity of ketene itself. Frequent reference will be made to the reactions of
disubstituted ketenes, since these more stable and easily handled compounds have been more intensively studied.

II. PATTERNS OF REACTIVITY

Ketene is polarized such that it is prone to nucleophilic attack at the central
carbon or, more rarely, to electrophilic reaction at a terminal atom [3,4].
A third mode of reactivity, the [2 + 2] cycloaddition to give four-membered-ring compounds, has been extensively studied by mechanistic and
synthetic chemists [5–7].

A. Nucleophilic Addition

Nucleophiles add to the central carbon of ketene to give derivatives of acetic acid. These reactions, although of interest, are generally not of practical utility, because the products are often more economically and practically made from acetic acid or anhydride. The reaction is straightforward and has been reviewed [3]. Wilsmore, in 1907, investigated the reaction of ketene with water to give acetic acid [8]. A second molecule of ketene may add to acetic acid to give the anhydride.

$$
\begin{array}{c}
\text{H}_2\text{O} \\
+ \\
\text{Ketene}
\end{array}
\longrightarrow
\underset{\text{CH}_3\text{COH}}{\overset{\overset{\text{O}}{\parallel}}{}}
\xrightarrow{\text{Ketene}}
\underset{\text{CH}_3\text{C O C CH}_3}{\overset{\overset{\text{O} \quad \text{O}}{\parallel \quad \parallel}}{}}
$$

Early mechanistic work on the addition of ketene to amines [9] established the role of steric and electronic factors that are important in all additions of this type. The following table shows how addition is slowed both by bulky substituents and by electron-withdrawing groups on the nucleophile. Other amines, such as hydrazines and the secondary amine of phenylhydrazone, add equally well.

$$
\text{Ketene} + \text{C}_6\text{H}_5\text{NH}_2 \longrightarrow \underset{\text{CH}_3\text{CNHC}_6\text{H}_5}{\overset{\overset{\text{O}}{\parallel}}{}}
$$

Amine	pKa of amine	Percent amine reacted after 1 h
Aniline	4.63	100
β-Naphthylamine	4.16	100
Ethylaniline	5.12	86
m-Nitroaniline	2.47	90
p-Nitroaniline	1.00	34
Diphenylamine	0.79	0.5

Alcohols afford esters, and again, the rates of reaction are subject to steric and electronic effects. Primary and secondary alcohols react readily with ketene, whereas tertiary alcohols and phenol are much slower. Thiols afford thioesters, whereas hydrogen sulfide yields a thioanhydride [10].

$$\underset{\text{Ketene}}{}\quad\xrightarrow{C_2H_5OH}\quad CH_3\overset{\displaystyle O}{\overset{\displaystyle \|}{C}}\,O\,Et$$

$$\xrightarrow{C_2H_5SH}\quad CH_3\overset{\displaystyle O}{\overset{\displaystyle \|}{C}}\,S\,C_2H_5$$

$$\xrightarrow{H_2S}\quad CH_3\overset{\displaystyle O}{\overset{\displaystyle \|}{C}}\,S\,\overset{\displaystyle O}{\overset{\displaystyle \|}{C}}\,CH_3$$

Unsymmetrical anhydrides are conveniently produced from the acid and a ketene [11].

$$(C_6H_6)_2\,C=C=O\ +\ C_6H_5\overset{\displaystyle O}{\overset{\displaystyle \|}{C}}OH\ \longrightarrow\ (C_6H_5)_2\,CH\,\overset{\displaystyle O}{\overset{\displaystyle \|}{C}}\,O\,\overset{\displaystyle O}{\overset{\displaystyle \|}{C}}\,C_6H_5$$

B. Addition to Single Bonds

Ketene inserts into single bonds [12]. In many cases, carbon-carbon bonds are formed under extraordinarily mild conditions—room temperature, high concentration, and a catalyst level of less than 5%. Unlike the nucleophilic products of ketene, these reactions are not easily duplicated with acetic acid and anhydride, and in some cases the products are difficult to obtain in any other way.

Ketene inserts into reactive carbon-halogen bonds. Generally, aliphatic chlorides like chloromethylmethyl ether require a Lewis acid catalyst, while trityl chloride affords a 1:1 adduct in a highly polar solvent, nitrobenzene, without catalyst [13].

$$Ketene\ +\ CH_3OCH_2Cl\ \longrightarrow\ CH_3OCH_2\,CH_2\overset{\displaystyle O}{\overset{\displaystyle \|}{C}}Cl$$

$$Ketene\ +\ (C_6H_5)_3\,CCl\ \longrightarrow\ (C_6H_5)_3\,CCH_2\overset{\displaystyle O}{\overset{\displaystyle \|}{C}}Cl$$

Acid halides react with ketene to give acid chlorides. For instance, phosgene reacts with 2 mol of ketene to give a potential raw material for citric acid [14,15], whereas thiophosgene affords a rare aliphatic thio acid chloride [16]. Ketenes also add to other acids chlorides to give β-ketoacid chlorides [17–19].

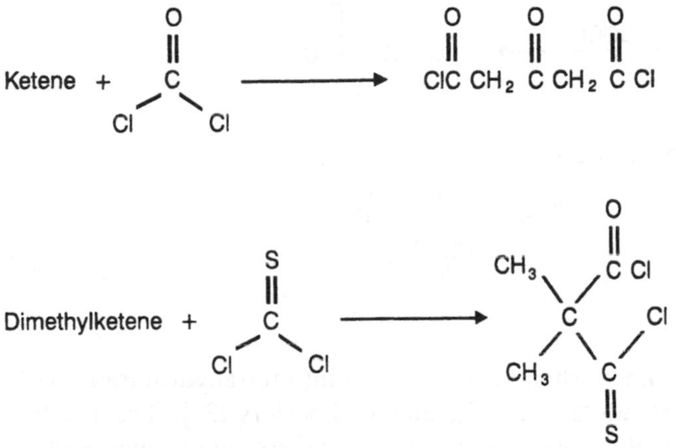
Ketene +
O
C
Cl
Cl
O O O
ClC CH2 C CH2 C Cl
Dimethylketene +
S
C
Cl
Cl
CH3
CH3
C
C Cl
O
Cl
C
S

It is not known if the reaction mechanism involves ionic intermediates, or if the bond making and breaking is concerted. The weak sulfur-chlorine bond is attacked without catalysts [20,21].

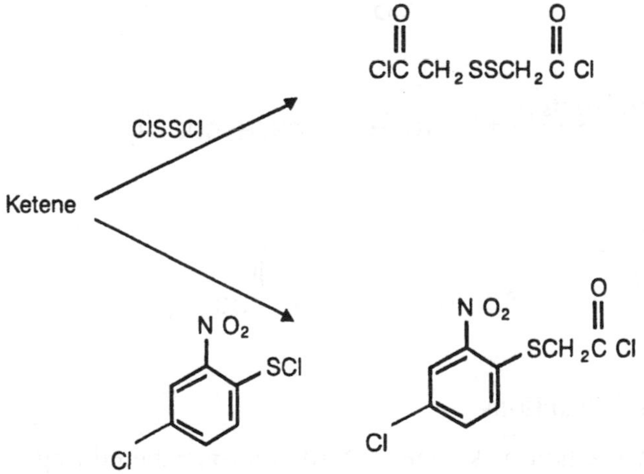
O O
ClC CH2SSCH2C Cl
ClSSCl
Ketene
N O2
SCl
Cl
N O2
SCH 2C Cl
O
Cl

Ketene inserts into hydrogen chloride [10] and bromide [10], as well as bromine [22], thionyl chloride [23], and sulfonyl chloride [23] in similar reactions, again giving acid halides.

$$CH_3CX \quad (X = Cl, Br)$$

Ketene $\xrightarrow{\text{HX}}$ (above)

Ketene $\xrightarrow{\text{SOCl}_2}$ $ClS \, CH_2 \, C \, Cl$ (with two C=O groups)

$\xrightarrow{\text{SO}_2Cl_2}$ $ClS \, CH_2 \, C \, Cl$ (with O above and below S, and C=O)

Esters are obtained when ketene inserts into tetrahydrofuran, diethoxymethane [25,28], acetals [26,27], and orthoesters [26]. The reaction is catalyzed by borontrifluoride etherate [24] and strong protonic acids [26].

Ketene $\xrightarrow{\text{CH(OC}_2\text{H}_5)_3}$ $(C_2H_5O)_2 \, CHCH_2 \, COC_2H_5$

$\xrightarrow{\text{C}_2\text{H}_5\text{OCH}_2\text{OC}_2\text{H}_5}$ $C_2H_5OCH_2 \, CH_2 \, COC_2H_5$ (with C=O)

C. Cycloaddition Reactions

The cycloaddition reaction of ketenes affords four-membered-ring, carbonyl-containing compounds with a high degree of peri-, regio-, and stereoselectivity. Mechanistically, the reaction has long been considered to be a concerted process, with both bonds being formed at nearly the same time [29]. Although more recent work has suggested that certain highly polar systems react via a process that almost certainly involves discrete, polar intermediates, recent theoretical calculations have upheld the concerted mechanism for nonpolar substrates [30]. The cycloaddition literature is vast [4–6, 31] and can only be summarized here.

1. [2 + 2] Cycloaddition

Ketenes add to olefins, as well as other double bonds such as C=N and C=O, in a [2 + 2] cycloaddition to give four-membered-ring carbonyl compounds. The reaction need not be catalyzed, although mild Lewis acids and bases and a polar solvent will often increase the rate. Solvent is generally not needed. An important competing reaction is dimerization of the ketene; this process precludes ketene itself from many (but not all) cycloadditions. More often, the more reactive dichloroketene will be reacted with the unsaturated substrate, followed by dechlorination. The regiochemistry of the cycloadduct is highly predictable, with the most electronegative terminus of the double bond bonded to the carbonyl of the ketene. Most often, the C=C of the ketene is attacked, although cycloaddition across the ketene carbonyl is not unknown [32]. The stereochemistry of the double bond is retained in the product. In the case of cyclopentadiene, [2 + 2] addition (rather than [4 + 2], Diels-Alder–type addition) occurs in a way that puts the larger substituent of the ketene adjacent to the cylopentene ring (endo addition) [33].

2. Dimerization

Ketene reacts with itself to give mostly "head to head" dimer, along with small amounts of the "head to tail" dimer, as well as higher oligomers [34]. If the cyclobutanedione product can enolize, the enol hydroxyl group is prone to further reaction with ketene. The monoalkyl ketenes give mainly the beta-lactone dimer [35], while dialkyl ketenes give only the 1,3-cyclobutanedione structure in most cases [36].

Ketene ⟶ (88 - 90%) + (4 - 5%)   R = H, COCH₃

Trimers are also known to form, especially in the presence of catalysts [37].

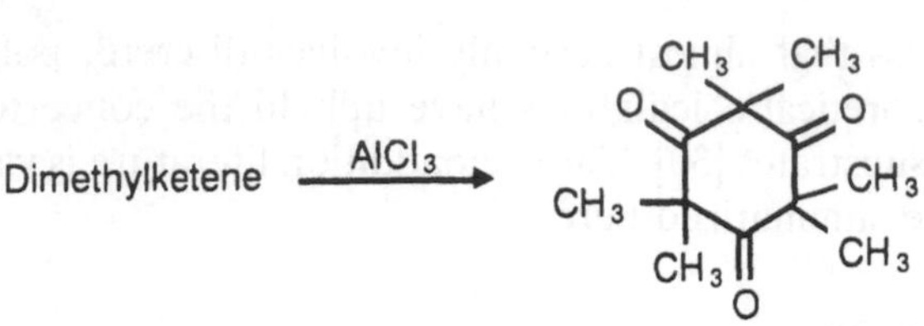
Dimethylketene
AlCl3
CH3
CH3
O
O
CH3
CH3
CH3
CH3
O

3. *Addition to Olefins*

Unsubstituted ketene does not add to unactivated olefins. Disubstituted ketenes form 1 : 1 adducts, with the success of the reaction depending mainly on the resistance of the ketene to self-condensation.

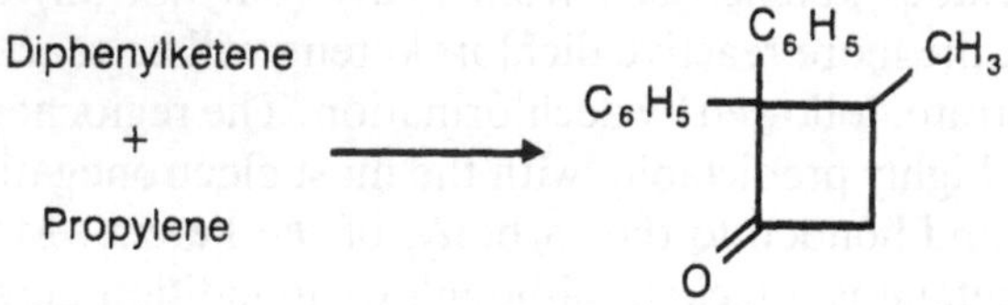
Diphenylketene
+
Propylene
C6H5
C6H5
CH3
O

Activation of the olefin with electron-donating groups increases the rate of the reaction, and cycloadducts can be isolated, even with the poorly reactive unsubstituted ketene [38,39].

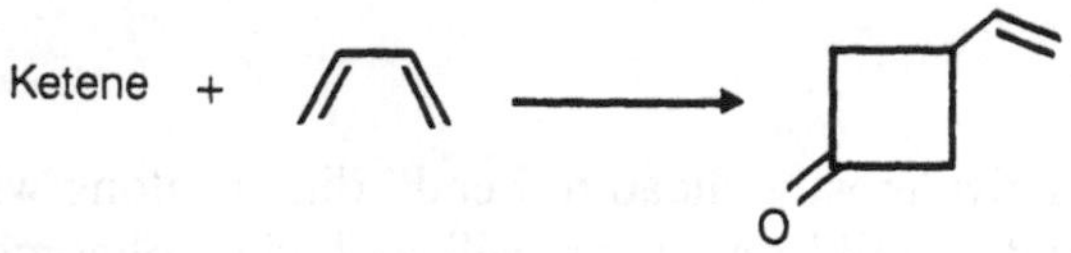
Ketene +
O

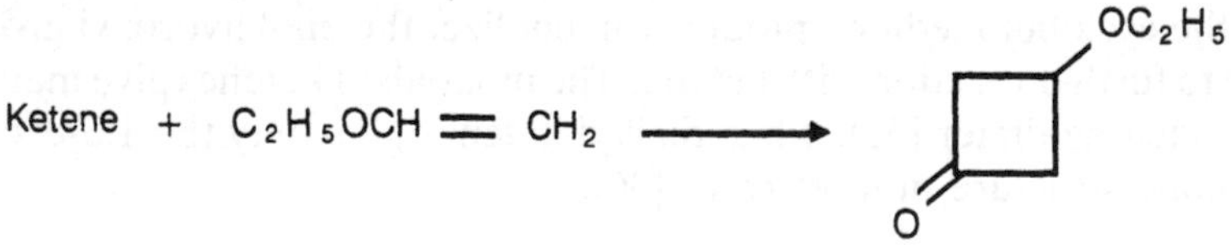
Ketene + C2H5OCH = CH2
OC2H5
O

Acetylenic compounds yield cyclobutenones [40].

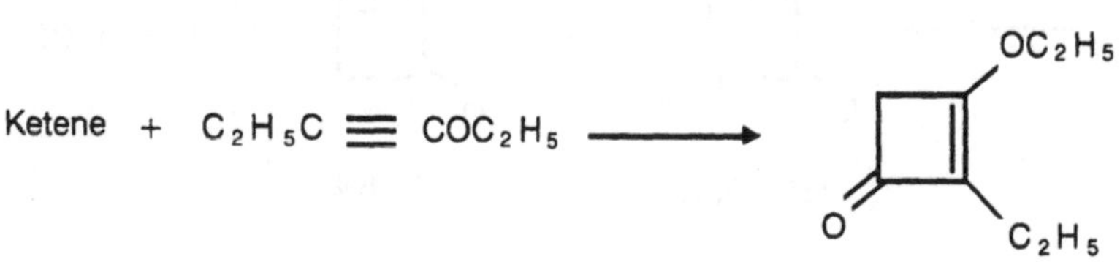
Ketene + C2H5C ≡ COC2H5
OC2H5
C2H5
O

Ketene reacts with carbonyl groups to give β-propiolactones, which often eliminate carbon dioxide under the reaction conditions to yield olefins

[41]. In the case of the reaction of ketene with crotonaldehyde, the conjectural propiolactone intermediate, which was neither isolated nor characterized, was hydrolyzed, yielding sorbic acid [42].

Ketene +

In a similar way, reaction can occur with unsaturated substituents containing nitrogen and sulfur to give a variety of four-membered-ring heterocycles.

Although the [2 + 2] mode of addition is most common, examples of [4 + 2] and, more rarely, [3 + 2] adducts are known [44].

III. SUMMARY

Ketene is highly reactive, readily available from C_2 feedstocks, and forms carbon-carbon bonds very efficiently. Commercial utility is limited by the propensity of ketene to react with itself and by the great care that must be exercised in designing a safe process. Nucleophilic attack on ketene yields products that are generally more readily accessible from acetic acid or anhydride. Insertion of ketene into single bonds, in what amounts to electrophilic attack on ketene, affords products that are not accessible from acetic acid or anhydride. Cycloaddition reactions occur readily and, as in the case of the dimerization of ketene, are of commercial importance.

REFERENCES

1. Williams, J. W., and Hurd, C. D., *J. Org. Chem.*, *5*, 122 (1940).
2. Andreades, S., and Carlson, H. D., *Org. Syn.*, *45*, 50 (1965).
3. Patai, S., ed., *The Chemistry of Ketenes, Allenes and Related Compounds*, Wiley, New York, 1980.
4. Staudinger, H., *Die Ketene*, Verlag von Ferdinand Enke, Stuttgart, 1912.
5. Ulrich, H., *Cycloaddition Reactions of Heterocumulenes*, Academic Press, New York, 1967.
6. Ghosez, L., and O'Donnell, M. J., in *Pericyclic Reactions*, Vol. II (A. P. Marchand and R. E. Lehr, eds.), Academic Press, New York, 1977.
7. Ghosez, L., *Stereoselective Synthesis of Natural Products* (W. Bartmann and E. Winterfeldt, eds.), Excerpta Medica, Amsterdam-Oxford, 1979.
8. Wilsmore, N. T. M., *J. Chem. Soc.*, *91*, 1939 (1907).
9. Lellmann, E., *Berichte*, *17*, 2719 (1884); Thiele, J., *Annalen*, *347*, 142 (1905); Ref. 3, p. 35.
10. Chick, F., and Wilsmore, N. T. M., *Proc. Chem. Soc.*, *24*, 77 (1909).
11. Staudinger, H., *Annalen*, *356*, 79 (1907).
12. Ref. 5, pp. 89–94.
13. Blomquist, A.T., Holley, R. W., and Sweeting, O. V., *J. Am. Chem. Soc.*, *69*, 2356 (1947).
14. Harmann, D., and Smith, C.W., British Patent 670,130 (1952); *Chem. Abstr.*, *47*, 5430 (1953).
15. Heyboer, N., U.S. Patent 3,963,775.
16. Martin, J. C., Gott, P. G., Meen, R. H., and Raynolds, P. W., *J. Org. Chem.*, *46*, 3911 (1981).
17. Beránek, J., Smrt, J., and Šorm, F., *Coll. Czech. Chem. Commun.*, *19*, 1231 (1954).
18. Šorm, F., Beránek, J., Smrt, J., and Sicher, J., *Coll. Czech. Chem. Commun.*, *20*, 593 (1955).
19. Smrt, J., Beránek, J., and Šorm, F., *Coll. Czech. Chem. Commun.*, *20*, 285 (1955).
20. Roe, A., and McGeehee, J. W., *J. Am. Chem. Soc.*, *70*, 1662 (1948).
21. Bohme, H., Bezzenberger, H., and Stachel, H. D., *Annalen*, *602*, 1 (1957).
22. Staudinger, H., and Klever, H. W., *Berichte*, *41*, 594 (1908).
23. Staudinger, H., Goring, O., and Scholler, M., *Berichte*, *47*, 40 (1914).
24. Grigsby, W. E., U.S. Patent 2,433,451 (1948); *Chem. Abstr.*, *42*, 7324 (1948).
25. Raynolds, P. W., and Jones, G. C., U.S. Patent 4,785,133.
26. Gresham, W. F., U.S. Patent 2,449,471 (1948); *Chem. Abstr.*, *43*, 1055 (1949).

27. Gresham, W. F., U.S. Patent 2,504,407 (1950); *Chem. Abstr.*, *44*, 5092 (1950).
28. Jones, G. C., Nottingham, D. W., and Raynolds, P. W., U.S. Patent 4,827,021.
29. Woodward, R. B., and Hoffmann, R., *The Conservation of Orbital Symmetry*, Verlag Chemie, Weinheim, 1970, p. 68.
30. Valenti, E., Pericas, M. A., and Moyano, A., *J. Org. Chem.*, *55*, 3582 (1990).
31. Holder, R. W., *J. Chem. Ed.*, *53*, 81 (1976).
32. Raynolds, P. W., and DeLoach, J. A., *J. Am. Chem. Soc.*, *106*, 4566 (1984).
33. Rey, M., Roberts, S., Dieffenbacher, A., and Dreiding, A. S., *Helv. Chim. Acta*, *57*, 417 (1970).
34. Tenud, L., Weilenmann, M., and Dallwigk, E., *Helv. Chim. Acta*, *60*, 975 (1977).
35. Farnum, D. G., Johnson, J. R., Hess, R. E., Marshall, T. B., and Webster, B., *J. Am. Chem. Soc.*, *87*, 5191 (1965).
36. Huisgen, R., and Otto, P., *J. Am. Chem. Soc.*, *90*, 5342 (1968).
37. Pregaglia, G. F., Mazzanti, G., and Binaghi, M., *Makromol. Chem.*, *48*, 234 (1966).
38. Vogel, E., and Müller, K., *Annalen*, *615*, 29 (1958).
39. Sieja, J. B., *J. Am. Chem. Soc.*, *93*, 130 (197).
40. Rosebeek, B., and Arens, J., *Rec. Trav. Chem.*, *81*, 549 (1962).
41. Vanderkooi, N., and Huthwaite, H., U.S. Patent 3,894,097 (1975).
42. Probst, O., Fernholz, H., Mundlos, E., Oheme, H., and Orth, A., U.S. Patent 3,311,149.
43. Gouesnard, J. P., *C. R. Acad. Sci.*, *Ser. C*, *277*, 883 (1973).
44. Kellogg, R. M., *J. Org. Chem.*, *38*, 844 (1973).

11

Diketene and Acetoacetates

Robert J. Clemens and J. Stewart Witzeman
Eastman Chemical Company, Kingsport, Tennessee

I. INTRODUCTION

A. General Overview

Diketene is an acetic acid derivative that is produced by dimerization of the ketene obtained from the pyrolysis of acetic anhydride or acetic acid [Eq. (1)]. Although it has no direct uses, the highly reactive diketene molecule is a valuable synthetic intermediate that is used to produce acetoacetic acid derivatives and heterocyclic compounds that find application in dyestuffs, agrichemicals, pharmaceuticals, and polymer intermediates.

Diketene provides the most economical and reactive source of the acetoacetyl moiety and is therefore the industrial precursor to commercially important acetoacetate esters and acetoacetamides. Diketene chemistry was comprehensively reviewed in 1940 [1], 1968 [2], and 1986 [3]; several other general review articles have also been published [4]. The current chapter provides a succinct overview of the chemistry of diketene and acetoacetic acid derivatives, particularly from an industrial perspective. It describes the manufacture of diketene, the fundamental chemistry of the diketene molecule, and then explores acetoacetylation and the preparation of heterocyclic molecules in greater detail. A section is also devoted to the increasingly important topic of acetoacetylated polymer molecules and their applications in coatings chemistry.

B. Manufacture

In 1907, Wilsmore prepared ketene by pyrolyzing acetic anhydride. He noted that ketene dimerized exothermically; distillation of this crude material afforded the first pure sample of the ketene dimer [5]. A recent nuclear magnetic resonance (NMR) study of secondary isotope effects during this dimerization is consistent with a [2 + 2] cycloaddition process [6]. Diketene is still prepared by the dimerization of ketene [7], which comes from the pyrolysis of acetic acid, acetic anhydride, or acetone at temperatures near 750°C (one industrial preparation of acetic anhydride is via the reaction of ketene and acetic acid; see Chapter 9). Acetic acid is the most common commercial precursor of diketene. A typical industrial diketene production facility is diagrammed in Figure 1. This plant consists of three major sections, which can be described as pyrolysis, dimerization (exothermic), and purification. The major technical concerns are maximizing throughput while avoiding the formation of large quantities of oligomers and tar. Also, the colorless, distilled product must be carefully stored and handled, as outlined below. For laboratory processes, however, acetic anhydride or acetone is generally preferred because they "crack" at lower temperatures. Diketene is generally purified by distillation at reduced pressure; ultra-high-purity diketene (99.99%) can be prepared by low-temperature recrystallization of the distilled material [8].

C. Properties and Handling of Diketene

Diketene is a colorless, lachrymatory liquid at room temperature. Its physical properties include:

Molecular weight:	84.07 g-mol^{-1}
Melting point:	-7.5°C
Boiling point:	127°C (760 torr)
	69–70°C (100 torr)
Density:	1.090 g-mL^{-1}
Refractive index:	1.4390 (20°C, Na-D line)
Dipole moment:	3.53 D (gas phase)

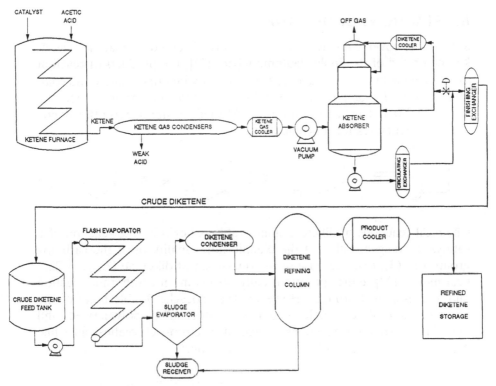

Figure 1 Production of diketene.

Additional physical analysis has included determination of the microwave [9], infrared [10], NMR [6,11], ultraviolet [12], and Raman [13] spectra; electron [14] and X-ray [15] diffraction studies have been reported, and the mass spectrum has been analyzed [16]. Several theoretical studies of diketene and other isomeric structures for a ketene dimer have been described [17].

Diketene must be carefully handled. It is a powerful lachrymator and can cause eye injury or burns to the skin or respiratory tract. Also, diketene will rapidly and exothermically self-condense in the presence of either acidic or basic catalysts; diketene must be kept free from contamination. Diketene is best stored as a solid, in which form it can be stored for many months without change. Liquid diketene undergoes gradual discoloration and decomposition with the liberation of carbon dioxide; it should not be stored in glass bottles. Copper sulfate [18] is reported to stabilize diketene; other processes utilize chemical reactions with water [19] or alcohols [20] to remove impurities.

II. REACTIONS OF DIKETENE

Since its discovery, diketene was frequently observed to react as acetylketene or one of its dipolar tautomers [Eq. (2)]; the products of reaction
of diketene with various reagents were correctly identified and found commercial uses. It was not until 1952, however, that the correct 4-methyle-
neoxetan-2-one structure was firmly established; earlier papers often depict
incorrect structures for diketene [21].

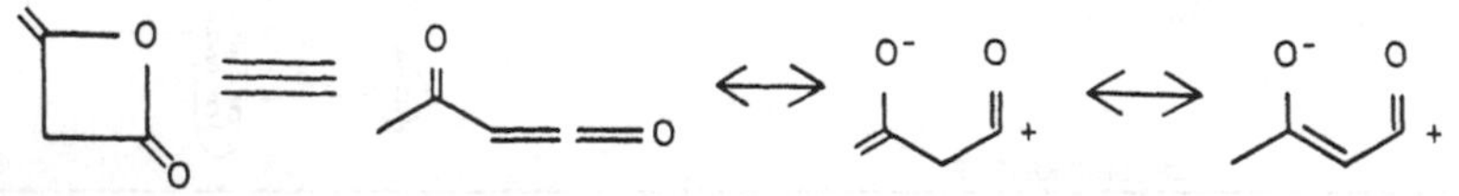

(2)

Chemical reactions of diketene almost always begin by the reaction of a
catalyst or reagent either at the olefinic functionality or at the carbonyl
group [Eq. (3), (4)]. Because of the ring-strain of diketene ($E_{strain} \sim 22.5$
kcal mol^{-1}) [22], either process usually results in an exothermic reaction
and a ring-opened product. However, the exocyclic double bond of the
diketene molecule can be hydrogenated, halogenated, polymerized, and
used in cycloaddition reactions typical of electron-rich olefins, all while
preserving the integrity of the beta-lactone ring.

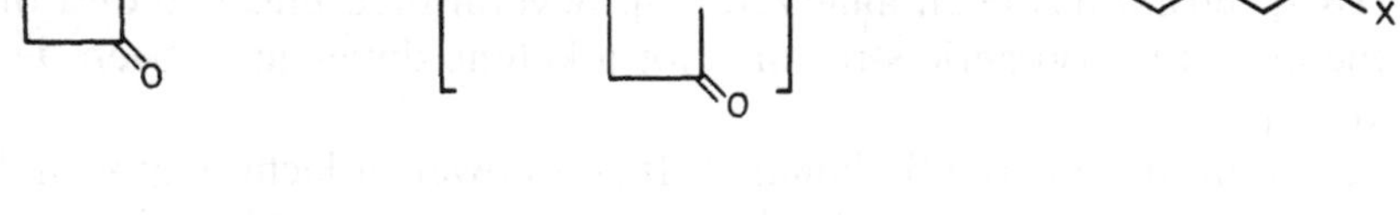

(3)

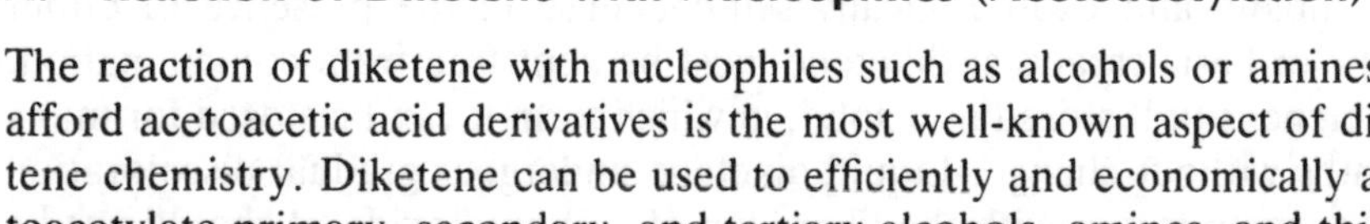

(4)

A.   Reaction of Diketene with Nucleophiles (Acetoacetylation)

The reaction of diketene with nucleophiles such as alcohols or amines to
afford acetoacetic acid derivatives is the most well-known aspect of diketene chemistry. Diketene can be used to efficiently and economically ace-
toacetylate primary, secondary, and tertiary alcohols, amines, and thiols.
The preparation and utility of acetoacetate esters and acetoacetamides are
described in detail in Section III. Diketene also acetoacetylates active
methylene groups, which is an important aspect of the heterocyclic chemistry of diketene and is discussed in Section V.

When exposed to water, diketene is slowly hydrolyzed to form aceto-acetic acid [23], which, although isolable under controlled conditions [24], gradually decarboxylates to provide carbon dioxide and acetone [Eq. (5)]. Sodium tetrachloropalladate catalyzes the hydrolysis of diketene; the palladium complex of acetoacetic acid is thus isolated [25]. The hydrolysis of diketene in 10% aqueous sodium hydroxide affords the more stable sodium acetoacetate, which can be used as an acetone enolate equivalent in Knoevenagel reactions [Eq. (6)] [26] and for alkylations such as in the preparation of allylacetone [Eq. (7)] [27]. Acetoacetic acid is also an intermediate in the formation of 2,6-heptanedione from aqueous formaldehyde and diketene [28] and in the formation of 2,4,6-heptanetrione from the hydrolysis of diketene in the presence of tertiary amines [29].

$$(5)$$

$$(6)$$

$$(7)$$

Mixing stoichiometric quantities of diketene and ammonia in an inert solvent, with cooling, provides acetoacetamide [Eq. (8)] [30]. The use of excess ammonia in the latter reaction results in the direct formation of beta-aminocrotonamide in yields of ~94% [31].

$$(8)$$

B. Reaction of Diketene with Hydrogen

Diketene can be selectively hydrogenated to provide beta-butyrolactone (Pd/C catalyst, ethyl acetate solvent, 0°C, 1 atm) [32] [Eq. (9)] or butyric acid (Raney nickel, 200°C, 150 atm). The isolation of beta-butyrolactone from the hydrogenation of diketene provided the first unequivocal proof of the correct structure for the ketene dimer [1,33].

$$\text{[structure]} \xrightarrow[\text{0°C} \quad 93\%]{\text{H}_2/\text{Pd/EtOAc}} \text{[structure]} \qquad (9)$$

C. Reaction of Diketene with Halogens and Hydrogen Halides

Acetoacetic acid (diketene) derivatives halogenated at either the 2- or the 4-positions are extremely important intermediates for the formation of heterocyclic molecules used in the pharmaceutical and agrichemical industries. Diketene reacts with hydrogen chloride in halogenated solvents to afford acetoacetyl chloride in excellent yield; sulfuric and acetic acids can be used as reaction catalysts [Eq. (10)] [34]. Acetoacetyl chloride is less stable than diketene and dimerizes to form dehydroacetic acid at 20°C. It is most commonly further halogenated with chlorine and subsequently treated with alcohols to form 2-chloroacetoacetate esters.

$$\text{[structure]} \xrightarrow{\text{HCl}} \text{[structure]} \longrightarrow \text{[structure]} \xrightarrow{\text{EtOH}} \text{[structure]} \qquad (10)$$

There are two other diketene-based routes to 2-chloroacetoacetate esters [Eq. (11)]. Most commonly, the preformed acetoacetic ester can be treated with sulfuryl chloride or chlorine gas [35]. Alternatively, the diketene-acetone adduct (vide infra) can be chlorinated to produce 5-chloro-2,6,6-trimethyl-1,3-4*H*-dioxin-4-one, which is then ring-opened with an alcohol (elevated temperatures) or its alkoxide salt (0°C) [Eq. (11)] [36]. The dioxinone-based approach enables the separation of overchlorinated products at an intermediate stage of the synthesis.

$$\text{[structure]} \xrightarrow{\text{SO}_2\text{Cl}} \text{[structure]} \xleftarrow{\text{EtONa}} \text{[structure]} \qquad (11)$$

Acetoacetamides can also be halogenated at the 2-position. The inclusion of alkali metal salts and urea in the monochlorination of *N*-methyl acetoacetamide is claimed to enhance yield and recovery of the desired chloro-*N*-methyl acetoacetamide [35e].

Reaction of elemental chlorine and diketene at low temperatures (ca. 0–10°C) produces 4-chloroacetoacetyl chloride, which can be subsequently reacted in situ with an alcohol to afford a 4-chloroacetoacetate ester [Eq.

(12)] [37]. This reaction is generally accomplished either via chlorination of diketene in an inert solvent such as methylene chloride followed by esterification of the intermediate acid chloride [37b] or via the continuous reaction of diketene spray with chlorine gas in a tube [37c] or thin-film reactor [37d], followed by down-line esterification.

$$\text{(12)}$$

$$\text{(13)}$$

$$\text{(14)}$$

4-Chloroacetyl chloride can be further treated with chlorine to afford 2,4-dichloroacetoacetyl chloride, which is also readily derivatized to give di-chloroacetoacetate esters [Eq. (14)] [38]. Hydrolysis and decarboxylation of these haloacylchlorides is an excellent method for the preparation of chloroacetone [39], dichloroacetone [40], and other halogenated acetones [Eq. (13), (14)].

4-Chloroacetoacetic acid derivatives have been used in a number of applications, including the preparation of alpha-hydroxycyclopentenones [41], quinacridone red pigments [37d], and 4-chloro-3-alkoxy-2-butenoate esters. The latter process involves reaction of the 4-chloroacetoacetyl chloride with an alcohol and a dialkyl sulfite to give an intermediate ketal ester (4-chloro-3,3-dialkoxybutyrate), which is subsequently converted to the corresponding olefin upon heating to ca. 130°C [Eq. (15)] [42a]. 4-Halo acetoacetates are also readily converted into the corresponding 4-alkoxy acetoacetates via nucleophilic displacement in aprotic solvents at 50–100°C [42b].

$$\text{(15)}$$

D. Cycloadditions to the Diketene Exocyclic Double Bond

The double bond of diketene participates in thermal [3 + 2] [43] and [4 + 2] (Diels-Alder) cycloaddition reactions as an electron-rich dienophile

[Eq. (16)] [44]. Diketene often acts as an allene equivalent in such reactions via the spontaneous decarboxylation and further reaction of the primary adduct.

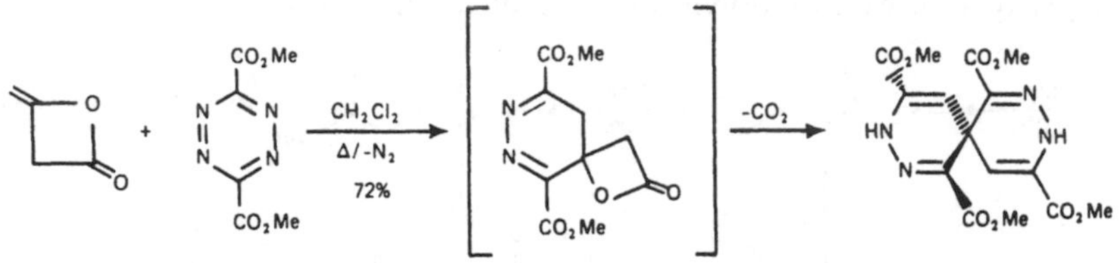

(16)

Diketene participates in photochemical [2 + 2] reactions with olefinic species such as maleic anhydrides [45a], maleimides [45b], and uracils [45c] to give mixtures of diastereomeric spirocyclobutanes [Eq. (17)]. In most cases, the product spirocyclobutanes can be ring-opened to give acetoacetate esters or pyrolyzed to provide methylenecyclobutanes [Eq. (18)] [45b]; ring-expanded products have also been isolated [46]. Carbonyl compounds such as benzaldehyde, benzophenone [47a], and p-benzoquinone [47b] also react with diketene in photochemical [2 + 2] reactions.

(17)

(18)

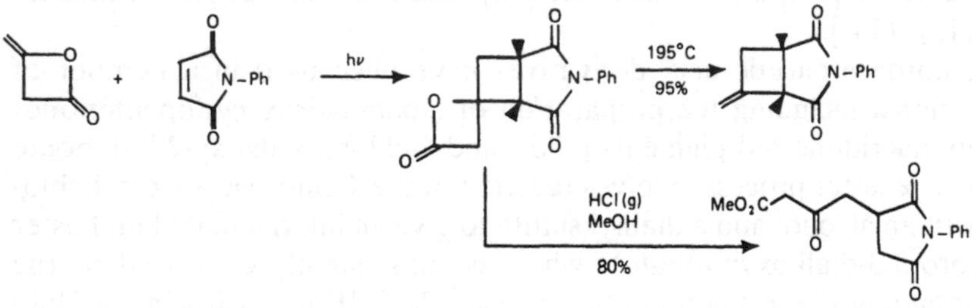

Carbenes, such as dichlorocarbene [48a], and carbenoids, such as carboethoxymethylene (generated in situ from the diazoacetate ester) [48b], add to the electron-rich double bond of diketene to give spirocyclopropanes, which are readily converted into cyclopropylacetate esters [48a,c,d]. The addition of carbenoids to diketene has been used in the synthesis of *cis*-jasmone [48e] and levulinic acid [48b,e,f].

Nitrenes generated from acyl azides or from ethyl azidoformate also add to the exocyclic double bond of diketene; the spirocyclic intermediates that are produced rapidly rearrange to afford N-acyltetramic acids [Eq. (19)] [49]. The yields are not good in this addition/rearrangement process, but it has proven useful in the complex total synthesis of althiomycin [49c].

$$\text{(19)}$$

E. Radical Additions to the Exocyclic Double Bond of Diketene

The double bond of diketene can be chlorinated with sulfuryl chloride in the presence of a radical initiator to give 4-chloro-4-(chloromethyl)oxetan-2-one [Eq. (20)] [50]. A number of compounds capable of undergoing homolytic cleavage and addition to olefins, including tetrahalomethanes, trichloroacetyl chloride, trichloroacetonitrile, and diethyl phosphite, will add to diketene under radical conditions. The resulting adducts can be ring-opened to provide 4-substituted acetoacetate derivatives. The radical addition products from diketene and secondary alcohols such as isopropanol can be further converted into conjugated dienoic acids and dihydropyrones [Eq. (21)] [51].

$$\text{(20)}$$

$$\text{(21)}$$

Aliphatic thiols undergo an exothermic, radical addition to diketene; the resulting sulfides precipitate from hydrocarbon solvents in high yield. These diketene/thiol adducts have been converted into alkenes, polymers, crotonate esters, and gamma-lactones [52].

F. Polymerization of Diketene

Under radical conditions, diketene readily copolymerizes with monomers such as vinyl acetate, vinyl chloride, vinylidene chloride, and acrylonitrile to afford polymers that contain beta-lactone rings (up to 20 mol%) in the monomer units [Eq. (22)] [53]. The beta-lactone rings can be treated with nucleophiles to provide more elaborately functionalized polymers [54]. A variety of industrial applications, especially in films, fibers [53b], pressure-sensitive adhesives [54b], and metal extraction [54c], have been described.

$$(22)$$

Diketene can be polymerized with $HgCl_2$ [55a,b], BF_3 [55c,d], or an ion-exchange resin [55e], to afford low-molecular-weight polymers (M.W. ~1630) containing unconjugated methylene groups in the form of enol ether linkages [Eq. (23)]. The polymers prepared from diketene and Lewis Acid catalysts such as BF_3 are claimed to give good organic polymeric semiconductors, particularly when doped with metals or iodine [55d].

$$(23)$$

G.   Pyrolysis of Diketene

Diketene can be "cracked" at temperatures above 400°C (E_{act} ~50 kcal-mol^{-1}) [56] to give two molecules of ketene in nearly quantitative yield. This pyrolysis can be performed in a "ketene lamp" with a glowing platinum filament, or in a hot tube, and is an excellent method of laboratory-scale ketene preparation [57].

Ketene is the kinetic product from pyrolysis of diketene; the thermodynamic products are allene and carbon dioxide [58]. Allene has occasionally been observed during pyrolysis of diketene [59], but it was not determined whether the allene was formed directly from diketene or from the reaction of ketene and the methylene that results from further pyrolytic decomposition of ketene. A careful study of the relative energies of the various possible isomeric structures for ketene dimers, such as 1,3-cyclobutanedione or 2,4-dimethylene-1,3-dioxetane, via semiempirical SINDO calculations has recently been performed [17,60].

Allene complexes are formed at ambient temperatures in the reaction of specific organometallic compounds with diketene [61]; these reactions may provide clues for those seeking to utilize diketene as a route to allene chemistry.

H. Reaction of Diketene Derivatives with Organometallic Reagents

Acetoacetate derivatives chelate with metal ions to provide enolate salts, which are an important aspect of acetoacetate chemistry that have industrial applications in the removal and separation of metal ions. Diketene itself reacts with organoboron, organoaluminum, organosilicon, and organotin compounds to afford acetoacetate complexes [Eq. (24)] [62]. In general, these reactions are fast, and both the conjugated (thermodynamic) and unconjugated (kinetic) enolates can be isolated; the reactions of diketene with (dimethylamino)trimethylsilane [63] or N-(trimethylstannyl)benzo-phenone imine [64] are representative of this chemistry [Eq. (25)]. Tri-methylsilyl iodide reacts with diketene to afford the trimethylsilyl enol ether of acetoacetyl iodide, which is reported to be a potent acetoacety-lating reagent for amides [65]. In a related reaction, trimethylsilyl azide reacts exothermically with diketene to afford a mixture of isocyanates [Eq. (25)] [66].

$$(24)$$

$$(25)$$

Diketene reacts with palladium(II) compounds and water or alcohols to form pi-allyl complexes of acetoacetic acid or acetoacetate esters [67]. The addition of pyridine to these latter complexes causes rearrangement of the pi-allyl complexes to the 4-sigma complexes [68], which can be carbonylated to afford esters of acetonedicarboxylic acid [Eq. (26)] [68b]. The direct conversion of diketene, carbon monoxide, and methanol into dimethyl acetonedicarboxylate in the presence of methyl nitrite and palladium(II) chloride has been effected by this latter process [69]. Diketene has been successfully carbonylated, albeit with low conversion, to provide methyl succinic anhydride [Eq. (27)] [70].

$$(26)$$

$$(27)$$

Diketene does not provide useful products upon treatment with normal Grignard reagents. However, in the presence of a nickel(II) catalyst, (trimethylsilyl)magnesium chloride reacts with diketene at the olefinic carbon-oxygen bond to give 3-methylene-4(trimethylsilyl)butyric acid [Eq. (28)], which has been used in the preparation of dihydropyrones and other synthetic intermediates [71]. A variety of other Grignard reagents add to diketene in the presence of cobalt(II) iodide to produce 3-methylene carboxylic acids such as have been used in the synthesis of several terpenoids and of the *Cecropia* juvenile hormone [Eq. (29)] [72].

$$(28)$$

$$(29)$$

I. Ozonolysis of Diketene

The ozonolysis of diketene at low temperatures produces formaldehyde and malonic anhydride [73], which subsequently decomposes into carbon dioxide and ketene [Eq. (31)]. The malonic anhydride can be converted into malonate derivatives in up to 76% yield by low-temperature (<0°C) reactions with simple nucleophiles such as aniline and ethanol [Eq. (30)].

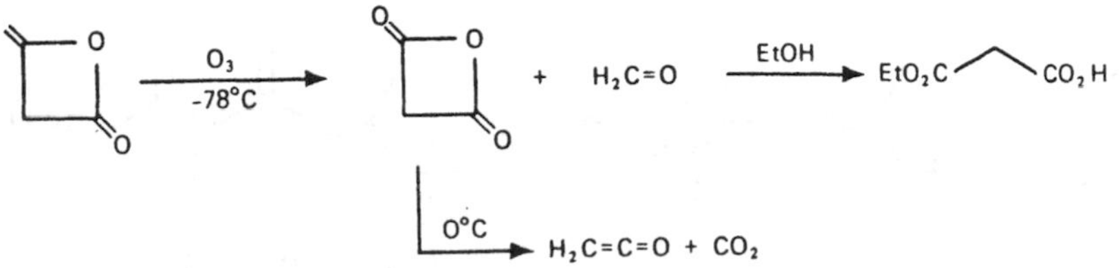

$$(30)$$

$$(31)$$

J. Miscellaneous Reactions of Diketene

Diketene reacts with acetals [74] and ketals [75] in the presence of TiCl$_4$ to give 4-substituted acetoacetates [Eq. (32)]. This reaction provides a convenient alternative to the dianion chemistry that is normally used to prepare such compounds. The resulting 4-substituted acetoacetates are readily cyclized under basic conditions to give 6-substituted 4-hydroxy-5,6-dihydro-2-pyrones [74]. These latter pyrones can also be prepared directly from diketene and ketones in the presence of dichlorodimethoxy zirconium(IV) [76]. These reactions of diketene with ketones have been used in the synthesis of kawains [Eq. (33)] [74], pestalotin [77], compactin [78], and a series of postemergent herbicides [79]. Surprisingly, the use of catalytic amounts of boron trifluoride in place of stoichiometric quantities of titanium tetrachloride in the above reaction yields 2-substituted acetoacetates [80].

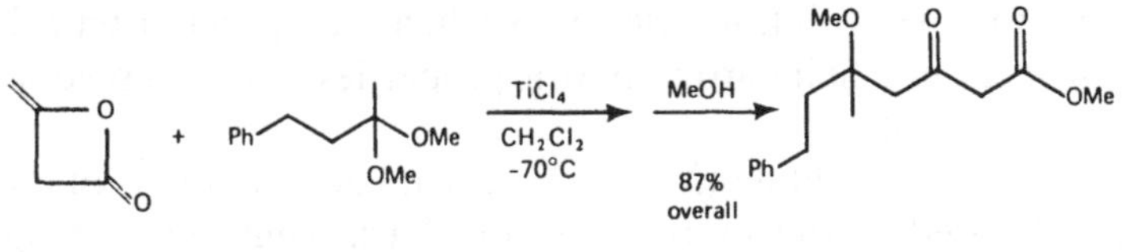

$$(32)$$

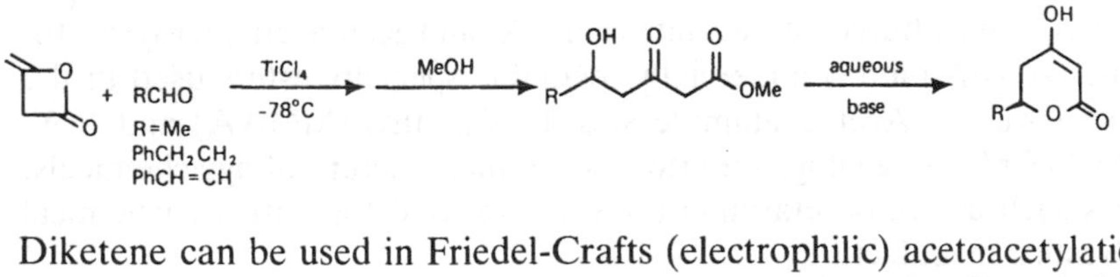

$$(33)$$

Diketene can be used in Friedel-Crafts (electrophilic) acetoacetylation at carbon to form beta-dicarbonyl compounds. Both olefins [Eq. (34)] [81] and aromatic compounds [Eq. (35)] [35,82] can be acetoacetylated in the presence of Lewis acids. Such acetoacetylations can also be run in hydrogen fluoride, in which acetoacetyl fluoride is the actual acetoacetylating reagent; this latter approach often provides products that result from the acid-catalyzed cyclization of the acetoacetyl group onto the activated aromatic ring [83]. Acetoacetylated aromatics have been used for the preparation of dyes [83b], for chelation with metals [84], and for the preparation of pharmaceutical polymers [85].

$$(34)$$

$$(35)$$

III. ACETOACETATES AND ACETOACETAMIDES

Acetoacetic acid derivatives such as acetoacetamides and acetoacetates have found numerous uses in chemical, pharmaceutical, agrichemical, dye-stuff, coating, and adhesive applications. The chemical properties associated with the beta-ketoester moiety also make these materials versatile synthetic building blocks. Examples of the chemical utility of this functional group include reactions such as condensation, halogenation, alkylation, diazotization, and metal chelation. The addition of organometallics, hydrides, or amines to the carbonyl group is also readily accomplished. Among the valued physical characteristics of acetoacetic acid derivatives are their ability to lower solution viscosities when incorporated onto hydroxylated polymers and their chromophoric properties when coupled with diazonium salts or metals.

Methyl and ethyl acetoacetate (MAA, EAA) are general items of commerce frequently used in the preparation of pharmaceuticals and agrichemicals, while *t*-butyl acetoacetate (t-BAA) and acetoacetoxyethyl methacrylate (AAEM) are commercially available specialty esters used in the coatings industry. Acetoacetamide and its *N*-methyl (MMAA) and *N,N*-dimethyl (DMAA) analogs are used in the manufacture of agrichemicals; arylides such as acetoacetarylide (AAA) are used for both agrichemical and dyestuff manufacture.

Simple acetoacetic acid derivatives are most readily prepared by reaction with diketene, although diketene-free approaches such as condensation, exchange with other acetoacetic acid derivatives (transacetoacetylation), and reaction with dioxinones can also be used. These alternative routes enable the preparation of materials for specialty applications where use of diketene is impractical because of manufacturing or raw material constraints.

A. Preparation of Acetoacetates and Acetoacetamides by Reaction with Diketene

The most direct and economical route for preparation of acetoacetates and acetoacetamides is reaction of the corresponding nucleophile with diketene [Eq. (36)]. This reaction is readily amenable to large-scale, continuous processes. When the nucleophile is an alcohol, the reaction requires the use of a catalyst. Catalysts used include acid catalysts, such as hydrochloric [86a] and sulfuric acids [86b]; tertiary amines, such as 1,4-diaza-[2.2.2]bicyclcoctane (DABCO) [86c], triethylamine, tributylamine, *N,N*-dimethyl ethanolamine [86d], *N,N,N'*-trimethyl hexanediamine [86d], or *N,N*-dimethyl-4-aminopyridine [86e,f]; salts, such as sodium acetate [86g,h], either alone or in combination with inorganic salts such as NaCl/NaBr/Na$_2$SO$_4$ [86i]; organophosphorus compounds, such as triphenyl phosphine [86j]; and organometallic compounds, such as dibutyltindilaurate [86k]. These reactions are usually carried out by continuous reaction of a diketene feed with a blend of alcohol and catalyst. One recently discovered aspect of acetoacetate ester preparation is that the reaction can be forced to provide a 2:1 alcohol:diketene adduct, an acetylketene acetal [87] under the proper conditions.

$$\text{(diketene)} \xrightarrow{\text{R X H}} \text{(acetoacetate product) X R} \qquad (36)$$

$$(X = O , S , N)$$

Figure 2 shows a schematic for a typical acetoacetate ester process. The facility involves a reaction zone where the two feeds are allowed to react and a continuous distillation train. Acetoacetamide facilities will differ slightly owing to the physical nature of the resulting product, which will determine the type of purification method used, and the more rapid rate of reaction, which will affect the reactor design.

The reaction of diketene with amines is much more exothermic than the corresponding reaction with alcohols and thus often does not necessitate

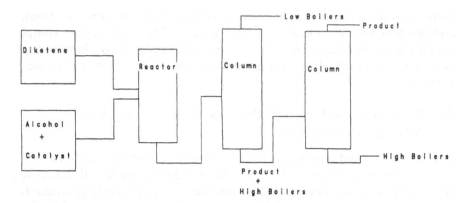

Figure 2   Schematic for a typical acetoacetate ester process.

the use of a catalyst. In some cases, these acetoacetamides can be prepared in aqueous solutions. Thus dimethyl acetoacetamide is prepared by reaction of aqueous dimethyl amine with diketene under conditions of high agitation.

Thiols are also readily acetoacetylated with diketene to afford thioacetoacetate esters, which have recently found considerable use in natural products and macrolide synthesis [88a]. Although mercaptans are readily acetoacetylated with diketene in the presence of basic catalysts, the reaction of diketene with thiols in sulfuric acid affords beta-alkylthiocrotonic acids [88b,c].

For companies with a captive source of diketene as well as the requisite acetic acid raw material base, the diketene-based approach to acetoacetic acid derivatives is the method of choice. The lachrymatory properties and reactivity characteristics of diketene have led to concerns regarding the shipping and handling of this material. These concerns, in turn, have led to restrictions on the shipping of diketene and thus have predicated the need for development of other, "diketene-free" acetoacetylation technologies. These methods also enable the preparation of specialized acetoacetates by companies that lack a diketene raw material position. Many of the following methods are described in the chemical literature, but only the exchange of acetoacetates with nucleophiles (transacetoacetylation) is used to any extent in industrial practice.

B.   Preparation of Acetoacetic Acid Derivatives by Condensation Reactions

The acetoacetic ester synthesis [89], in which two acetate esters are condensed to yield an acetoacetate [Eq. (37)], is a quintessential name reaction in organic chemistry. Although this process has been used extensively in laboratory applications, the requirement for stoichiometric quantities of base in this reaction makes the process economically unattractive for the preparation of simple acetoacetates. This route becomes more economically attractive for more specialized materials such as highly functionalized beta-ketoesters. Thus, ethyl-4,4,4,-trifluoroacetoacetate [90], which is used as an intermediate in the preparation of pharmaceuticals and agrichemicals, is prepared by reaction of ethyl acetate and trifluoro ethyl acetate with sodium alkoxide base.

$$\text{(37)}$$

A number of other condensation reactions have been used to prepare acetoacetate derivatives. Although these methods can be used to prepare simple acetoacetic acid derivatives, they are generally of much greater utility in the preparation of highly functionalized beta-ketoesters. One such reaction involves the reaction of dithioglutaramides with acetate ylides [91a], which produce the beta-ketoester upon hydrolysis. The acid-catalyzed addition of ketene to ketene acetals [91b] has been used to prepare simple acetoacetates as well as 2-chloroacetoacetates. Alpha diazo carbonyl compounds produce acetoacetic acid derivatives either via addition of an ester enolate with a functionalized alpha-diazoketone [92a], or, alternatively, by reaction of an alpha-diazo ester with an aldehyde [92b]. Yet another condensation-like approach to beta-ketoesters involves the carbonylation of mono-haloacetones with lower alcohols using cobalt catalysts [93].

C. Preparation of Acetoacetates from Dioxinones or Exchange with Other Acetoacetates

The previously mentioned condensative syntheses of acetoacetates were based on precursors derived from sources other than diketene. In the following approaches, the essential $C_4H_4O_2$ unit is derived from diketene and used as a "diketene equivalent." These intermediate, nontoxic, storage-stable diketene equivalents make possible the preparation of various specialty diketene derivatives without the capital requirements associated with construction of diketene facilities.

One such "diketene-free" approach involves use of the dioxinone shown in Eq. (38), which is prepared from the acid-catalyzed reaction of diketene with acetone. This diketene-acetone adduct (2,2,6-trimethyl-4H-1,3-dioxin-4-one, TKD) can be used to prepare a wide variety of acetoacetates, acetoacetamides, and thioacetoacetates [94]. Kinetic and spectroscopic studies have shown that this reaction proceeds via the intermediary of acetylketene, which is formed in a rate-determining unimolecular decomposition [94e].

$$\text{(38)}$$

Industrial application of the diketene-acetone adduct has been limited by its availability. A second diketene-free approach to acetoacetic acid derivatives involves the exchange reaction of a nucleophile with an acetoacetate [Eq. (39)] [95]. This approach is particularly attractive in industrial settings owing to the similarity of this reaction to the transesterification of simple esters. (See Chapter 14 for a discussion of this reaction.) Despite the apparent similarities between the exchange reactions of acetoacetates and those of simple esters, the mechanism of reaction of acetoacetates (transacetoacetylation reaction) is quite different from that of typical transesterification processes [96]. As with the reactions of the diketene-acetone adduct, the transacetoacetylation reaction occurs via a rate-determining, unimolecular decomposition to produce an acetylketene intermediate that subsequently reacts with the nucleophile [96a,b]. This decomposition presumably occurs via an intermediate enol, which can subsequently decompose, via a cyclic transition state, to the ketene intermediate [96c]. One of the most remarkable aspects of the transacetoacetylation reaction is the effect of the ester component on the rate of reaction [96b]. Thus, tertiary esters such as *t*-butyl acetoacetate (t-BAA) or *t*-amyl acetoacetate react 10- to 20-fold faster than the less sterically encumbered methyl or ethyl analogs. Not only does the nature of the ester component affect the rate of transacetoacetylation, but it also affects the overall direction of the equilibrium. For example, reaction of ethyl or methyl acetoacetates with hindered alcohols such as 2-methyl-3-buten-2-ol gives none of the desired acetoacetate [95c], while the corresponding reaction with *t*-butyl acetoacetate produces the corresponding acetoacetate in 72% yield [95g,h]. Other hindered alcohols such as methylcyclohexanol and isopinocampheol can similarly be acetoacetylated with t-BAA.

$$\text{(39)}$$

The transacetoacetylation reaction can also be used to prepare acetoacetamides. This exchange reaction can be complicated by formation of the enamine, particularly when unhindered primary alkyl amines are used. Formation of these enamines can be minimized by proper choice of reaction conditions such as dilution and temperature.

D. Uses of Acetoacetates

The acetoacetyl moiety can undergo a variety of synthetic transformations, including reactions at the active methylene group, such as condensation, alkylation, and halogenation; reactions at the ketone carbonyl group, such as addition of nucleophiles and reduction; reactions involving the ester moiety, such as decarboxylation and the previously mentioned exchange reactions; as well as complexation/chelation processes with metals in which the acetoacetate reacts as a 1,3-diketone. These reactions can be used to prepare a variety of materials for use in agrichemicals, pharmaceuticals, chemicals, paints, and plastics.

Methyl acetoacetate can be used in an economical, large-scale preparation of valproic acid, a key intermediate in the preparation of an important antiepileptic drug [97]. MAA is also used in the preparation of geranyl acetone, which is used in the preparation of vitamin E [98]. The latter process involves the reaction of MAA with myrcene using a meta-trisulfonated triphenylphosphine rhodium salt as a catalyst.

MAA and EAA are also used as amine-protecting agents during the synthesis of beta-lactam antibiotics such as amoxicillin and cephalexin, via enamine formation (vide infra), and to produce the isoxazole sidechain for oxacillin antibiotics, via reaction of EAA with an aryl oximinoyl chloride.

E. Carroll Rearrangement

The Carroll rearrangement is formally a 3,3 sigmatropic shift of the enol form of a vinylogous acetoacetate, derived from diketene and an allylic or propargylic alcohol, followed by decarboxylation [Eq. (40)] [99]. This reaction provides a versatile method for preparing unsaturated methyl ketones, which are important to the flavor and fragrance industry. For example, 3-buten-2-ols can be acetoacetylated with diketene and then pyrolyzed to provide 5-hepten-2-ones via a rearrangement and decarboxylation of the intermediate beta-keto acid [100]. The Carroll rearrangement frequently proceeds in good yield and is often run as a one-pot acetoacetylation/rearrangement sequence. Thus, the simultaneous addition of diketene and the unsaturated alcohol to a high-boiling solvent that contains an amine catalyst often provides the rearranged product directly. Aluminum isopropoxide often catalyzes the Carroll rearrangement [101]. Kinetic studies have shown that the rate of this reaction can be correlated with the degree of substitution of the resulting olefin, with the more highly substituted material reacting faster.

$$\text{(structure)} \quad (40)$$

Acetoacetate esters of propargylic alcohols rearrange to afford conjugated dienones [Eq. (41)] [102].

$$\text{(structure)} \quad (41)$$

The Carroll reaction has been used to prepare a number of mono-, di-, and trisubstituted olefins [103] and especially to prepare terpene and steroid derivatives, which are important to the perfume and pharmaceutical industries, respectively [104]. Thus, geranylacetone has been prepared in 87% yield by the 150°C, aluminum isopropoxide–catalyzed Carroll rearrangement of linalool [105]: farnesylacetone has been similarly prepared from nerolidol in 85% yield.

In an important variation of the Carroll rearrangement, the dianions of allylic acetoacetates were found to undergo Carroll rearrangement at low temperature [86f]. This work has been elaborated into a technique for the synthesis of contiguous quaternary centers, via the diastereoselective rearrangement of a *bis*-silylated ketene acetal intermediate [106].

F. Uses of Acetoacetamides

Several acetoacetamides that are produced from small aliphatic amines are commodity chemicals which are heavily used in the preparation of insecticides such as monocrotophos (Azodrin), dicrotophos, phosphamidon, and oxamyl. The artificial sweetener Acesulfame K can be derived from acetoacetamide or simple acetoacetamides [107]. *N*-Hydroxyethyl acetoacetamide is used in the preparation of olaquindox, an animal feed additive.

Simple aliphatic acetoacetamides are often used as copromotors in the curing of unsaturated polyesters and alkyd coatings. Also, acetoacetamides of cyclic amines have recently found use as accelerators in the curing of vinyl ester resins [108b]. This latter application circumvents the use of the more commonly used (and toxic) N,N-dimethyl aniline. Acetoacetates such as MAA and EAA have also found use in this application. The acetoacetate or acetoacetamide presumably functions by complexation with the polyvalent metal free radical promoter [108a].

Acetoacetamides that are produced from aromatic amines (acetoacetarylides) are widely used in dye and colorants applications. These dye applications involve the coupling of the acetoacetarylide with a diazonium salt to produce the corresponding diazocompound. Pigment Yellow 12, one of the largest volume yellow pigments sold in the United States, is produced in this fashion by reaction of acetoacetanilide (AAA) and the *bis*(diazo) salt of 3,3'-dichlorobenzidine. Orange pigments can be similarly produced from the corresponding reaction of AAA and diazonium salt of 3,3'-dimethoxybenzidine. A variety of substituted analogs of AAA have been used to prepare pigments with improved properties. For example, the pigments produced from coupling of *bis*(azo) salts with acetoacetamides such as acetoacet-2,4-xylidide (AAMX), acetoacet-2-toluidide (AAOT), and acetoacet-2,5-dimethoxy-4-chloroaniline are of commercial interest as pigments for offset printing [109]. Yellow pigments have also been recently described that are claimed to have improved heat stability and excellent compatibility with numerous resins and polymers [110a]. A novel family of pigments that contain pendant polyoxypropylene or polyoxyethylene groups have been prepared by reaction of monofunctional amine-terminated ethylene oxide or propylene oxide polymers with an acetoacetarylide such as AAA, AAOT, or acetoacet-2,5-dimethoxy-4-chloroaniline to give the corresponding enamine (see Section III.H). These adducts can be diazotized or tetra-azotized with various amines and diamines to give pigments that are claimed to have utility as colorants in solvent and water-borne applications, such as inks, paints, and plastics [110c,d].

The fungicide carboxin, which is widely used for seed treatment, is prepared from acetoacetanilide, via 2-chloroacetoacetanilide [111]. Further chemistry of acetoacetarylides has been extensively reviewed elsewhere [112].

G. Substituted Acetoacetates

Acetoacetates can be readily alkylated at the 2-position [Eq. (42)]. This reaction can be efficiently carried out either under phase-transfer conditions [113a] or by preforming the corresponding enolate with bases such as sodium hydride or alkyl lithiums prior to alkylation. The O-alkylated congeners can also be formed, particularly when combinations of alkoxide bases and alkylating reagent are employed [113b,c]. The formation of the 2,2-dialkylated species can also complicate such alkylations, although preformation of the enolate usually minimizes this problem. An alternate method of preparation of 2-allyl acetoacetates is by reaction of a beta-ketoester with an allylic alcohol in the presence of a te-trakis(triphenylphosphine)palladium(O) catalyst [114a]. A similar approach involves the palladium-catalyzed telomerization of 1,3-dienes with beta ketoesters [114b]. Thus, reaction of butadiene with ethyl acetoacetate using a [1,2-*bis*(dialkylphosphino)-ethane] palladium complexes gives the 2-butenylacetoacetate ester.

$$\text{(structure)} \quad \xrightarrow[\text{Base}]{\text{R}'-\text{X}} \quad \text{(structure)} \tag{42}$$

2-Alkylated acetoacetates can be decarboxylated with strong acids to produce substituted 2-alkanones. Thus, reaction of MAA with 1,3-dibromo-propane followed by hydrolysis/decarboxylation gives the 6-bromohexan-2-one in good yield [115a]. When combined with a halogenation reaction, this chemistry can be used to prepare 3-chloro-2-alkanones from 2-chloro-2-alkyl-acetoacetates [115b].

Acetoacetates can be similarly acylated to produce the 2-acylated materials [Eq. (43), (44)]. Treatment of these materials with acid produces the substituted diketones, while basic hydrolysis produces substituted 1,3-beta-ketoesters [116].

$$\text{(structure)} \quad \xrightarrow[\text{Base}]{\text{R}'\text{COCl}} \quad \text{(structure)} \tag{43}$$

Acid/Water   Base/Water

$$\text{(structure)} \tag{44}$$

Alkylation of acetoacetates at the 4-position is more difficult owing to the much lower basicity of the protons at C-4 relative to those at the 2-position. The 4-position can be functionalized by preparation of the corresponding dianion [Eq. (45)]. This is usually accomplished by treatment of the ace-toacetate with an equivalent of sodium hydride followed by a second equiv-alent of strong base such as alkyl lithium or lithium diisopropyl amide. The necessity for use of strong bases in this process has limited the industrial practice of this chemistry [117]. Alternate approaches to acetoacetate esters alkylated at C-4 were discussed in Section II.J.

$$\text{(45)}$$

H. Formation of Enamines (Beta-Aminocrotonamide) from Acetoacetates

The reaction of acetoacetates and/or acetoacetamides with amines provides enamides (beta-crotonamides) via addition of the amine to the acetoacetyl ketone moiety followed by loss of water [Eq. (46)]. The conjugation in-herent in these products makes these beta-aminocrotonamides more stable than most enamines.

$$+ \ H_2O \quad \text{(46)}$$

Ethyl beta-anilinocrotonate, a typical beta-aminocrotonate, has been pre-pared by a number of methods, such as by heating ethyl acetoacetate and aniline in benzene with an acetic acid catalyst [118a] or by reacting the amine and acetoacetate in the absence of a catalyst either at room tem-perature or with mild (steam bath) heating [118b,c,d]. Aniline hydrochlo-ride and iodine have also been used as catalysts in this latter process [118e].

Methyl 3-aminocrotonate is useful in the preparation of various heter-ocycles. It can be prepared by reacting methyl acetoacetate with gaseous ammonia in solvents such as ether or toluene, using catalysts such as am-monium nitrate or toluenesulfonic acid [119a,b]. More recently, it has been noted that beta-aminocrotonates can be readily prepared by mixing am-monium hydroxide with a neat acetoacetate ester at such a rate as to keep the temperature of the highly exothermic reaction between 40 and 45°C [119c]. These conditions are claimed to enable preparation of the amino-crotonate in high yield and substantially free of acetoacetamide by-prod-ucts.

This formation of beta-amino crotonic acid derivatives can compete with acetoacetylation reactions (see Section III.C). For example, transaceto-acetylation of *N*-heptylamine with t-BAA can result in the synthesis of enamine (crotonamide) by-products in addition to the desired *n*-heptyl acetoacetamide [95g]. Reaction conditions that involve high dilution and slow addition of the amine to a preheated solution of t-BAA were found to minimize these by-products. Similarly, reaction conditions that include extended reaction times and the use of excess amine can result in formation of the amino-crotonamides resulting from enamine formation from the product acetoacetamide [118].

A model study on the reaction of ethyl acetoacetate (EAA) and morpholine has been described in conjunction with work on the preparation of novel acetoacetamides for use as fuel dispersants [120]. It is noted in these studies that reaction of morpholine with the acetoacetate at room temperature favors production of the enamine, while reaction temperatures between 120 and 180°C give the corresponding acetoacetamide.

The use of both acetoacetamide and 3-aminocrotonamide as intermediates for heterocyclic synthesis has been reviewed [121], and the mechanism of the formation of 2,6-dimethyl-3H-4-pyrimidone by pyrolysis of 3-aminocrotonamide has recently been clarified [122].

I. Hantzsch Pyridine Synthesis

The Hantzsch pyridine synthesis is a preparation of 1,4-dihydropyridines from ammonia, an aldehyde, and a beta-keto ester or a related derivative [Eq. (47)] [123]. Symmetrical 1,4-dihydropyridines are easily prepared by mixing (cautiously) the three reagents.

$$(47)$$

Unsymmetrical Hantzsch dihydropyridines are best prepared by condensation of a 3-aminocrotonate ester with a 2-alkylidene acetoacetate ester [124]. Many dihydropyridines, such as nifedipine [125], nimodipine [126], and nicardipine [127], act as vasodilators and antihypertensives [128]. This area of medicinal chemistry is quite active, and many analogous 1,4-dihydropyridines have been patented.

Dihydropyridines unsubstituted in the 4-position can be prepared by using formaldehyde or hexamethylenetetramine in the Hantzsch reaction [129]; many of these dihydropyridines are effective antioxidants because they are readily oxidized to pyridines under a variety of conditions.

IV. ACETOACETYLATED POLYMERS AND RESINS

Polymers and resins that contain pendant acetoacetyl groups have been the subject of considerable interest over the last few years, owing to both the physical and the chemical properties of the resulting acetoacetylated polymers. In particular, it has been noted that many acetoacetylated polymers have much lower solution viscosities than the corresponding hydroxyl bearing polymers. These viscosity characteristics result in a greater ease of processing and thus enable use of lower amounts solvents in the preparation of polymer solutions [130]. In coatings applications, these viscosity characteristics make possible the preparation of formulations with lower volatile organic contents (VOCs) and thus are particularly attractive. In addition to these physical properties, the chemical reactivity of the acetoacetyl functional group makes a variety of cross-linking reactions possible for these polymeric and oligomeric species. The combination of these chemical and physical properties has led to the use of acetoacetylated materials in coatings, adhesive, and general polymer applications.

A. Preparation of Acetoacetylated Polymers and Resins

In general, acetoacetylated polymers can be prepared either by use of an acetoacetylated monomer or by acetoacetylation of a resin or polymer. The method of preparation of a given acetoacetylated polymer or resin will depend on the nature and composition of the desired material. For addition polymers such as acrylics or methacrylics, the acetoacetyl moiety can be incorporated into the polymer backbone by use of the commercially available monomer acetoacetoxyethyl methacrylate (AAEM) [130b,e,f]. This material can be copolymerized under standard free-radical polymerization conditions with a variety of acrylic, methacrylic, and vinyl monomers. An acetoacetylated acrylamide monomer has also been described [131].

A second general approach toward the preparation of acetoacetylated acrylics involves the reaction of the hydroxylated polymer or resin with an acetoacetylating reagent such as diketene, the diketene-acetone adduct, or an acetoacetate such as MAA, EAA, or t-BAA (see Section III.C). The limited availability of the diketene-acetone adduct and concerns regarding the safety and handling of diketene often make transacetoacetylation the most industrially attractive route for acetoacetylating polymeric hydroxyl groups. The higher reactivity of t-BAA relative to MAA and EAA makes it the reagent of choice for these exchange reactions [95h,130g,h]. Acetoacetylation of polymer-bound hydroxyl groups with t-BAA allows control of the extent of acetoacetylation of a given polymer merely by adjusting the ratio of acetoacetylating reagent to polymer [130g,h].

Recent patents have described the use of mercaptoalkyl acetoacetates in the preparation of acetoacetylated addition polymers. These compounds function as chain transfer agents and thus allow control of molecular weight of the polymer while also enabling incorporation of the acetoacetyl moiety [132].

For condensation polymers and resins such as polyesters, polyether polyols, and polyurethanes, the use of acetoacetylated monomers during the polycondensation process is not feasible because the acetoacetyl groups cannot tolerate the high reaction temperatures used in these processes. In these cases, the acetoacetyl group can be efficiently incorporated into the polymer backbone by subsequent acetoacetylation of the pendant hydroxyl groups on the polymer following the resin synthesis, as described above.

B. Cross-Linking Reactions of Acetoacetylated Polymers and Resins

Acetoacetylated polymers and resins are capable of undergoing a large number of cross-linking reactions. These materials can react via the active methylene of the acetoacetyl group with commonly used cross-linkers such as alkyl methylol melamines, isocyanates, and epoxies [130]. The unique chemical nature of the acetoacetyl group enables the use of nontraditional cross-linking reactions such as condensation of the acetoacetyl methylene with activated olefins (Michael reaction) [130a,b,j], reaction of the acetoacetyl ketone group with amines (the enamine reaction) [130i,k], and chelation/complexation of the beta-ketoester with metals (chelation reaction).

Melamine and isocyanate cross-linkers that react with hydroxylated polymers can also cross-link with acetoacetylated polymers. This cross-linking reaction with melamines is thought to occur via the active methylene group [Eq. (48)], and is catalyzed by acids such as p-toluenesulfonic acid. Divalent metals with easily displaced ligands, such as $Zn(AcAc)_2$ and similar complexes of Ni and Cu, are useful catalysts in the acetoacetate-isocyanate reaction, whereas traditional urethane catalysts such as tin compounds do not effectively catalyze this process. The fact that these cross-linkers can react with either hydroxyl or acetoacetyl groups provides considerable latitude in the formulation of coatings employing acetoacetylated resins and polymers.

(48)

In addition to the cross-linking reactions discussed above, acetoacetylated polymers and resins can be used in conjunction with materials such as trimethylolpropane triacrylate to produce cross-linked films at ambient temperatures via the Michael reaction [130a,b,d,e]. This process has applications as varied as the preparation of paints capable of curing at ambient temperatures [132] and the preparation of novel acrylic polymers for flooring materials [133]. The effect of various catalysts on this cross-linking reaction has been studied; a good combination of pot-life and catalytic activity are obtained if catalysts with a pKb of ca. 12.5–13.6 are used [130a,d,j]. Examples of such catalysts include 1,8-diazabicyclo(5.4.0)undec-7-ene (DBU; pK$_b$ 12.5), 1,5-diazabicyclo(4.3.0)non-5-ene (DBN; pK$_b$ 12.7), and tetramethyl guanidine (TMG; pK$_b$ 13.6). Catalysts such as sodium hydroxide or tetrabutyl ammonium hydroxide often do not provide a sufficiently long pot-life to the liquid paint. Another approach to effectively catalyzing the Michael reaction while also giving a workable pot-life involves the use of a two-part catalyst system, which is a combination of a tertiary amine and an epoxy compound [134]. The actual catalytic species in this application is presumably a zwitterionic species produced from the addition of the amine to the epoxide. The pot-life of the formulation is thus controlled by the induction time required to accumulate an appreciable concentration of the catalytic species.

Another cross-linking reaction that can be used with acetoacetylated materials is via complexation with metals. This reaction takes advantage of the ability of the beta-ketoester to chelate with polyvalent metals via its enol (or enolate) form. One example of the reaction is the cross-linking of an acetoacetylated epoxy resin with group I, VI, or Ia metals. Reaction of the epoxy backbone with common epoxy cross-linkers, such as polyamines, polyacids, or polyanhydrides, can be carried out to impart additional properties to the film. This approach is claimed to produce exceptionally hard films [135] and can also be used to prepare primers for fluorocarbon topcoats. Another application of this cross-linking reaction involves the use of metal chelates (for example, Al, Zr, Ti) of acetoacetates with a terminal oxysilane functional group in conjunction with epoxy or anhydride functional polymers [136].

Monomeric acetoacetates that are chelated to metals have also been used as ligands to impart characteristics such as solubility or reactivity to the metallic reagent. These applications typically involve the use of titanium or aluminum chelates, which can function as cross-linkers with various materials such as polysiloxanes [137].

Polyacetoacetates can also react with polyfunctional amines to produce cross-linked networks via formation of the enamine. This approach is useful for the preparation of coating formulations that cure under ambient conditions. The pot-life of such formulations can be extended either by blocking the amine with ketones or aldehydes or, in some cases, merely by the use of ketone solvents [138]. The type of amine used has a dramatic effect on the overall rate of this reaction, with primary amines being ca. 100-fold faster than secondary amines. The use of this enamine reaction in combination with epoxy cross-linking has recently been described in automotive refinish applications [139a]. Another related approach involves the use of acetoacetylated diols or triols in epoxy/amine cross-linked films [139b]. This approach enables reduction of solvent levels owing to the fact that the polyacetoacetate can function as both a solvent and a participant in the cross-linking chemistry.

C. Uses/Applications of Acetoacetylated Acrylic Polymers

Acetoacetylated acrylic polymers have found a variety of uses in coating and adhesive applications. These applications take advantage of the ability of these polymers to undergo cross-linking reactions under ambient conditions, as well as the ultimate properties obtained from the cured material. This chemistry has been used in pressure-sensitive adhesives by use of a terpolymer containing 50% butyl acrylate, 37–39% 2-ethylhexyl acrylate, 10% vinyl acetate, and 1–3% acetoacetoxyethyl methacrylate (AAEM). This acrylic residue is cross-linked with polyisocyanates, epoxy resins, aldehydes, diamines, and metals such as zirconium or zinc to produce adhesives with a good combination of strength, pot-life tack, adhesion, and cohesion [140]. Room-temperature-curable adhesives with excellent water and boiling-water resistance have been described which use an emulsion prepared from acetoacetylated polyvinyl alcohol (see below) and a copolymer containing 1–20% AAEM. Combination of this emulsion with a resorcinol precondensate and hexamethylene tetraamine produces the adhesive composition [141].

Acetoacetylated acrylic polymers have also found uses in numerous coatings applications. These applications can involve either a room temperature (ambient) cure process or an oven-bake (thermosetting) process. The resulting coatings have been noted to provide a number of desirable characteristics. For example, acetoacetylated latex acrylic polymers (prepared from AAEM) have been cross-linked with urea or urea/formaldehyde to give films with excellent wet tensile strength for paper or textile coating applications [142]. Acetoacetylated acrylic polymers that contain glycidyl and cyclocarbonate groups have been used as pigment dispersants [143]. These latter polymers are used in conjunction with isocyanate-terminated polyesters and polyamines to produce materials that enable efficient dispersion of pigments in paints without agglomeration. The use of acetoacetylated materials in water-borne coating formulations has also been examined [138a]. Acetoacetylated polymers can be emulsified and used in conjunction with polyamines to give water-borne systems that can be cured at ambient temperatures. The use of emulsifying agents such as nitroalkanes, lower formate esters, and/or enolizable materials with a pKa of 6.5–9, such as 2,4-pentanedione or methylcyanoacetate, are noted to be critical to obtaining useful emulsions in these systems.

In addition to coating and adhesive applications, acetoacetylated acrylic polymers have also found several other applications. Thus AAEM and its acrylic analogs have been grafted onto ABS (acrylonitrile, butadiene, styrene polymers) and/or ABS-polycarbonate blends to impart improved flame resistance to molding compounds [144]. Similarly, novel abrasives compositions have been prepared by electron beam irradiation of an aluminum-cross-linked acetoacetylated acrylic/epoxy copolymer [144c]. This material is claimed to have excellent water resistance and to be easier to process than other commercial systems.

D. Acetoacetylated Polyvinyl Alcohol

Acetoacetylated polyvinyl alcohol (PVOAcAc) has been the subject of a great deal of patent activity, particularly in the unexamined Japanese patent literature [145]. This material was first disclosed by Staudinger and Haberle [145a], and detailed studies of its preparation in solvents such as DMF (dimethylformamide) have been published [145b]. The industrial method of choice for preparation of this material involves spraying the solid polyvinyl alcohol, which has been swollen by addition of small amounts of water, with diketene, and subsequently heating the heterogenous material [145c].

PVOAcAc is claimed to be of use as a protective colloid in the emulsion polymerization of various materials for use in paints, textile sizings, adhesives, and paper coatings; as a freeze-thaw stabilizer for various formulations [145d]; as a binder for gels such as are used in deodorants; and to enable lower formaldehyde levels in adhesives [145e]. Another application of this material is in thermally sensitive processes such as facsimile machines [145f].

V. HETEROCYCLES

Because it contains both electrophilic and nucleophilic functionality, diketene reacts with a variety of substrates to provide heterocyclic molecules. A number of these compounds are of commercial importance, particularly in the pharmaceutical and agrichemical industries. The commercial insecticide Diazinon, the fungicide Vitavax, the antisecretory drug Cimetidine, and many beta-lactam antibiotics are derived from diketene.

Heterocyclizations that involve diketene usually begin with acetoacetylation of a substrate, followed by an intramolecular condensation reaction; this latter ring-closure reaction is often an equilibrium process. One can frequently predict the product of heterocyclization reactions of diketene from the initial position of acetoacetylation. For example, when a substrate is initially acetoacetylated on carbon, a six-membered ring system containing one heteroatom is usually formed. The simplest example of this type of reaction is the dimerization of diketene to give dehydroacetic acid.

A. Six-Membered Ring Heterocycles with One Heteroatom Prepared from Diketene

Diketene readily self-condenses to form dehydroacetic acid (DHA) [Eq. (49)], which, along with its sodium salt, is an important fungicide and food preservative. DHA is frequently a by-product of processes that utilize diketene.

$$
\text{DHA} \quad \xleftarrow[\text{(Cat)}]{\text{DABCO}} \quad \xrightarrow[\substack{\text{KF, PhMe} \\ 93\%}]{\text{Aliquat 336}}
\tag{49}
$$

$$
\xleftarrow[\substack{\text{TsOH} \\ 50 \text{ C}}]{\text{MeCN}} \quad \xrightarrow[63\%]{98\% \text{ HF}} \quad \xrightarrow{\text{Heat}}
\tag{50}
$$

At room temperature, diketene is very slowly converted into DHA, but catalysts can accelerate this reaction into a rapid and exothermic process. Such catalysts include pyridine [5,146], imidazole [147], and other tertiary amines [148]. The catalyst of choice for the industrial preparation of DHA from diketene appears to be diazabicyclo[2.2.2]octane (DABCO); yields of over 95% have been realized under mild reaction conditions (40–50°C, 15 min) [149].

The self-condensation of diketene in the presence of larger quantities of a tertiary amine results in the formation of several products resulting from further condensation [150]. The use of a strongly basic ion-exchange resin as a catalyst for the self-condensation of diketene resulted in the formation of one coumarin in excellent yield [55e]. Like DHA, this latter compound is sometimes found as an insoluble by-product of diketene reactions.

Under acidic conditions, diketene self-condenses into dimeric and trimeric species [Eqn. (50)]. Two isomeric spirocyclic diketene trimers, which appear to result from further reaction of DHA, have been isolated from the treatment of diketene with *p*-toluenesulfonic acid in acetonitrile [151]. Under more vigorous conditions, in the presence of water or alcohols, 2,6-dimethyl-4-pyrone is generally formed. Note that, once again, the initial acetoacetylation occurs on carbon, but the acidic condition favors ring closure viathe enolic hydroxyl group instead of the carboxylic acid (which is protonated under the acidic reaction conditions).

Several types of pyrones can be prepared via acetoacetylation of active methylene groups with diketene and subsequent intramolecular condensations [Eq. (51), (52)]. Thus, the sodium or thallium salts of beta-keto esters react with diketene to provide orsellinate esters and 5-substituted resorcinols [152]. The use of strong bases in these latter reactions favors ring closure via the unconjugated enolate anion to provide benzene derivatives via carbon-carbon bond formation; the use of other catalysts, such as triethylamine [153], Aliquat 336 [55e], or sulfuric acid [154], affords 4-pyrones. Other compounds with active methylene groups that can be acetoacetylated with diketene to provide heterocycles include sodium diethylmalonate [155], malononitrile [156], and 1,3-dimethylbarbituric acid [157].

$$ (51) $$

R = Me (60%)

$$ (52) $$

Diketene self-condenses in the presence of amines to provide pyridones, in a manner analogous to the preparation of pyrones from diketene [Eq. (53), (54)]. Thus, 2 equivalents of diketene and ammonia combine at 20°C to give 3-acetyl-4-hydroxy-6-methyl-2-pyridone [Eq. (53), R = H] in excellent yield, presumably via C-acetoacetylation of a molecule of acetoacetamide [158]. Unhindered, primary aliphatic amines react with 2 equivalents of diketene to give 2-pyridones, which can also be obtained by reacting 1 equivalent of diketene with an N-alkylacetoacetamide. Sterically hindered aliphatic amines, however, give 4-pyridones [Eq. (54)] [158]; steric hindrance of the amide nitrogen appears to be a major factor in controlling the mode of ring closure, as established by a study on the reaction of diketene with aminocrotonamides [159]. Diketene reacts with anilines to afford 1-aryl-2-pyridones; several of these pyridones exhibit fungicidal activity [160].

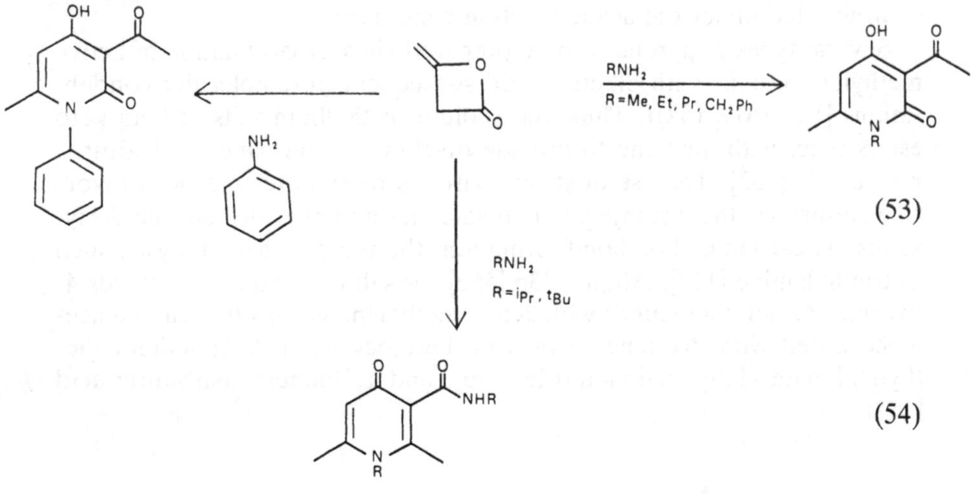

(53)

(54)

Diketene reacts with enamines to provide six-membered heterocycles containing one ring heteroatom, as would be expected from an initial acetoacetylation on carbon [Eq. (55)–(57)]. These reactions are frequently described as cycloaddition reactions even though polar intermediates are involved. In part, this description is used because acetylketene has been considered an intermediate in uncatalyzed, thermal heterocyclization reactions of diketene. The reaction of diketene with enamines derived from secondary amines gives fused gamma-pyrones, following elimination of the amine from the intermediate "cycloadduct" [161]. N,N-Dimethylhydrazones of acetoacetanilides also react with diketene as the enamine, with subsequent elimination of dimethylhydrazine, to afford pyrones [162].

Yneamines [163] and ketene acetals and their congeners [164] undergo analogous polar cycloaddition reactions with diketene to afford substituted 4-pyrones.

$$(55)$$

$$(56)$$

$$(57)$$

B. Six-Membered Ring Heterocycles with Two Heteroatoms Prepared from Diketene

Another major type of heterocyclization that involves diketene is that in which diketene initially *N*-acetoacetylates a substrate containing a C—N or C—O multiple bond to afford a six-membered ring containing two heteroatoms. The resulting product can be predicted by anticipating the reaction of acetylketene and the imine or nitrile. Thus, many imines [165] and *N*-acyl imines [166] react with diketene under basic conditions to afford 1,3-oxazin-4-ones, as illustrated by the synthesis of the tranquilizer Ketazolam from diazepam [Eq. (58)] [165c]. Diketene can also react with imines to afford azetidinone (beta-lactam) products that would be expected from [2 + 2] reactions of imines and acetylketene (see Section V.G).

$$(58)$$

Imidate esters react with diketene to provide oxazinones via intermediate 2-alkoxy-1,3-oxazin-4-ones [Eq. (59)–(63)] [167]. Thus, the interaction of diketene, ethyl isobutyrimidate, and ammonia [Eq. (59)] provides 4-hydroxypyrimidine (Oxy-P) [168], which is widely used in the manufacture of the insecticide Diazinon [Eq. (61)]. Oxy-P has also been prepared from diketene and either isobutyramidine [169], isobutyronitrile [170], isobutyrate esters [171], or isobutyramide [172], providing an excellent demonstration of the versatility of diketene in heterocyclic synthesis.

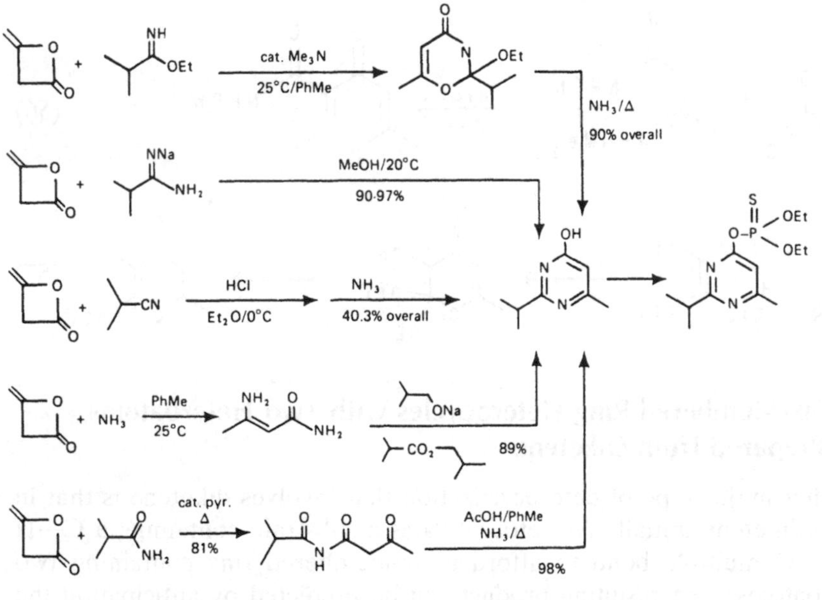

$$
\text{(59)}
$$

$$
\text{(60)}
$$

$$
\text{(61)}
$$

$$
\text{(62)}
$$

$$
\text{(63)}
$$

Diketene reacts with carbonyl compounds in the presence of Arrhenius acids to produce 1,3-dioxin-4-ones [Eq. (64)]. Thus, in the presence of p-toluenesulfonic acid, diketene reacts with acetone to afford 2,2,6-trimethyl-4H-1,3-dioxin-4-one [94f]. Analogous 1,3-dioxin-4-ones have been similarly prepared from other ketones and aldehydes [94g]. Palladium(II) complexes [173] and a quaternary amine resin [174] have also been used to catalyze the formation of 1,3-dioxinones from diketene.

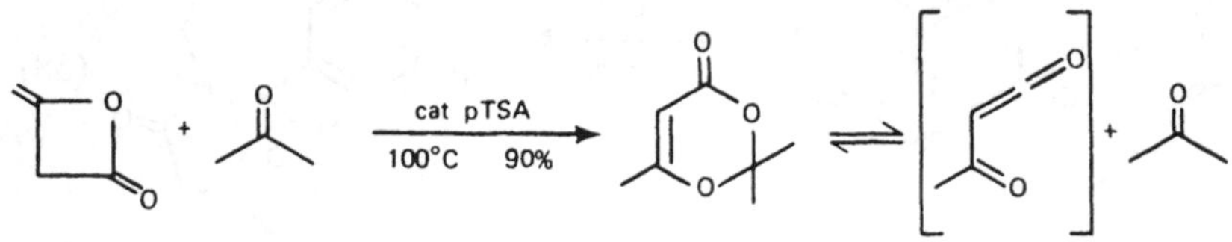

$$
\text{(64)}
$$

Thermolysis of the diketene-acetone adduct provides acetylketene [94e], a reactive intermediate that is useful for acetoacetylation (see Section III.C) and cycloaddition reactions [Eq. (65)–(67)] [94j]. Other 1,3-dioxin-4-ones have also been used as convenient sources of acylketenes [175].

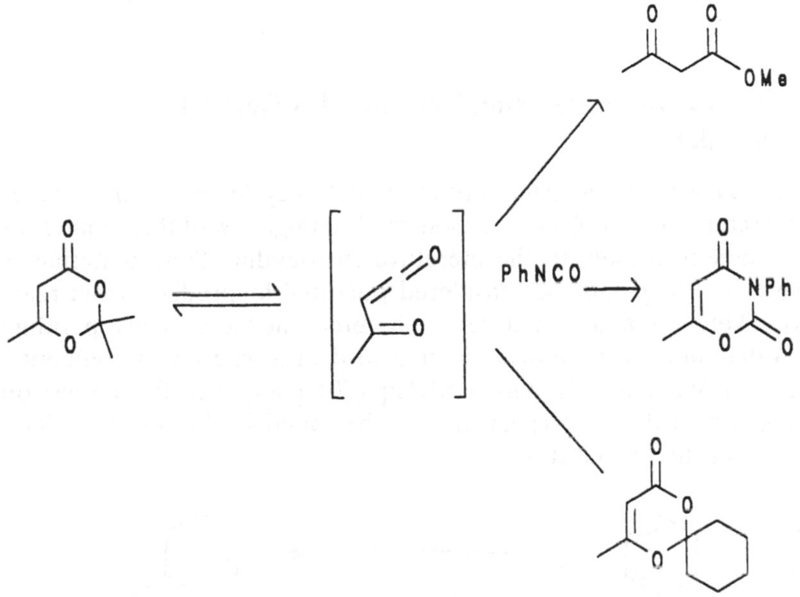

(65)

(66)

(67)

Diketene reacts with urea to give 6-methyluracil [Eq. (68)] [176]; the yield can be greatly improved with a catalyst such as pyridine [177] or diazabicyclo[2.2.2]octane (DABCO) in an acetic acid/acetic anhydride solvent system [178]. In pyridine at 20°C, diketene acetoacetylates *N*-alkylureas almost exclusively on the unsubstituted amino group, and the resulting acetoacetimides can be cyclized to give 1-alkyl-6-methyluracils [179]. Many diketene-derived uracils exhibit herbicidal activity, including terbacil and bromacil [Eq. (69)]. 6-Methyluracil is also used for the preparation of the analgesic mepirizole and as an intermediate in the preparation of the vasodilator dipyridamol (via 5-nitroorotic acid), which is used for drug-induced stress tests.

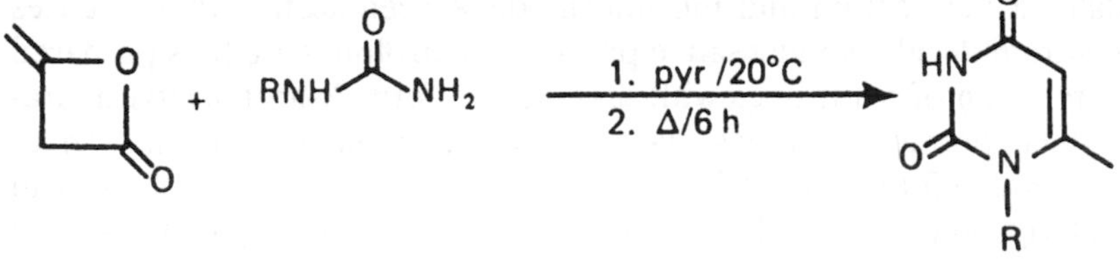

(68)

(69)

C. Other Six-Membered Ring Heterocycles Derived from Diketene

Diketene reacts readily with many nitrogen heterocycles and their *N*-oxides to give discrete, fused polycyclic products. Although few of these materials are of commercial interest, the literature surrounding these materials is extensive, and they must be considered potential by-products from reactions of diketene that are catalyzed by heterocyclic bases. Thus, pyridine reacts with either ketene or diketene to afford a tricyclic system frequently referred to as Wollenberg's compound [Eq. (70)] [180]. A further discussion of this aspect of diketene chemistry can be found in the review articles referenced in the Introduction.

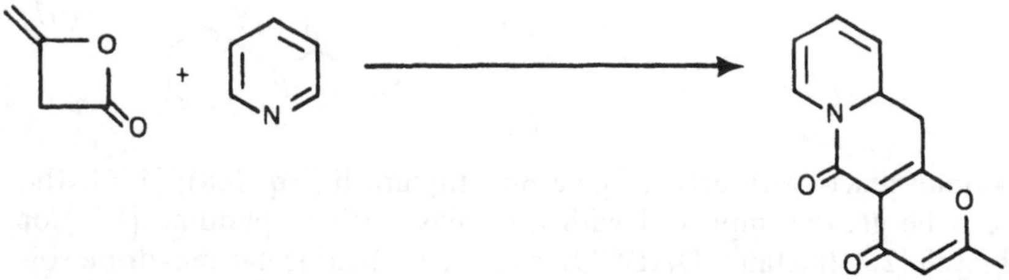

(70)

The preparation of the 1,2,3-oxathiazin-4-one ring system present in the artificial sweetener Acesulfame K was discussed in Section III.F as a reaction of acetoacetamides.

D. Five-Membered Heterocycles Derived from Hydrazines and Diketene

Five-membered heterocycles are generally formed when diketene acetoacetylates a heteroatom and the intermediate acetoacetate then cyclizes via an intramolecular condensative process. Pyrazolones are thus prepared by the reaction of hydrazines with diketene to form *N*-acetoacetyl hydrazine, which promptly cyclizes via intramolecular enamine formation to afford 5-methyl-3-pyrazolone [Eq. (71)] [181]; other alkyl pyrazolones can be similarly prepared [182]. 5-Methyl-3-pyrazolone readily couples with diazo compounds to form intermediates of great importance in dyestuff manufacture.

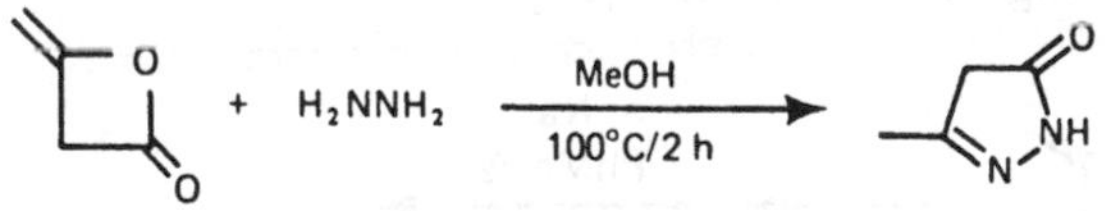

(71)

Phenylhydrazine reacts with diketene to give *N'*-phenylacetoacetohydrazide phenylhydrazone [Eq. (72) and (73)] [5], which provides 2-phenyl-3-pyrazolone upon heating with an additional equivalent of diketene. If the intermediate *N*-phenylacetoacetohydrizide is heated in hydrochloric acid without the addition of diketene, 1-phenyl-3-pyrazolone is isolated [183]. It is therefore possible to selectively prepare either of the two 3-pyrazolone isomers from a common intermediate. A mild, one-pot reaction is used to prepare commercially desirable 5-methyl-2-aryl-3-pyrazolones [184], which are coupled with diazonium salts to form colorants, such as the pyrazolone pigment shown in Eq. (74), which is known as Pigment Yellow 10.

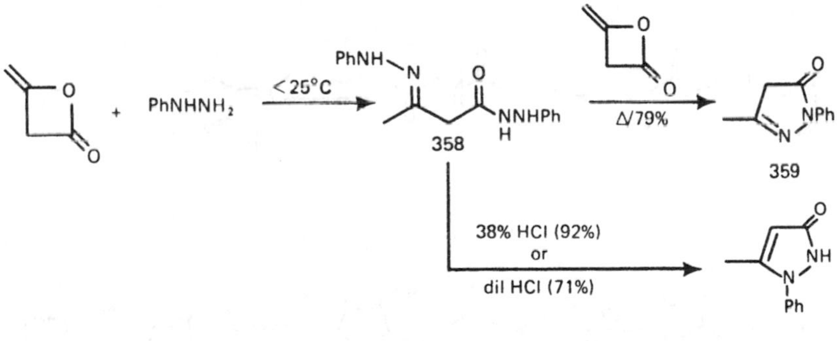

(72)

(73)

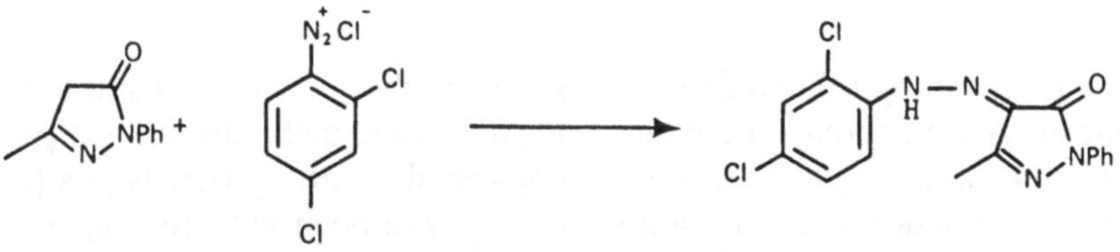

(74)

E. Other Five-Membered Heterocycles Derived from Diketene

Hydroxy ketones undergo a sequential acetoacetylation/condensation reaction to give five-membered butenolide ring systems [Eq. (75)] [185]. These butenolides are easily converted into furan-3-carboxylic acids. Alpha-hydroxy acids react similarly to provide furanones [186], which are important intermediates in the preparation of a series of natural products [187]. Tetrabutylammonium fluoride has recently been shown to be an effective catalyst for effecting this cyclization [88].

$$\text{(75)}$$

The analogous reaction of diketene and alpha-amino ketones provides a route to pyrrolidinones [Eq. (76), (77)] [188]; alpha-amino acids and their esters react

$$\text{(76)}$$

$$\text{(77)}$$

similarly to afford pyrrolidinediones [189]. A number of these substituted tetramic acids have interesting biological properties, including the strep-tolydigin and tirandamycin series of antimicrobial natural products [190]. Such acetoacetylation/condensation sequence have been used to prepare the antibiotic holomycin from diketene and S-benzyl-L-cysteine ethyl ester [191] and the mycotoxic cyclopiazonic acids [192].

As would be expected, alpha-amino amides and alpha-amino nitriles can also be used to prepare pyrrolidinone derivatives. In the case of the alpha-amino amides, the product is a function of the substituent on nitrogen [Eq. (78), (79)] [193]. Alpha-amino nitriles can be used to prepare ami-nopyrrolinones [194].

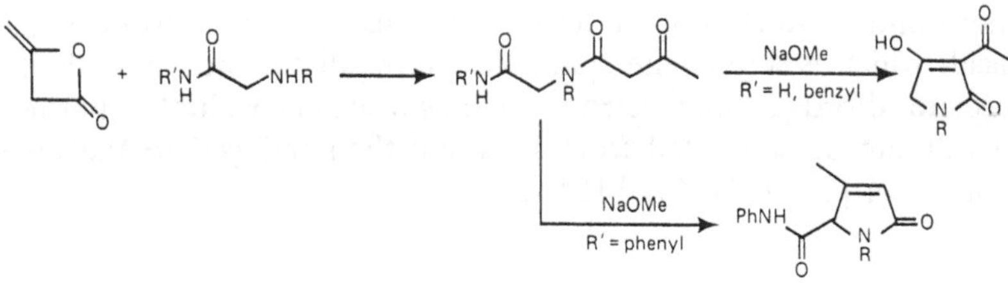

(78)

(79)

F. Five-Membered Heterocycles Derived from Haloacetoacetate Esters

Halogenated acetoacetic acid derivatives have at least three functional groups available for heterocyclization, which means that (1) several types of heterocycles should be accessible from these derivatives, and (2) the resulting products will still have functionality remaining for subsequent reactions.

2-Chloroacetoacetic acid derivatives are commonly used intermediates for the preparation of five-membered heterocycles, such as oxazoles, imidazoles, and thiazoles. Usually, heterocyclizations occurs via displacement of the halogen at C2, followed by reaction of the carbonyl group at C3. Thus, the oxazole used for preparing pyridoxine can be made from 2-chloroacetoacetamide in formic acid/formamide at moderate temperatures [195a]; further heating results in the formation of imidazoles [Eq. (80)–(82)]. This latter imidazole synthesis has been used in the preparation of antisecretory compounds, such as cimetidine [195b,c]. Likewise, thiazoles can be prepared from 2-chloroacetoacetate esters and thioformamide [196].

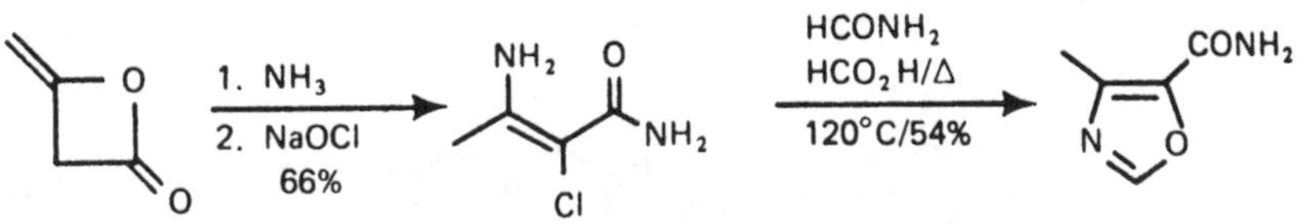

(80)

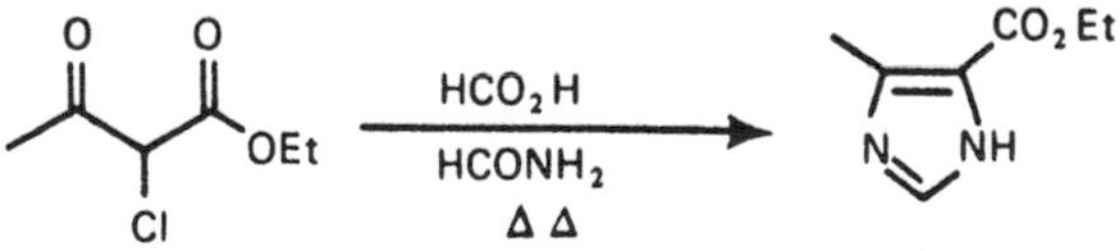

(81)

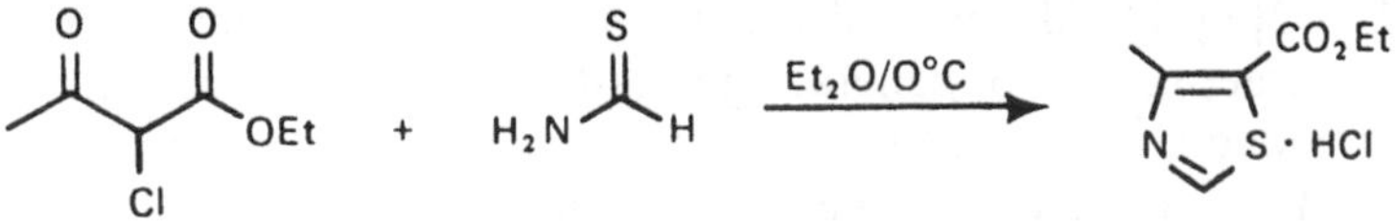

(82)

4-Chloro and 2,4-dichloroacetoacetates are used as precursors to te-
tronic acid, which is a versatile synthetic intermediate; the latter dihal-
oacetoacetate affords 3-chlorotetronic acid as a water-insoluble intermediate, which can be separated from salts and then reduced to the final
product in water [Eq. (83), (84)] [197].

$$ \text{(83)} $$

$$ \text{(84)} $$

Displacement of the halogen of 4-haloacetoacetate esters, followed by ring
closure on the carbonyl group (C3), provides five-membered ring heterocycles such as are found in the substituted acetoacetamido side chains for
beta-lactam antibiotics. The aminothiazoleacetic acid side chain that is
present on many of the current "third generation" cephalosporin antibiotics
is prepared from a 4-chloroacetoacetic acid derivative and thiourea, as
shown in Eq. (85) and (86) [198].

$$ \text{(85)} $$

$$ \text{(86)} $$

G. Other Heterocycles Derived from Diketene

The [2 + 2] cycloaddition of diketene to imines has recently been shown
to be an effective way of generating beta-lactams [Eq. (87)]; chiral imines
have provided sufficient chiral induction to provide optically active intermediates for making carbapenem antibiotics [199].

$$R-CH(PhOMe)_2$$

$$3.3 \quad : \quad 1 \qquad (87)$$

VI. CONCLUSION

The chemistry of diketene and its derivatives is extensive and diverse. Diketene is readily prepared from an inexpensive and readily available feedstock and should therefore continue to be a valuable synthetic intermediate for a variety of polymer intermediates, agrichemicals, dyes, and pharmaceuticals. Furthermore, diketene chemistry still offers ample opportunity for providing new discoveries in the future. It is hoped that this chapter has provided a background in the chemistry of diketene and acetoacetic acid derivatives, and has provided some new ideas.

ACKNOWLEDGMENTS

The authors thank Ms. D. Davis-Waltermirc for providing the flow diagram illustrating the manufacture of diketene (Fig. 1).

REFERENCES

1. Boese, A. B., Jr., *Ind. Chem. Eng.*, *32*, 16 (1940).
2. Borrmann, D., in *Methoden der Organischen Chemie (Houben-Weyl)*, 4th ed., Thieme Verlag, Stuttgart, 1968, Vol. 7, Part 4, p. 53.
3. Clemens, R. J., *Chem. Rev.*, *86*, 241 (1986).
4. (a) Kato, T., *Acc. Chem. Res.*, *7*, 265 (1974). (b) Kato, T., *Lect. Heterocycl. Chem.*, *6*, 105 (1982). (c) Lacey, R. N., in *Advances in Organic Chemistry*, Interscience, New York, 1960, p. 214. (d) Hasek, R. H., in *Kirk-Othmer: Encyclopedia of Chemical Technology*, 3rd ed., Wiley, New York, 1982, Vol. 13, p. 874.
5. (a) Chick, F., and Wilsmore, N. T. M., *J. Chem. Soc.*, *93*, 946 (1908). (b) Chick, F., and Wilsmore, N. T. M., *Proc. Chem. Soc.*, *24*, 100 (1908).
6. Pascal, R. A., Jr., Baum, M. W., Wagner, C. K., Rodgers, L. R., and Huang, D. S., *J. Am. Chem. Soc.*, *108*, 6477 (1986).
7. (a) Hanford, W. E., and Sauer, J. C., *Org. React.*, *3*, 108 (1946). (b) William, J. W., and Krynitsky, J. A., *Org. Synth.*, *21*, 13 (1941). (c) Gibaud, A., and Willemart, A., *Bull. Soc. Chim. Fr.*, 620 (1955).
8. (a) Lonza, Br. Patent 852,865 (1960); *CA*, *55*, 12299 (1961). (b) Keller, H. (Lonza), Swiss Patent 423,754 (1967); *CA*, *68*, 21825.

9.   Moenning, F., Dreizler, H., and Rudolph, H. D., *Z. Naturforsch.*, *22*, 1471 (1967).
10. (a) Whiffen, D. H., and Thompson, H. W., *J. Chem. Soc.*, 1005 1946). (b) Miller, F. A., and Carlson, C. L., *J. Am. Chem. Soc.*, *79*, 3995 (1957). (c) Carreira, L. A., and Lord, R. C., *J. Chem. Phys.*, *51*, 3225 (1969).
11. (a) Ford, P. T., and Richards, R. E., *Discuss. Faraday Soc.*, *19*, 230 1955). (b) Bader, A. R., Gutowsky, H. S., Williams, H. A., and Yankwich, P. E., *J. Am. Chem. Soc.*, *78*, 2385 (1956). (c) Brookes, D., Sternhell, S., Tidd, B. K., and Turner, W., *Aust. J. Chem.*, *18*, 373 (1965). (d) Matter, V. E., Pascual, C., Pretsch, E., Pross, A., Simon, W., and Sternhell, S., *Tetrahedron*, *25*, 2023 (1969).
12. Calvin, M., Magel, T. T., and Hurd, C. D., *J. Am. Chem. Soc.*, *63*, 2174 (1941).
13. (a) Angus, W. R., Leckie, A. H., LeFevre, C. G., LeFevre, R. J., and Wassermann, A., *J. Chem. Soc.*, 1751 (1935). (b) Kohlrausch, K. W. F., and Skrabal, R., *Proc. Indian Acad. Sci.*, *Sect. A*, *8A*, 424 (1938). (c) Taufen, H. J., and Murray, M. J., *J. Am. Chem. Soc.*, *67*, 754 (1945).
14. Bregman, J., and Bauer, S. H., *J. Am. Chem. Soc.*, *77*, 1955 (1955).
15. (a) Katz, L., and Lipscomb, W. N., *J. Org. Chem.*, *17*, 515 (1952). (b) Kay, I., and Katz, L., *Acta. Crystallogr.*, *11*, 897 (1958).
16. (a) Long, F. A., and Friedman, L. J., *J. Am. Chem. Soc.*, *75*, 2837 (1953). (b) Olivares, J. A., Flesch, G. D., and Svec, H. J., *Int. J. Mass. Spectrom. Ion Processes*, *56*, 293 (1984).
17. (a) Seidl, E. T., and Schaefer, H. F., III, *J. Am. Chem. Soc.*, *112*, 1493 (1990). (b) Dwivedi, C. P. D., *Pramana (J. Phys.)*, *25*, 547 (1985).
18. Hitzler, F. (Bayer), Ger. Offen. 890,341 (1953); *CA*, *52*, 14658 (1958).
19. Jacobs, M. L., and Higdon, B. W. (Celanese), U.S. Patent 3,759,955 (1973).
20. Bergamin, R., and Quittmann, W., (Lonza), Eur. Appl. 287,894 (1988).
21. See Ref. 3 for a detailed discussion of this structure elucidation.
22. Mansson, M., Nakase, Y., and Sunner, S., *Acta. Chem. Scand.*, *22*, 171 (1968).
23. (a) Briody, J. K., and Satchell, D. P. N., *J. Chem. Soc.*, 3778 (1965). (b) Briody, J. M., and Satchell, D. P. N., *Chem. Ind. (Lond.)*, 893 (1964).
24. Eck, H. (Wacker), Br. Patent 1,346,701 (1974); Ger. Offen. 2,154,875 (1973); *CA*, *79*, 18109 (1973).
25. Okeya, S., and Kawaguchi, S., *Bull. Chem. Soc. Jpn.*, *57*, 1217 (1984).
26. Kaku, T., Katsuura, K., and Sawaki, M. (Nippon Soda), U.S. Patent 4,335,184 (1982).
27. Moulin, F. (Lonza), Swiss Patent 647,495 (1985); *CA*, *102*, 203605 (1985).
28. Micheli, R. A., Hajos, Z. G., Cohen, N., Parish, D., Partland, L. A., Sciamanna, W., Scott, M. A., and Wehrli, P. A., *J. Org. Chem.*, *40*, 675 (1975).
29. Marcus, E., Chan, J. K., and Strow, C. B., *J. Org. Chem.*, *31*, 1369 (1966).
30. (a) Chick, F., and Wilsmore, N. T. M., *J. Chem. Soc.*, *97*, 1982 (1910). (b) Kato, T., Yamanaka, H., and Shibata, T., *Chem. Pharm. Bull.*, *15*, 921

(1967). (c) Kuenstle, G., and Jung, H. (Wacker), Ger. Offen. 3,101,650 (1982); *CA*, *97* 144409 (1982).

31. (a) Inoi, T., Sueyoshi, K., Fuji, M., Shoji, K., Shudo, A., Yoshizaki, T., and Matsuo, Y. (Chisso), U.S. Patent 3,703,518 (1971). (b) Pressler, W., and Meidert, H. (Hoechst), Ger. Offen. 2,842,149 (1980).
32. Sixt, J. (Wacker), U.S. Patent 2,763,664 (1956); *CA*, *31*, 5115 (1957).
33. Hurd, C. D., and Kelso, C. D., *J. Am. Chem. Soc.*, *62*: 1548 (1940).
34. Blum, R., and Tenud, L. (Lonza), Eur. Appl. 37015 (1981); *CA*, *96*, 19682 (1982).
35. (a) DeKimpe, N., Brunet, P., *Synthesis*, 595 (1990). (b) Rigamont, F., Roghunandan, P., US Patent 4,992,585 (1991).
36. Clemens, R. J., U.S. Patents 4,582,913 and 4,633,013 (1986).
37. (a) Hurd, C. D., and Abernethy, J. L., *J. Am. Chem. Soc.*, *62*: 1147 (1940). (b) Gross, M. (Lonza), U.S. Patent 4,473,508, (1984); Eur. Appl. 28709, (1981). (c) Dianippon Ink and Chemicals, Inc., Japan Kokai Tokkyo Koho 83-157747 (1983); *CA*, *100*, 51103 (1984). (d) VanSickle, D. E., Newland, G. C., Siirola, J. J., and Cook, S. L. (Kodak), U.S. Patent 4,468,356 (1984); *CA*, *101*, 191166 (1984).
38. Boosen, K. (Lonza), U.S. Patent 3,950,412 (1976), *CA*, *85*, 20628 (1976).
39. Boese, A. B., Jr. (Carbide), U.S. Patent 2,209,683 (1930); *CA*, *35*, 139 (1941).
40. Nollett, A. J. H., Ladage, J. W., and Mijs, W. J., *Recl. J. R. Neth. Chem. Soc.*, *94*, 59 (1975).
41. Wild, H. J. (Givaudan), U.S. Patent 4,892,966 (1990).
42. (a) Duc, L. (Lonza), EP 346,852 (1989). (b) Abacherli, C. (Lonza), U.S. Patent 4,565,696 (1986).
43. (a) d'Alcontres, G. S., Cum, G., and Gattuso, M., *Ric. Sci.*, *37*, 750 (1967); *CA*, *68*, 95733 (1968). (b) Kurita, K., Hirakawa, N., Dobashi, T., and Iwakura, Y., *J. Polym. Sci.*, *Polym. Chem. Ed.*, *17*, 2567 (1979).
44. (a) Clemens, R. J. (Eastman Chemical Company), unpublished results. (b) Seitz, V. G., Dhar, R., and Mohr, R., *Chem.-Ztg.*, *107*, 172 (1983).
45. (a) Kato, T., Chiba, T., and Tsuchiya, S., *Chem. Pharm. Bull.*, *28*, 327 (1980). (b) Chiba, T., Tsuchiya, S., and Kato, T., *Chem. Pharm. Bull.*, *29*, 3715 (1981). (c) Chiba, T., Takahashi, H., and Kato, T., *Heterocycles*, *19*, 703 (1982). (d) Chiba, T., Takahashi, H., Kato, T., Yoshido, A., and Moroi, R., *Chem. Pharm. Bull.*, *30*, 544 (1982). (e) Kato, T., Sato, M., and Kitagawa, Y., *J. Chem. Soc.*, *Perkin Trans.*, *1*, 352 (1918). (f) Chiba, T., Kato, T., Yoshida, A., Moroi, R., Shimomura, N., Momose, Y., Naito, T., and Kaneko, C., *Chem. Pharm. Bull.*, *32*, 4707 (1984).
46. Kato, T., Sato, M., and Kitagawa, Y., *J. Chem. Soc.*, *Perkin Trans.*, *1*, 352 (1978).
47. (a) Kato, T., Sato, M., and Kitagawa, Y., *Chem. Pharm. Bull.*, *23*, 365 (1975). (b) Adam, W., Kleim, U., and Lucchini, V., *Liebigs Ann. Chem.*, 869 (1988).
48. (a) Kato, T., Chiba, T., Sato, R., and Yashima, T., *J. Org. Chem.*, *45*, 2020 (1980). (b) Kato, T., and Katagiri, N., *Chem. Pharm. Bull.*, *21*, 729 (1973).

(c) Kato, T., Katagiri, N., and Sato, R., *Chem. Pharm. Bull.*, 27, 1176 (1979). (d) Kato, T., Katagiri, N., and Sato, R., *J. Org. Chem.*, 45, 2587 (1980). (e) Takano, S., Sugahara, T., Ishiguro, M., and Ogasawara, K., *Heterocycles*, 6, 1141 (1977). (f) Kato, T., Katagiri, N., and Sato, R., *J. Chem. Soc., Perkin Trans.*, 1, 525 (1979).

49. (a) Kato, T., Suzuki, Y., and Sato, M., *Chem. Lett.*, 697 (1978). (b) Kato, T., Suzuki, Y., and Sato, M., *Chem. Pharm. Bull.*, 27, 1181 (1979). (c) Inami, K., Saito, Y., and Shiba, T., *Chem. Pharm. Bull.*, 28, 327 (1980).

50. Dingwall, J. G., and Tuck, B., *J. Chem. Soc., Perkin Trans.*, 1, 2081 (1986).

51. Kato, T., Sato, M., Kitagawa, Y., and Sato, R., *Chem. Pharm. Bull.*, 29, 1624 (1981).

52. (a) Hull, G. A., Daniher, F. A., and Conway, T. F., *J. Org. Chem.*, 37, 1837 (1972). (b) Theobald, C. W. (DuPont), U.S. Patent 2,675,392 (1954); *CA*, 49, 4722 (1955).

53. (a) Coffmann, D. D. (DuPont), U.S. Patent 2,585,537 (1962). (b) Gabrielyan, G. A., Stanchenko, G. I., and Rogovin, Z. A., *Khim. Volokna*, 13 (1965); *CA*, 64, 9865 (1966).

54. (a) Gray, H. W., and Theobald, C. W. (DuPont), U.S. Patent 2,653,146 (1953); *CA*, 48, 1068 (1954). (b) Franzen, K., and Lehmann, H. (P. Beiersdorf and Co.), U.S. Patent 3,189,480 (1963). (c) Werntz, J. H. (DuPont), U.S. Patent 2,722,981 (1955); *CA*, 50, 4760 (1956).

55. (a) Furukawa, J., Saegusa, T., Mise, N., and Kawasaki, A., *Makromol. Chem.*, 39, 243 (1960). (b) Kawasaki, A., Furukawa, J., Saegusa, T., Mise, M., and Tsuruta, T., *Makromol. Chem.*, 42, 25 (1960). (c) Oda, R., Munemiya, S., and Okano, M., *Makromol. Chem.*, 43, 149 (1961). (d) Okamoto, Y., and Hwang, E. F. (W.R. Grace), U.S. Patent 4,588,792, (1986). (e) Dehmlow, E. V., and Shamout, A. R., *Ann. Chem.*, 2062 (1982).

56. (a) Chickos, J. S., *J. Org. Chem.*, 41, 3176 (1976). (b) Chickos, J. S., Sherwood, D. E., Jr., and Jug, K., *J. Org. Chem.*, 43, 1146 (1978).

57. Andreades, S., and Carlson, H. D., *Org. Synth.*, 45, 50 (1965).

58. Rice, F. O., and Roberts, R. J., *J. Am. Chem. Soc.*, 65, 1677 (1943).

59. (a) Fitzpatrick, J. T., *J. Am. Chem. Soc.*, 69, 2236 (1947). (b) Conley, R. T., and Rutledge, T. F. (Air Reduction Co.), U.S. Patent 2,818,456 (1957); *CA*, 52, 6391 (1958).

60. Jug, K., Dwivedi, C. P. D., and Chickos, J. S., *Theor. Chim. Acta*, 49, 249 (1978).

61. (a) Herrmann, W. A., Weichmann, J., Ziegler, M. L., and Pfisterer, H., *Angew. Chem. Int. Ed. Engl.*, 21, 551 (1982). (b) Arce, A. J., and Deeming, A. J., *J. Chem. Soc., Chem. Commun.*, 364 (1982). (c) Yamamoto, T., Ishizu, J., and Yamamoto, I., *Bull. Chem. Soc. Jpn.*, 55, 623 (1982).

62. (a) Horder, J. R., and Lappert, M. F., *J. Chem. Soc., Chem. Commun.*, 485 (1967). (b) Horder, J. R., and Lappert, M. F., *J. Chem. Soc.*, A, 173 (1969). (c) Urata, K., Itoh, K., and Ishii, Y. J., *Organomet. Chem.*, 63, 11 (1973). (d) Bodesheim, F., Hahn, G., and Nischk, G. (Bayer), Ger. Offen. 1,223,103 (1966); *CA*, 65, 17112 (1966).

63. Sheludyakov, V. D., Kozyukov, V. P., Petrovskaya, L. I., and Mironov, V. F., *Khim. Geterotsikl. Soedin.*, 185 (1967).
64. Suzuki, H., Matsuda, I., Itoh, K., and Ishii, Y., *Bull. Chem. Soc. Jpn.*, *47*, 2736 (1974).
65. Yamamoto, Y., Ohnishi, S., and Azuma, Y., *Synthesis*, 22 (1981).
66. Kricheldorf, H. R., *Chem. Ber.*, *106*, 3765 (1973).
67. Okeya, S., Ogura, T., and Kawaguchi, S., *Inorg. Nucl. Chem. Lett.*, *5*, 713 (1969).
68. (a) Baba, S., Sobata, T., Ogura, T., and Kawaguchi, S., *Bull. Chem. Soc. Jpn.*, *47*, 2792 (1974). (b) White, A. W. (Eastman Chemical Co.), unpublished results.
69. Uchiumi, S., and Ataka, K. (Ube Industries), Eur. Appl. 83-110,641 (1984); *CA*, *101*, 90431 (1985).
70. Pino, P., and VonBezard, D., U.S. Patent 4,186,140 (1978); Ger. Offen. 2,807,251 (1978); *CA*, *90*, 6235 (1979).
71. (a) Itoh, K., Fukui, M., and Kurachi, Y., *J. Chem. Soc., Chem. Commun.*, 500 (1977). (b) Armstrong, R. J., and Weiler, L., *Can. J. Chem.*, *61*, 2530 (1983).
72. Fujisawa, T., Sato, T., Gotoh, Y., Kawashima, M., and Kawara, T., *Bull. Chem. Soc. Jpn.*, 3555 (1982).
73. (a) Hurd, C. D., and Blanchard, C. A., *J. Am. Chem. Soc.*, *72*, 1461 (1950). (b) Perrin, C. L., and Arrhenius, T., *J. Am. Chem. Soc.*, *100*, 5249 (1978).
74. Izawa, T., and Mukaiyama, T., *Chem. Lett.*, 161 (1975).
75. (a) Izawa, T., and Mukaiyama, T., *Chem. Lett.*, 1189 (1974). (b) Ishikawa, T., and Yamamoto, M., *Chem. Pharm. Bull.*, *30*, 1594 (1982).
76. Garnero, J., Caperos, J., Anwander, A., and Jacot-Guillarmod, A., in *Proceedings of the 10th International Congress of Essential Oils*, Elsevier, Amsterdam, 1988.
77. Izawa, T., and Mukaiyama, T., *Chem. Lett.*, 409 (1978).
78. Girotra, N. N., and Wendler, N. L., *Tetrahedron Lett.*, *23*, 5501 (1982).
79. Watson, K. G. (ICI), U.K. Patent 2,140,803 (1984); *CA*, *102*, 184973 (1985).
80. Yufit, S. S., and Kucherov, V. F., *Izv. Akad. Nauk SSSR, Ser. Khim.*, 1658 (1960); *CA*, *55*, 9273 (1961).
81. Byrns, A. C. (Union Oil), U.S. Patent 2,453,619 (1948); *CA*, *43*, 2634 (1949).
82. Estes, R. R., and Tockman, A., *Trans. Ky. Acad. Sci.*, *13*, 265 (1952).
83. (a) Olah, G. A., and Kuhn, S. J., *J. Org. Chem.*, *26*, 225 (1961). (b) Eiglmeier, K., U.S. Patent 3,937,737 (1975); U.S. Patent 4,082,807 (1975); *CA*, *83*, 78865 (1975). (c) Eiglmeier, K., U.S. Patent 3,865,879 (1974).
84. Neunteufel, R. A., and Pfueller, P. (Hoechst), Ger. Offen. 2,631,526 (1978); *CA*, *88*, 139983 (1978).
85. Wang, P. C., and Dawson, D. J. (Dynapol), U.S. Patent 4,489,197 (1984).
86. (a) Lonza, French Patent 1,575,144 (1969); *CA*, *72*, 132082z. (b) Marti, O., and Zimmerli, W. (Lonza), U.S. Patent 3,651,130 (1972); *CA*, *76*, 139949d. (c) Marcus, E. (Union Carbide), U.S. Patent 3,513,189 (1970); *CA*, *73*, 34818v. (d) Netherlands Patent Appl. 75/8253 (1975); *CA*, *86*, 89198s. (e)

Nudelman, A., Kelner, R., Broida, N., and Gottlieb, H. E., *Synth.*, 387 (1989). (f) Wilson, S. R., and Price, M. F., *J. Org. Chem.*, *49*, 722 (1984). (g) Wacker, French Patent 1,550,747 (1968); *CA*, 72, 31251y. (h) *Organic Synthesis Col.*, *5*, 155 (1973). (i) Kasper, W., Matthias, G., and Schulz, G. (BASF), Ger. Offen. DE 1,912,406; *CA*, *74*, 31553q (1970). (j) Hoechst, French Patent 1,549,825 (1968); *CA*, 72, 12134w. (k) Heckles, J. S., (Armstrong) US Patent 4,217,396 (1980).

87. Guenther, K. (Hoechst), German Patent 3,818,244 (1989).

88. (a) Booth, P. M., Fox, C. M. J., and Ley, S. V., *J. Chem. Soc.*, *Perkin Trans.*, *1*, 121 (1987). (b) Yaggi, N. F., and Douglas, K. T., *J. Chem. Soc.*, *Chem. Commun.*, 609 (1977). (c) Nakazumi, H., Asada, A., and Kitao, T., *Bull. Chem. Soc. Jpn.*, *53*, 2046 (1980).

89. (a) Grigorescu, R., Sbiera, A., and Florescu, C. (Centrala Industruala de Medicamente si Coloranti), Romanian Patent 57,247 (1974); *CA*, *83*, 96456. (b) Libkin, M. A., and Markevich, V. S., Russian Patent 1,214,654 (1986); *CA*, *105*, 190603. (c) Hauser, C. R., and Hudson, B. E., *Org. React.*, *1*, 266 (1942).

90. Amiet, L., and Longlois, B. (Rhone-Poulenc), U.S. Patent 4,883,904 (1989).

91. (a) Gossauer, A., Roessler, F., Zilch, H., and Ernst, L., *Liebis Ann. Chem.*, 1309 (1979); *CA*, *92*, 6379b. (b) Kato, T., Yamamoto, Y., and Takeda, S., *Yakugaku Zasshi*, *94*, 884 (1974); *CA*, *81*:104677q.

92. (a) Schoellkopf, U., Frasnell, H., Banhidai, B., and Meyer, R. (BASF), German Patent 2,246,095 (1974). (b) Wenkert, E., and McPherson, C. A., *J. Am. Chem. Soc.*, *94*, 8084 (1972).

93. (a) Kawatetsu Kagaku Co., Jpn. Kokai Tokkyo Koho JP 58/154535 (1983); *CA*, *100*, 34166g. (b) Suzuki, T., Matsuki, T., Kudo, K., and Sugita, N., *Nippon Kagaku Kaishi*, *10*, 1482 (1983); *CA*, *100*, 33956c.

94. (a) Clemens, R. J., and Hyatt, J. A., *J. Org. Chem.*, *50*, 2431 (1985). (b) Hyatt, J. A., Feldman, P. L., and Clemens, R. J., *J. Org. Chem.*, *49*, 5105 (1984). (c) Sato, M., Kanuma, N., and Kato, T., *Chem. Pharm. Bull.*, *30*, 1315 (1982). (d) Sato, M., Ogasawara, H., and Yohizumi, E., *Chem. Pharm. Bull.*, *31*, 1902 (1983). (e) Clemens, R. J., and Witzeman, J. S., *J. Am. Chem. Soc.*, *111*, 2186 (1989). (f) Carroll, M. F., and Bader, A. R., *J. Am. Chem. Soc.*, *75*, 5400 (1953). (g) Carroll, M. F. (A. Boake, Roberts), Br. Patent 749,749 (1956); *CA*, *51*, 7415 (1957). (h) Pietsch, S., and Schaeffer, G. (Hoechst), Ger. Offen. 2,149,650 (1973); *CA*, *79*, 126507 (1973). (i) Sato, M., Ogasawara, H., Komatsu, S., and Kato, T., *Chem. Pharm. Bull.*, *32*, 3848 (1984). (j) Coleman, R. S., and Grant, E. B., *Tetrahedron Lett.*, *26*, 3677 (1990). (k) Kato, T., Chiba, T., and Sato, M., *Chem. Pharm. Bull.*, *26*, 3877 (1978).

95. (a) Bader, A. R., Cummings, L. O., and Vogel, H. A., *J. Am. Chem. Soc.*, *74*, 3992, (1952). (b) Bader, A. R., and Vogel, H. A., *J. Am. Chem. Soc.*, *74*, 3992 (1952). (c) Taber, D. F., Amdeio, J. C., Jr., and Patel, Y. K., *J. Org. Chem.*, *50*, 3618 (1985). (d) Gilbert, J. C., and Kelly, T. A., *J. Org. Chem.*, *53*, 449 (1988). (e) Carroll, M. F., *Proc. XIth Intern. Congr. Pure Appl. Chem.*, *2*, 39 (1947); *CA*, *45*, 7015e (1951). (f) Cossy, J., and Thelland, A., *Synthesis*, 753 (1989). (g) Witzeman, J. S., and Nottingham, W. D., *J.*

Org. Chem., *56*, 1713 (1991). (h) Witzeman, J. S., and Nottingham, W. D. (Eastman Kodak), U.S. Patent 5,051,529 (1991). (i) Witzeman, J. S., and Nottingham, W. D., *ChemSpec*, *89*, 17 (1989).

96. (a) Campbell, D. S., and Lawrie, C. W., *J. Chem. Soc.*, *Chem. Commun.*, 355 (1971). (b) Witzeman, J. S., *Tetrahedron Lett.*, *31*, 1401 (1990). (c) Wentrup, C., and Friermilk, B., *J. Org. Chem.*, *56*, 2286 (1991).
97. *Comline Biotechnol. Med.*, 2 (July 13, 1988); *Jpn. Chem. Week*, 4 (June 30, 1988).
98. *Chem. Eng. News*, 17 (Dec. 19, 1988).
99. Carroll, M. F., *J. Chem. Soc.*, 704 (1940); 1266 (1940); 507 (1941).
100. (a) Kimel, W. (Hoffmann LaRoche), U.S. Patent 2,638,484 (1953); *CA*, *48*, 2763 (1954). (b) Cane, D. E., and Thomas, P. J., *J. Am. Chem. Soc.*, *106*, 5295 (1984).
101. (a) Yamamoto, A., and Ishihara, T. (Shin-Etsu) Japan Kokai 77-49445 (1977); *CA*, *88*, 169600 (1978), *84*, 179676 (1976). (b) Kitagaki, T., Yamamoto, A., and Ishihara, T. (Shin-Etsu), U.S. Patent 3,975,446 (1975); Ger. Offen. 2,430,192 (1975); *CA*, *82*, 124771 (1975).
102. (a) Lacey, R. N., *J. Chem. Soc.*, 827 (1954). (b) Carroll, M. F. (A. Boake, Roberts), Br. Patent 762,656 (1956); *CA*, *51*, 12143 (1957). (c) Kato, T., and Chiba, T., *Chem. Pharm. Bull.* *23*, 2263 (1975). (d) Kimel, W. (Hoffmann LaRoche), U.S. Patent 2,812,353 (1957); *CA*, *52*, 3860 (1958).
103. Hoffmann, W., and Baumann, M. (BASF), Ger. Offen. 2,532,799 (1977); *CA*, *86*, 155831 (1977).
104. (a) Fujita, Y., Ohnishi, T., Nishida, T., Omura, Y., and Mori, F. (Kuraray), Japan Kokai Tokkyo Koho 78-105407 (1978); *CA*, *90*, 23332 (1979). (b) Ohnishi, T., Fujita, Y., Nishida, T., Ishiguro, M., and Hosogai, T. (Kuraray), Japan Kokai Tokkyo Koho 79-14905 (1979); *CA*, *91*, 5383 (1979).
105. Ninagawa, Y., Nakahara, F., Nakamoto, T., and Kawaguchi, T. (Kuraray), Japan Kokai 77-68115 (1977); *CA*, *87*, 184730 (1977).
106. Gilbert, J. C., and Kelley, T. A., *Tetrahedron*, *44*, 7587 (1988).
107. (a) Linkies, A., and Reuschling, D. B., *Synthesis*, 405 (1990). (b) Weber, E., Reuschling, D. B., and Linkies, A. H. (Hoechst), German Patent 3,531,357 (1987). (c) Schuetz, J., and Schweikert, O. (Hoechst), German Patent 3,527,070 (1987).
108. (a) Waters, W. D., U.S. Patent 4,931,514 (1990). (b) Takiyama, E., Yokoyama, A., Ishihara, K., Abe, H., and Igarashi, Y., Japan Kokai Tokkyo Koho, 1,254,720 (1989).
109. Blackburn, J. B., and Hamilton, A. (Ciba-Geigy), U.S. Patent 4,885,033 (1989).
110. (a) Kuhne, R., and Hamel, H. (Hoechst), U.S. Patent 4,870,164 (1989). (b) Schwartz, R. J., and Gregario, M. Z. (Sun Chemical), U.S. Patent 4,946,508 (1990). (c) Schwartz, R. J., Gregario, M. Z., and Zwirgzdas, A. C. (Sun Chemical), U.S. Patent 4,946,509 (1990). (d) Schwartz, R. J., and Gregario, M. Z. (Sun Chemical), U.S. Patent 5,024,698 (1991).
111. *The Merck Index*, 9th ed., Merck, Rahway, NJ, 1976, p. 1834 and references therein.

112. Hussain, S. M., El-Reedy, S. A., and El-Sherabasy, S. A., *Heterocycles*, 9, 25 (1988).

113. (a) Burgstahler, A. W., Sandes, M. E., Shaefer, C. G., and Weigel, L. O., *Synthesis*, 405 (1977). (b) leNoble, W. J., and Morris, H. F., *J. Org. Chem.*, 34, 1969 (1969); leNoble, W. J., and Puerta, J. E., *Tetrahedron Lett.*, 1087 (1966). (c) Brieger, G., and Pelltier, W. M., *Tetrahedron Lett.*, 3555 (1965).

114. (a) Bergbreiter, D. E., and Weatherford, D. A., *J. Chem. Soc.*, *Chem. Commun.*, 883 (1989). (b) Jolly, P. W., and Kokel, N., *Synthesis*, 771 (1990).

115. (a) Bellas, T. E., Brownlee, R. G., and Silverson, R. M., *Tetrahedron*, 25, 5149, (1969). (b) De Kimpe, N., and Brunet, P., *Synthesis*, 595 (1990).

116. (a) McElvain, S. M., and Weber, K. H., *Org. Syn. Coll.*, *III*, 379, (1955). (b) Bowman, R. E., and Fordham, W. D., *J. Chem. Soc.*, 2758 (1951). (c) Rathke, M. W., and Cowan, P. J., *J. Org. Chem.*, 50, 2622 (1985).

117. (a) Sum, P-E., and Weiler, L., *Can. J. Chem.*, 55, 996 (1977). (b) Hukin, S. N., and Weiler, L., *J. Am. Chem. Soc.*, 96, 1082 (1974). (c) Yamaguchi, M., Shibuto, K., and Hirao, I., *Chem. Lett.*, 1145 (1985).

118. (a) Reynolds, G. A., and Hauser, C. R., *Org. Syn. Coll.*, *III*, (1955) 374. (b) Knorr, H., *Chem. Ber.*, 16, 2593 (1883). (c) Limpach, W., *Chem. Ber.*, 64, 969 (1931). (d) Conrad, C. and Limpach, W., *Chem. Ber.*, 20, 944 (1887). (e) Coffey, S., Thomson, J. K. and Wilson, F. J., *J. Chem. Soc.*, 856 (1936).

119. (a) Conrad, M., and Epstein, W., *Chem. Ber.*, 20, 3054 (1887). (b) Meyer, H., *Arzn. Forsch*, 31, 409 (1981). (c) Gyogyszergyar, E. GB Patent 2,219,294 (1989).

120. (a) Gutierrez, A., and Kleist, R. A. (Exxon), U.S. Patent 4,839,070 (1989). (b) Gutierrez, A., Lundberg, R. D., Kleist, R. A., and Bloch, R., (Exxon), U.S. Patent 4,906,252 (1990).

121. Kato, T., Katagiri, N., and Daneshtalab, M., *Heterocycles*, 413 (1975).

122. (a) Shim, S. C., Kim, D. W., Moon, S. S., and Chae, Y. B., *J. Org. Chem.*, 49, 1449 (1984). (b) Kato, T., Yamanaka, H., Kawamata, J., and Shibata, T., *Chem. Pharm. Bull*, 16, 1835 (1968). (c) Kato, T., Yamanaka, H., and Shibata, T., *Tetrahedron*, 23, 2965 (1967).

123. Hantzsch, A., *Ann. Chem.*, 215, 1 (1882).

124. Tacke, R., Bentlage, A., Sheldrick, W. S., Ernst, L., Towart, R., and Stoepel, K. *Z. Naturforsch. B: Anorg. Chem.*, *Org. Chem.*, 37B, 443 (1982).

125. (a) *USAN and USP Dictionary of Drug Names*, USP, Rockville, MD, 1981, p. 259. (b) Bossert, F., Horstmann, H., Meyer, H., and Vater, W., *Arzneim-Forsch.*, 29, 226 (1979); *CA*, 90, 179914 (1979).

126. *USAN and USP Dictionary of Drug Names*, USP, Rockville, MD, 1981, p. 261.

127. (a) *USAN and USP Dictionary of Drug Names*, USP, Rockville, MD, 1981, p. 258. (b) Iwanami, M., Shibanuma, T., Fujimoto, M., Kawai, R., Tamazawa, K., Takenaka, T., Takahashi, K., and Murakami, M., *Chem. Pharm. Bull.*, 27, 1426 (1979).

128. Tacke, R., Bentlage, A., Towart, R., Meyer, H., Bossert, F., Vater, W., and Stoepel, K., *Z. Naturforsch. B: Anorg. Chem.*, *Org. Chem.*, 35B, 494 (1980).

129. (a) Chekavichus, B. S., Sausinsh, A. E., and Dubur, G. Y., *Khim. Geterotsikl. Soedin.*, 1238 (1975); *CA*, *84*, 17091 (1976). (b) Uldrikis, Y. R., Dubur, G. Y., Dipan, I. V., and Chekavichus, B. S., *Khim. Geterotsikl. Soedin.*, 1230 (1975).

130. (a) Clemens, R. J., *Proc. 15th Water-Borne and Higher-Solids Coatings Symp.*, 55 (1988). (b) Rector, F. D., Blount, W. W., and Leonard, D. R., *Proc. 15th Water-Borne and Higher-Solids Coatings Symp.*, 68 (1988). (c) Vogel, H. A., and Bader, A. R. (PPG), U.S. Patent 2,730,517 (1956). (d) Clemens, R. J., and Rector, F. D., *J. Coat. Technol.*, *61*(770), 83 (1989). (e) Rector, F. D., Blount, W. W., and Leonard, D. R., *J. Coat. Technol.*, *61* (771), 31, (1989). (f) Rector, F. D., and Witzeman, J. S., *Surface Coat. Austr.*, *26*, 6 (1989). (g) Witzeman, J. S., Nottingham, W. D., and Rector, F. D., *J. Coat. Technol.*, *62* (789), 101 (1990). (h) Witzeman, J. S., Rector, F. D., and Nottingham, W. D., *Proc. 16th Water-Borne and Higher-Solids Coatings Symp.*, 400 (1989). (i) Witzeman, J. S., Clemens, R. J., Blount, W. W., and Rector, F. D., *Polym. Mat. Sci. Eng. Preprints*, *63*, 1000 (1990). (j) Clemens, R. J. (Eastman Kodak), U.S. Patent 5,107,649 (1991). (k) Clemens, R. J., Witzeman, J. S., and Rector, F. D., *Proc. XVI Intern. Conf. Organic Coatings Technol.*, Athens, Greece, 1990, p. 127. (And references cited therein.)

131. (a) Fong, D., and Cramm, J. R. (Nalco), U.S. Patent 4,743,668 (1988). (b) Clemens, R. J., U.S. Patent 4,992,584 (1991).

132. Bors, D. A., and Emmons, W. D. (Rohm and Haas), EP 367,508 (1990); U.S. Patent 4,960,924 (1990).

133. Heckles, H. S. (Armstrong Cork), U.S. Patent 4,217,396 (1980).

134. (a) Schindler, F. J., Hurwitz, M. J., Feely, W. E., and Fulton, R. L., Jr. (Rohm and Haas), EP Appl. 326,723 (1989). (b) Say, T. E. [RTZ (now Rhone-Poulenc)], U.S. Patent 4,906,684 (1990).

135. Ritz, J., and Reese, J. (Hoechst), U.S. Patent 4,016,141 (1977).

136. (a) Mizuguchi, K., Okude, Y., Miwa, H., and Okude, H., EP 295,657 (1988). (b) Mizuguchi, K., Okude, Y., Miwa, H., and Okude, H., EP 336,428 (1989).

137. (a) Love, D. J., *Proc. 17th Water-Borne and Higher Solids Coatings Symp.*, 1, (1990). (b) Smith, S. D., and Hamilton, S. B. (G.E.), U.S. Patent 3,779,986 (1973). (c) Barfurth, D., and Nestler, H. (Huels), EP 316,893 (1989). (d) Barfurth, D., and Nestler, H. (Huels), U.S. Patent 4,924,016 (1990).

138. (a) Noomen, A., Vandervoorde, P. M., and Akkerman, H. (AKZO), U.S. Patent 4,772,680 (1988). (b) Hoy, K. L., and Carder, C. H., *J. Paint Technol.*, *46*, (591), 70 (1974). (c) Hoy, K. L., and Milligan, C. L. (Union Carbide), U.S. Patent 3,668,183 (1972).

139. (a) Den Hartog, H. C., and Palmer, G. T. (DuPont), U.S. Patent 4,987,177 (1991). (b) Stark, C. J., and Pietruszka, R. D. (Shell), U.S. Patent 5,021,537 (1991).

140. U.K. Patent Application 2,213,157 (1989).

141. Takeshi, K., and Makoto, Y. (National Starch), U.S. Patent 4,687,809 (1987).

142. Kissel, C. L. (Union Oil of California), EP 357,287 (1990).
143. Yamamoto, T., Matsukara, Y., Ohe, O., Ogawa, H., Ishidoya, M., and Matsubara, Y. (Nippon Oil & Fat), EP 358,358 (1990).
144. (a) Bauer, W., Lindner, C., Muller, F., Kress, H. J., Zabrocki, K., Bueckers, J. (Bayer), U.S. Patent 4,548,987 (1985). (b) Zabrocki, K., Lindner, C., Muller, F., and Doring, J. (Bayer), U.S. Patent 4,600,747 (1986). (c) Hesse, W., Ritz, J., and Tescher, E. (Hoechst), U.S. Patent 4,047,903 (1977).
145. (a) Staudinger, H., and Haberle, M. *Makromol. Chem.*, *9*, 52 (1953). (b) Arranz, F., Sanchez-Chavez, M., and Jarrin, C. M., *Angew. Makromol. Chem.*, *143*, 101 (1986). (c) Kobayshi, K., Matsuhiro, K., and Ito, Y. (Nippon Synthetic Chemical Industry Co.), Japanese Patent 1-34245 (1989). (d) Shimokawa, W., Ito, Y., Kobayashi, K., Fukumori, K., and Iwase, N. (Nippon Gohsei), U.S. Patent 4,350,788 (1982). (e) Shimokawa, W., and Fukamori, K. (Hoechst Gohsei), U.S. Patent 4,708,821 (1987). (f) Takayama, Y., Noguchi, A., and Matsushita, T. (Kazaki Paper), U.S. Patent 4,513,301 (1985). These are just a few of the many patents on this subject. The unexamined Japanese patent literature is particularly full of references to uses of this material.
146. Boese, A. B., Jr. (Carbide), Canadian Pat. 384,872 (1939); *CA*, *34*, 1335 (1940).
147. Marcus, E., and Chan, J. K. F. (Union Carbide), U.S. Patent 3,592,826 (1971); *CA*, *75*, 5701 (1971). (b) Steele, A. B., Boese, A. B., Jr., and Dull, M. F., *J. Org. Chem.*, *14*, 460 (1949).
148. Branch, S. J. (Distillers Co.), U.S. Patent 2,849,456 (1958); *CA*, *53*, 4306 (1959).
149. (a) Minamidate, M., and Veda, I. (Nippon Synthetic Chemical), Japan Kokai 74-31688 (1974); *CA*, *81*, 25554 (1971). (b) Friedl, Z., and Dofek, R., Czech. Patent 156,584 (1975) and 165,170 (1976); *CA*, *85*, 46390 (1976) and *CA*, *86*, 139858 (1977). (c) Kortvelyessy, G., Matolcsy, K., Sziliagyi, J., Wurdits, I., and Hartmann, J., Ger. Offen. 2,254,516 (1973); *CA*, *79*, 66173 (1973).
150. Marcus, E., and Chan, J. K., *J. Org. Chem.*, *32*, 2881 (1967).
151. Ernest, I., Fritz, H., and Rihs, G., *Helv. Chim. Acta.*, *70*, 203 (1987).
152. (a) Kato, T., and Hozumi, T., *Chem. Pharm. Bull.*, *20*, 1574 (1972). (b) Hase, T. A., Suokas, E., and McCoy, K., *Acta Chem. Scand.*, *Ser. B*, *B32*, 701 (1978).
153. Kato, T., Yamamoto, Y., and Hozumi, T., *Chem. Pharm. Bull.*, *21*, 1840 (1973).
154. Hamamoto, K., Isoshima, T., and Yoshioka, M., *Nippon Kagaku Zasshi*, *79*, 840 (1958); *CA*, *54*, 4552 (1960).
155. Suzuki, E., Sekizaki, H., and Inoue, S., *Synthesis*, 652 (1975).
156. (a) Kato, T., Kubota, Y., Tanaka, M., Takahashi, H., and Chiba, T., *Heterocycles*, *9*, 841 (1978). (b) Pietsch, H. (Hoechst), Ger. Offen. 2,726,685 (1979); *CA*, *90*, 137676 (1979).
157. Shoji, N., Kondo, Y., and Takemoto, T., *Chem. Pharm. Bull.*, *21*, 2639 (1973).
158. Kato, T., and Kubota, Y., *Yakugaku Zasshi*, *89*, 1477 (1970).

159. (a) Goto, Y., Masamoto, K., Arai, Y., and Ueda, Y., *Chem. Express*, *4*, 523 (1989). (b) Pierce, J. B., Ariyan, Z. S., and Ovenden, G. S., *J. Med. Chem.*, *25*, 131 (1982). (c) Kato, T., Chiba, T., Sasaki, M., and Kamo, M., *Yakugaku Zasshi*, *101*, 40 (1981); *CA*, *95*, 7179 (1981).

160. (a) Nagai, S., Yorie, T., Hirota, Y., Hibi, T., Sato, K., Yamamura, H., Wada, T., and Aoi, I. (Hokko), Japan Kokai 77-66630 (1977); *CA*, *88*, 1599 (1978). (b) Kato, T., and Kubota, Y., *Yakugaku Zasshi*, *101*, 47 (1967).

161. (a) Hunig, S., Benzing, E., and Hubner, K., *Chem. Ber.*, *94*, 486 (1961). (b) Eiden, F., and Wanner, K. T., *Arch. Pharm.*, *317*, 958 (1984). (c) Millward, B. B., *J. Chem. Soc.*, 26 (1960). (d) Lounasmaa, M., Langenskiold, T., and Holmberg, C., *Tetrahedron Lett.*, *22*, 5179 (1981).

162. Daicel, German Patent 3,544,850 (1986).

163. Ficini, J., and Genet, J. P., *Bull. Soc. Chim. Fr.*, 2086 (1974).

164. (a) Gompper, R., and Stetter, J., *Tetrahedron Lett.*, 233 (1973). (b) Kato, T., Yamamoto, Y., and Takeda, S., *Chem. Pharm. Bull.*, *21*, 1047 (1973).

165. (a) Kato, T., and Sakamoto, T., *Yakugaku Zasshi*, *87*, 1322 (1967); *CA*, *68*, 114524 (1968). (b) Maujean, A., and Chuche, J., *Tetrahedron Lett.*, 2905 (1976). (c) Hach, V. (Delmar Chemicals), Br. Patent 1,479,202 (1976) and Ger. Offen. 2,512,092 (1975); *CA*, *84*, 17444 (1976). (d) Hsi, R. S. P., *J. Labelled Compd.*, 341 (1974). (d) Matsuda, I., Yamamoto, S., and Ishii, Y., *J. Chem. Soc.*, *Perkin Trans.*, *1*, 1523, 1528 (1976).

166. Matsuda, I., Yamamoto, S., and Ishii, Y., *J. Chem. Soc.*, *Perkin Trans.*, *1*, 1523, 1528 (1976).

167. Kato, T., and Kondo, M., *Chem. Pharm. Bull.*, *24*, 356 (1976).

168. Blackwell, J. T., Gupton, J. T., Miyazaki, T. U., Nabors, J. B., and Pociask, J. R. (Ciba-Geigy), U.S. Patent 4,163,848 (1978); *CA*, *92*, 58804 (1980); *CA*, *94*, 65717 (1981).

169. Omori, M., Miyazaki, H., Hasegawa, K., Watanabe, Y., and Utsugi, Y. (Nippon Kayaku), Japan Kokai 73-39943 (1973); *CA*, *80*, 108569 (1974).

170. Pociask, J. R. (Ciba-Geigy), U.S. Patent 4,052,396 (1977); *CA*, *87*, 201581 (1977).

171. (a) Blackwell, J. T., Gupton, J. T., and Nabors, J. B. (Ciba-Geigy), U.S. Patent 4,052,397 (1977); *CA*, *87* 201580 (1977). (b) Kato, T., Yamanaka, H., and Konno, S., *Yakugaku Zasshi*, *90*, 509 (1970).

172. Gupton, J. T., III, Jelenevsky, A. M., Miyazaki, T. U., and Petree, H. E. (Ciba-Geigy), U.S. Patent 4,018,771 (1977).

173. Okeya, S., Ogura, T., and Kawaguchi, S., *Kogyo Kagaku Zasshi*, *72*, 1656 (1969); *CA*, *72*, 30858 (1970).

174. Dehmlow, E. V., and Shamout, A. K., *Ann. Chem.*, 1753 (1982).

175. Sakaki, J., Kobayashi, S., Sato, M., and Kaneko, C., *Chem. Pharm. Bull.*, *37*, 2952 (1989).

176. (a) Boese, A. B., U.S. Patent 2,138,756 (1938). (b) Gleason, A. H. (Standard Oil), U.S. Patent 2,174,239 (1939); *CA*, *34*, 450 (1940).

177. (a) Khromov-Borisov, N. V., and Karlinskaya, R. S., *Zh. Obsch. Khim.*, *26*, 1728 (1956); *CA*, *51*:2809 (1957). (b) Yeksler, M. A., and Vitvitskaya, A. S., *Khim-Farm. Zh.*, *16*, 727 (1982); *CA*, *97*, 55772 (1982).

178. Eck, H., and Spes, H. (Wacker), Br. Patent 1,371,155 (1973); *CA*, *78*, 124622 (1973).
179. Gunar, V. I., and Zavyalov, S. I., *Dokl. Akad. Nauk SSSR*, *158*, 1358 (1964).
180. Wollenberg, W. O., *Chem. Ber.*, *67*, 1675 (1934).
181. Ube Industries, Ltd., Japan Kokai Tokkyo Koho 80-108856 (1980); *CA*, *94*, 175114 (1981).
182. Kato, T., Yamanaka, H., and Hamaguchi, F., *Yakugaku Zasshi*, *83*, 741 (1963); *CA*, *59*, 13964 (1963).
183. Lecher, H. Z., Parker, R. P., and Conn, R. C., *J. Am. Chem. Soc.*, *66*, 1959 (1944).
184. (a) Levin, P. A., *Zh. Obsch. Khim.*, *27*, 2864 (1957). (b) Skowronski, R., and Szalecki, W., Polish Patent 105501 (1980); *CA*, *93*, 132486 (1980). (c) Johnston, F., U.S. Patent 2,017,815 (1933). (d) Hoechst, German Patent 3,416,205 (1985).
185. (a) Lacey, R. N., *J. Chem. Soc.*, 822, 850 (1954). (b) Falsone, G., and Hundt, B., *Arch. Pharm.*, *318*, 190 (1985).
186. (a) Lacey, R. N., *J. Chem. Soc.*, 832 (1954). (b) Bloomer, J. L., and Kappler, F. E., *J. Org. Chem.*, *39*, 113 (1974).
187. Ager, D. J., and Mole, S. J., *Tetrahedron Lett.*, *29*, 4807 (1988).
188. Kato, T., Sato, M., and Yoshida, T., *Chem. Pharm. Bull.*, *19*, 292 (1971).
189. (a) Lacey, R. N., *J. Chem. Soc.*, 850 (1954). (b) Schmidlin, T., and Tamm, C., *Helv. Chim. Acta*, *63*, 121 (1980). (c) Matsuo, K., Kimura, M., Kinuta, T., Takai, N., and Tanaka, K., *Chem. Pharm. Bull.*, *32*, 4197 (1984).
190. (a) Jones, R. C. F., and Bates, A. D., *Tetrahedron Lett.*, *28*, 1565 (1987). (b) Rosen, T., Fernandes, P. B., Marovich, M. A., Shen, L., Mao, J., and Pernet, A. G., *J. Med. Chem.*, *32*, 1062 (1989) and references therein.
191. Buchi, G., and Luka, G., *J. Am. Chem. Soc.*, *86*, 5654 (1964).
192. (a) Holzapfel, C. W., and Gildenhuys, P. J. W., *Afr. J. Chem.*, *30*, 125 (1977). (b) Kozikowski, A. P., Greco, M. N., and Springer, J. P., *J. Am. Chem. Soc.*, *106*, 6873 (1984).
193. Kato, T., and Sato, M., *Yakugaku Zasshi*, *92*, 1507 (1972); *CA*, *78*, 84162 (1973).
194. Kato, T., and Sato, M., *Yakugaku Zasshi*, *92*, 1515 (1972); *CA*, *78*, 97451 (1973).
195. (a) Coffen, D. L. (Hoffmann La Roche), U.S. Patent 4,093,654 (1978); *CA*, *89*, 179983 (1978). (b) Durant, G. J., Emmett, J. C., and Ganellin, C. R. (SKF), U.S. Patent 3,876,647 (1975). (c) Erlenmeyer, H., *Helv. Chim. Acta*, *31*, 32 (1948).
196. (a) Clarke, H. T., and Gurin, S., *J. Am. Chem. Soc.*, *57*, 1876 (1935). (b) Kawasaki, T., Osaka, Y., and Tsuchiya, T. (Kureha Chemical Industry Co), U.S. Patent 4,363,813 (1981); *CA*, *95*, 7265 (1981).
197. Meul, T., Miller, R., and Tenud, L., *Chimia*, *41*, 73 (1987).
198. Huweiler, A., and Tenud, L. (Lonza), U.S. Patent 4,391,979 (1983).
199. Kawabata, T., Kimura, Y., Ito, Y., Terashima, S., Sasaki, A., and Sunagawa, M., *Tetrahedron*, *44*, 2149 (1988).

12

Vinyl Acetate

Charles E. Sumner and Joseph R. Zoeller
Eastman Chemical Company, Kingsport, Tennessee

I. INTRODUCTION

Vinyl acetate represents the single largest use of acetic acid, with a worldwide production of about 2.7 billion kg/year in 1989. Several processes have been used to generate this material on commercial scales. These processes have included the addition of acetic acid to acetylene [Eq. (1)], the addition of acetic acid to ethylene [Eq. (2)], and the sequential addition of acetic anhydride to acetaldehyde to form ethylidene diacetate (1,1-diacetoxyethane) and subsequent cracking of ethylidene diacetate to vinyl acetate and acetic acid [Eq. (3) and (4)].

Acetylene-based route:

$$C_2H_2 + AcOH \longrightarrow (AcO)CH=CH_2 \tag{1}$$

Ethylene-based route:

$$C_2H_4 + AcOH \longrightarrow (AcO)CH=CH_2 \tag{2}$$

Acetaldehyde/acetic anhydride–based routes:

$$AcH + Ac_2O \longrightarrow (AcO)_2CHCH_3 \tag{3}$$

$$(AcO)_2CHCH_3 \longrightarrow (AcO)CH=CH_2 + AcOH \tag{4}$$

225

The predominant route today, which was first introduced in the late 1960s, is the addition of acetic acid to ethylene using a heterogeneous catalyst. However, preexisting plants for the corresponding homogeneous process and for the addition of acetic acid to acetylene are still operating.

The last process, involving the addition of acetic anhydride to acetaldehyde and subsequent cracking of ethylidene diacetate, has been practiced on a limited basis by Celanese and is still practiced on a small scale by Celmex. When this process was initially introduced, acetic acid (needed to make acetic anhydride) and acetaldehyde were derived either from acetylene or ethylene. Therefore, the development of a direct method for the addition of acetic acid to ethylene rapidly displaced this process.

However, in the early 1970s and 1980s, the long-term availability and cost of petroleum came under question, and the chemical industry turned its research efforts to using synthesis gas as an alternative feedstock. The successful development of laboratory and pilot plant–scale processes for the generation of acetaldehyde, acetic anhydride, and ethylidene diacetate from synthesis gas has led to a renewed interest in this last process.

In this chapter, we will discuss the uses of this very important product and will survey the different processes involved in its manufacture. The existing acetylene- and ethylene-based routes have been reviewed [1–3], and we will present a brief survey of these processes. Although it is of lesser importance at present, a strong emphasis will be placed on examining the potential synthesis gas–based route as it has not yet been extensively described in the literature.

II. USES FOR VINYL ACETATE

Vinyl acetate–derived polymers are ubiquitous in modern society. They may have been used in binding this book, producing the glossy cover, for the adhesive on the stamp and mailing label, and to glue the cardboard box in which it arrived. Probably, the walls in the room where the reader is sitting were painted (or finished in the case of wood paneling) using a vinyl acetate–derived polymer.

These important polymers are generally generated via free radical polymerization. Commercially, vinyl acetate polymerized with itself to generate the homopolymer polyvinyl acetate (PVAc) or copolymerized with other olefins. The most important copolymer is the copolymer with ethylene (ethylene vinyl acetate copolymer, EVA), although the copolymer with vinyl chloride has some utility as well. PVAc and EVA are either used directly or reacted with methanol to form methyl acetate and polyvinyl alcohol (PVA) or ethylene vinyl alcohol (EVOH), respectively. (The methyl acetate generated in these processes is generally hydrolyzed to methanol and acetic acid, which are recycled through the process.) Polyvinyl alcohol can be further reacted in the presence of an acidic catalyst with either butyraldehyde or formaldehyde (or its precursor) to form polyvinyl butyral (PVB) or polyvinyl formal (PVF), respectively.

Vinyl acetate is also hydrogenated to ethyl acetate on a limited commercial basis. The major classes of commercial products that flow from vinyl acetate are summarized in schematic form in Figure 1.

The applications of these polymers are incredibly diverse. Most of these polymers are sold commercially as emulsions or resins, although a number of the materials, particularly the EVA, PVB, and PVF polymers, are more commonly sold as film or sheet. These materials are compounded in a variety of ways and are subsequently used in the assembly of consumer products or sold directly to the consumer. Table 1 summarizes the *major* uses of these materials. It is not complete as these polymers are found in so many applications, it would require a separate book to describe all the products and processes. Good reviews of the methods used to generate and apply these various polymers are available [1,2,4,5].

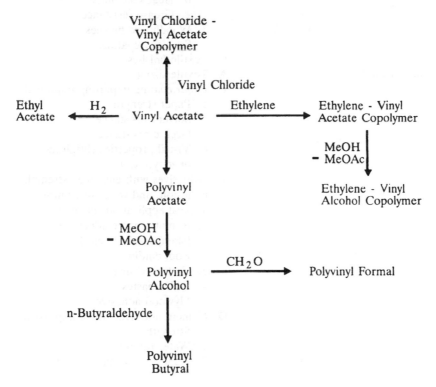

Figure 1   Product flow chart for vinyl acetate derivatives.

Table 1 Typical Uses of Vinyl Acetate–Derived Polymers

Product group	Commercial applications and properties

Polyvinyl acetate	A. Adhesives used in:
	   1. Packaging
	   2. Labeling (labels, stamps, etc.)
	   3. "White" glue
	   4. Construction
	   5. Wallboard laminate
	   6. Smoothing compound for wall joints
	   7. Specialty adhesives
	B. Coatings
	   1. Architectural coatings (latex paints)
	   2. High-quality paper coatings and paperboard, which allow:
	      a. Printable surfaces
	      b. Heat sealability
	      c. Grease resistance
	   3. Woodboard finishes
	   4. Specialty coatings
	C. Textile finishes

Polyvinyl alcohol	A. Textile sizing
	B. Paper coating imparting improved:
	   1. Paper strength
	   2. Solvent resistance
	   3. Grease resistance
	   4. Visual properties (brightness, opacity, etc.)
	C. Adhesives with enhanced strength, adhesion, and water resistance. Typical applications include:
	   1. Remoistenable adhesives (stamps, labels, etc.)
	   2. Bookbinding
	   3. Carton sealing
	   4. Office pastes
	   5. Plywood adhesives
	D. Cement additive, which improves:
	   1. Strength
	   2. Water retention
	   3. Adhesion to other surfaces

Table 1 Continued

Product group	Commercial applications and properties
Ethylene–vinyl acetate copolymer	A. Film, generally for packaging. Generally displays enhanced: 1. "Cling" properties 2. Toughness/strength 3. Low-temperature utility 4. Clarity 5. Heat sealability B. Hot-melt adhesives. Particularly useful in: 1. Cardboard box construction 2. Paper bags 3. Bookbinding 4. Furniture assembly C. Barrier coatings D. Wire and cable coverings
Ethylene–vinyl alcohol copolymer	A. Packaging films B. Packaging containers 1. Flexible tubes (e.g., toothpaste containers) 2. Bottles C. Thermofilm or sheet D. May have future application as a gas barrier polymer
Polyvinyl butyral	A. Laminated safety glass used in: 1. Automotive applications 2. Architectural applications B. Hot-melt adhesives
Polyvinyl formal	A. Magnet/wire insulation B. Hot-melt adhesives
Vinyl chloride–vinyl acetate copolymer	A. Protective surface coatings

III. ETHYLENE-BASED PROCESSES FOR VINYL ACETATE

At present, the majority of vinyl acetate manufactured in the world is generated from ethylene. Both the homogeneous and heterogeneous processes for accomplishing the transformation of acetic acid and ethylene to vinyl acetate utilize a palladium-based catalyst. In these processes a great deal of care must be exercised to maintain the ethylene-oxygen ratio below the explosive limits. The homogeneous and hetereogeneous processes have both been described previously [1,2] and will be discussed separately in this chapter.

A. Processes Using a Homogeneous Catalyst

The homogeneous catalysts for the oxidative acetoxylation of ethylene were the first to be commercialized. The process used a catalyst consisting of a palladium salt, generally palladium chloride (concentration approximately 50–80 mg/L), and a copper salt (Cu:Pd > 100:1) and was usually run with excess chloride present. (Halide-free systems have been reported, although no known commercialization of these halide-free systems is known to the authors.) The addition of an alkali metal salt has been reported to accelerate the reaction by creating a more nucleophilic acetate species [2]. The commercial chloride-containing processes were very corrosive and required the utilization of highly corrosion-resistant, very expensive materials of construction.

The reported pressures vary widely and depend on the exact nature of the catalyst. However, for commercial processes, the pressure is believed to be about 3 MPa (30 atm) with operating temperatures of 100–130°C. The oxidant is oxygen and the oxygen is maintained at less than 5.5% to maintain the mixture below the ignition limits. This limits the ethylene conversion per pass to about 2–3% [1]. The overall yields for the process are reported to be about 90% based on ethylene and 95% based on acetic acid.

These processes were often integrated to coproduce acetaldehyde by the addition of water to the reaction. The acetaldehyde would be oxidized separately to provide a source of acetic acid for the vinyl acetate process.

The similarity to the Wacker acetaldehyde synthesis is obvious and the role of the catalyst components is similar [6]. In the homogeneous oxidative acetoxylation, palladium, present as Pd^{2+}, coordinates ethylene, which activates the ethylene to nucleophilic attack by acetate forming a β-acetoxypalladium complex. The β-acetoxypalladium complex undergoes β-elimination of hydrogen to generate vinyl acetate and a hydridopalladium complex. Subsequent elimination of HCl generates palladium (0), which must be reoxidized.

Unfortunately, palladium (0) is well known to be unreactive with air. The reaction is thermodynamically favorable but kinetically slow. This problem is circumvented by the use of copper as an oxygen shuttle.

Cu^+ is readily oxidized by oxygen in chloride-containing solutions to form cupric chloride. Fortunately, cupric chloride is an efficient oxidant for the oxidation of palladium (0) to palladium (II) in HCl solutions forming cuprous chloride as the reduced species, thus providing the necessary means

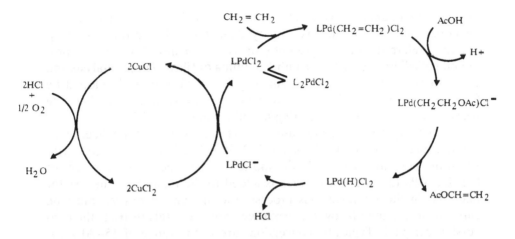

L = ligand (e.g. ethylene, Cl⁻)

Figure 2 Proposed mechanism for the homogeneous oxidative acetoxylation of ethylene to vinyl acetate. L = ligand (e.g., ethylene, Cl⁻).

of overcoming palladium's refractory behavior toward oxygen. A mechanism for the process is shown in Figure 2.

Because of its highly corrosive nature and the large capital required, this process was displaced by the more appealing, less capital-intensive vapor phase process. Since the homogeneous process has been displaced and represents antiquated technology, the plant schematics for this process have not been reproduced in this chapter; however, they are readily available [1,7].

B. Processes Using a Heterogeneous Catalyst

The most widely used and economically most attractive process is the heterogeneous process for the oxidative acetoxylation of ethylene. The process was first reported by National Distillers [8], which was later known as USI and, more recently, as Quantum. The catalyst was a supported palladium system, but the most useful catalysts were not apparent from the patent disclosure.

However, Hoescht and Bayer later described their joint development of a vapor phase process [9], which was based on 0.1–2% palladium catalyst supported on either alumina or silica. This catalyst has been the subject of enormous amounts of research, and innumerable patents have appeared. The catalysts have been reported to be enhanced by the addition of other noble metals, such as rhodium [1,7,9]. Particularly interesting promotors included vanadium (10), gold (11), and cadmium (11). The initial catalysts introduced are believed to have employed alkali metals as promotors [1,7], but the preferred catalysts presently in use are believed to be gold-promoted palladium supported on either alumina or silica. These catalysts and the associated manufacturing technology for vinyl acetate developed by Quantum, Hoescht, and Bayer have been extensively licensed and are the basis of most of the manufacturing facilities built since 1970.

The heterogeneous reaction uses a reactor consisting of a collection of catalyst tubes that are maintained at 150–200°C. Ethylene and acetic acid vapor are preheated to 120–160°C and fed to the reactor at a pressure of 0.6–1.1 MPa (5–10 atm). Oxygen is added to the mixture at the reactor inlet. As in the homogeneous process, the amount of oxygen must be carefully maintained below the explosive limits, and this restricts the conversion per pass. Typically, conversions are in the range of 15–30% for acetic acid, 10–15% for ethylene, and 60–90% for oxygen. The yield, based on ethylene, is generally about 90–95%, with most of the yield loss due to CO_2 generation. Approximately 1% of the ethylene is converted to acetaldehyde. The space time yield for the process is about 200 g vinyl acetate/L of catalyst.

This vapor phase ethylene process is reported to require 50% more capital than the corresponding acetylene processes, which will be described

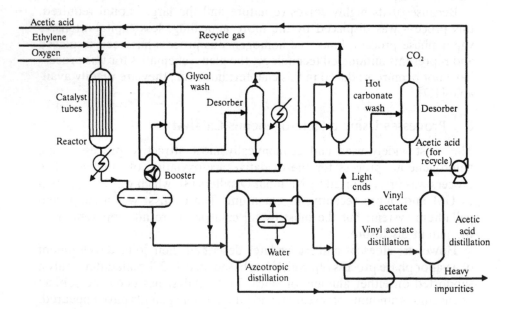

Figure 3   Acetic acid to ethylene vapor phase process. (Reprinted with permission from Ref. 7.)

later. However, the larger capital investment is offset by the much lower cost of ethylene. A schematic plant diagram appears in Figure 3.

IV. ACETYLENE-BASED PROCESSES

The addition of acetic acid to acetylene is the oldest process for the commercial generation of vinyl acetate. Although most capacity is presently ethylene based and nearly all expansions since the 1960s have been ethylene based, the older plants, particularly those in Germany, still contribute to the overall worldwide production of vinyl acetate.

The very first processes were homogeneous processes, and a variety of catalysts and catalyst combinations were used commercially. The best processes were based on a mercury catalyst in the presence of a strong acid [1–3]. The operation is completely obsolete and will not be discussed further.

The vapor phase process, which is still in practice in several locations, is based on a zinc acetate catalyst impregnated in a carbon support. These processes are operated at only slightly above atmospheric pressure at a temperature of 180–210°C. The reactants are preheated, and conversions of acetic acid are generally about 60–80%. The yields for this process are also high, generally in the range of 92–98% for acetylene and 95–99% based on acetic acid [1].

Although the acetylene process appears to be simpler to operate and cheaper to build, its use will continue to decline because these plants are dependent on having a source of acetylene at reasonable cost. The generation of acetylene is expected to be considerably more expensive than the cost of obtaining ethylene for the foreseeable future, and this ties the use of this process to preexisting acetylene sources in which the cost of acetylene generation was incurred prior to development of the ethylene-based processes.

If a competitive process for the generation of acetylene (as compared to ethylene) were to become available, this process would likely be of interest again because it is a simpler, less capital intensive process. Since this process represents disappearing technology, we have not included a process flowsheet, although process flowsheets are readily available [1].

V. SYNTHESIS GAS–BASED ROUTES TO VINYL ACETATE

As mentioned earlier, synthesis gas–based alternatives to the now well-established oxidative acetoxylation of ethylene have been proposed and may be of significant importance in the future. Several process variants have been proposed [12–14]. All the proposed processes start with methanol, or a downstream methanol derivative, and mimic the very old Celanese process [shown by Eq. (3) and (4)] in that the synthesis gas–based processes generate acetaldehyde as an intermediate that is either isolated or reacted in situ with acetic anhydride to generate ethylidene diacetate [(AcO)$_2$CHCH$_3$, EDA]. The method of cracking ethylidene diacetate to vinyl acetate, which involves an acidic catalyst, preferably a sulfonic acid [4], is unchanged from the early Celanese process. The various alternatives are summarized in Figure 4.

The reason these new variants of the older Celanese-type process have received renewed interest is that now all the key ingredients, namely acetic anhydride and acetaldehyde, can be derived from synthesis gas as the sole carbon source. Thus, in the event that ethylene prices rise again, these synthesis gas–based chemical processes could be implemented as a replacement process for vinyl acetate.

The generation of acetic anhydride from synthesis gas is discussed in Chapter 9. Therefore, this discussion will be limited to describing numerous ways in which acetaldehyde is generated from synthesis gas and will explore the different schemes whereby it can be converted to ethylidene diacetate.

For the sake of discussion, it is convenient to first look at synthesis gas–based methods for the generation of acetaldehyde, all of which can be integrated into a synthesis gas–based process by reaction with acetic anhydride. These processes have generally been based on methanol, a methyl ester, or acetic anhydride.

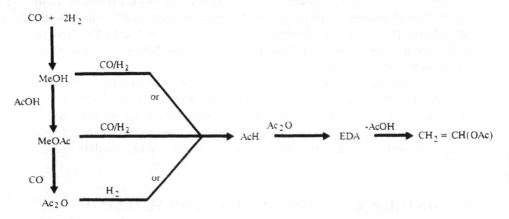

Figure 4 Alternative pathways for the generation of vinyl acetate from synthesis gas.

A number of methods for the generation of acetaldehyde from methanol have been demonstrated [16]. [See Eq. (5).] Most methods used cobalt or cobalt plus an additional transition metal and were not particularly selective for acetaldehyde, with ethanol and/or methyl acetate generally representing major by-products [16–25]. The lack of selectivity demonstrated by the methanol reactions shifted the attention of several workers away from methanol toward methanol derivatives such as methyl ketals [17] and methyl esters [13,17,26–29].

$$MeOH + CO + H_2 \longrightarrow AcH + H_2O \qquad (5)$$

The generation of acetaldehyde from methyl esters has proven to be more useful. A very efficient cobalt catalyst, which is composed of a mixture of CoI_2, LiI, and NPh_3 and operates at 170°C and 5000 psig (340 atm), has been described [17,26]. However, the more commonly described catalysts are homogeneous palladium processes in which iodine promotors, particularly methyl iodide, and a cation, such as lithium, sodium, quarternary ammonium, or quarternary phosphonium, are added [13,27,28]. These palladium-based processes appear to be slightly less selective but operate at comparable temperatures (150–175°C) and considerably lower pressures (ca. 1000 psi; 68 atm).

There is one important complication in the later palladium-based processes. Although it is generally not comparable to rhodium for the carbonylation of methyl acetate to acetic anhydride, palladium is capable of catalyzing the carbonylation of methyl acetate to acetic anhydride. Therefore, in the reductive carbonylation of methyl acetate, acetic anhydride is also formed and is probably a requisite intermediate.

As we pointed out earlier, acetic anhydride reacts with acetaldehyde to generate ethylidene diacetate. [See Reaction (3).] However, although strongly favoring the formation of ethylidene diacetate, the reaction of acetic anhydride with acetaldehyde, as shown in Eq. (3), is reversible and actually represents an equilibrium process. To generate acetaldehyde instead of ethylidene diacetate in these processes, the reaction vessel is operated by continuously removing the most volatile component, acetaldehyde, from the reactor as a means of driving the equilibrium toward acetaldehyde.

At this point, it is reasonable to ask whether isolation of the acetaldehyde is actually desirable since ethylidene diacetate is the ultimately desired intermediate. If ethylidene diacetate can be generated directly from methyl acetate, this would represent a shorter synthesis. [The shorter route to ethylidene diacetate is represented in Eq. (6).]

$$2MeOAc + 2CO + H_2 \longrightarrow CH_3CH(OAc)_2 + AcOH \qquad (6)$$

This possibility was obvious to the majority of the research groups investigating the generation of vinyl acetate from synthesis gas. As a result, a large body of information on processes that involve the direct generation of ethylidene diacetate from methyl acetate or dimethyl ether is available in the patent literature. Homogeneous carbonylation catalysts that have shown some activity in the carbonylation of methyl acetate were obvious choices. Following this expected pattern, the catalyst systems investigated were based on rhodium [13,14,30–33], palladium [13,14,32–37], nickel [32,38,39], and cobalt [32,39,40]. However, rhenium [41,42] and manganese [42] were also reported to catalyze this process.

The major drawback to the combined process for the direct conversion of methyl acetate or dimethyl ether to ethylidene diacetate is selectivity. The carbonylation process for the conversion of methyl acetate is fairly complex, requiring several simultaneous reactions to occur. It is very difficult to attain both an efficient carbonylation and an efficient hydrogenation and not get one in excess. Several mixed catalysts, including combinations of Rh and Pd [43,44], Ni and Pd [45], and Ni and Ru [46], have been examined to get around this problem. However, the net result is that, in practice, one must cogenerate acetic anhydride, ethylidene diacetate, and acetic acid if this process is employed.

This cogeneration requires large distillations and recycles. As a result, the additional synthetic step incurred when acetaldehyde is isolated represents a negligible economic penalty because it requires a simpler distillation process, and processes that generate ethylidene directly and processes in which acetaldehyde is isolated are likely equivalent economically.

Other esters have proven useful in the generation of acetaldehyde. Methyl formate, which can be synthesized by the base-catalyzed addition of methanol to CO, has also been useful in the generation of acetaldehyde [29]. However, the process utilizes different chemistry than the earlier processes. The methyl formate is reacted at 180°C under a partial CO pressure of 80 bar (79 atm) in the presence of a Rh-LiI catalyst in an N-methyl pyrrolidinone solvent. Rather than using hydrogen, the reducing agent is CO, which is converted to CO_2. [The balanced equation for this process is shown in Eq. (7).] This last process has not been well examined and has several drawbacks compared to the methyl acetate processes, particularly the use of a very expensive catalyst and considerably more dilute solutions.

$$HCO_2Me + CO \longrightarrow CH_3CHO + CO_2 \qquad (7)$$

Acetic anhydride, which can be derived from methyl acetate or dimethyl ether via carbonylation, as described in some detail in Chapter 9, represents another desirable starting point, particularly since it is a requisite material in the generation of ethylidene diacetate and must be generated anyway. Generally, the acetaldehyde generated from the reduction of acetic anhydride [Eq. (8)], is allowed to react in situ with excess acetic anhydride starting material generating ethylidene diacetate, as shown in Eq. (3). A number of catalysts for the hydrogenation of acetic anhydride to ethylidene diacetate are known [12–14,47–63]. However, the most common and useful catalysts have been based on rhodium [12–14,47–55], palladium [12–14,47–50,55–58], and ruthenium [12,47–51,53,54,59,60]. Cobalt [12–14,47–50,61], nickel [12–14,47–50,62], and rhenium [63] have also been reported as catalysts, but appear to be less active. The rhodium catalysts are generally homogeneous and used in the presence of some carbon monoxide- and iodine-containing components. The palladium catalysts and ruthenium catalysts generally applied to this reduction have been either heterogeneous or homogeneous catalysts, and catalysis is most efficient when a strongly acidic cocatalyst, including HI, sulfonic acids, or phosphoric acid, is included.

$$Ac_2O \longrightarrow AcH + AcOH \tag{8}$$

As in the reductive carbonylation of methyl acetate, acetaldehyde can be isolated from the hydrogenation of acetic anhydride by taking advantage of the equilibrium between ethylidene diacetate and acetaldehyde [12–14]. By removing the low-boiling acetaldehyde component either as the reaction proceeds or from an actively equilibrating ethylidene diacetate/acetaldehyde/acetic anhydride product solution, the less favored acetaldehyde is obtained.

Alternatively, Reaction (8) can be driven very close to completion, leaving little or no acetic anhydride to participate in the generation of ethylidene diacetate represented by the equilibrium reaction shown in Reaction (3). Catalysts that have proven useful in attaining the very high levels of conversion required for this approach to acetaldehyde are palladium [13,64,65] and cobalt [66]. In the liquid phase variants [13,64,66], it is advisable to continuously remove the acetaldehyde because acetaldehyde undergoes self-condensation under the reaction conditions. The self-condensation represents a significant source of yield loss if the acetaldehyde concentration reaches or exceeds about 10%. The vapor phase process [13,65] does not encounter this problem, but must still be run at near 100% conversion to avert the generation of ethylidene diacetate. The isolation of acetaldehyde from an acetic anhydride has no significant advantage over the in situ generation of ethylidene diacetate and would be useful only if by-product was problematic in the in situ process. As mentioned earlier, the acid-catalyzed cracking of ethylidene diacetate to vinyl acetate and acetic acid has been known for a long time [12–15] and is best catalyzed by sulfonic acid catalysts.

Combining the Eastman/Halcon technologies for acetic anhydride with their own in-house processes, Halcon International is known to have successfully piloted a number of these processes. (This technology is now completely owned by the Eastman Chemical Co.) Mitsubishi Gas Chemical (and possibly several other companies) have also piloted these processes.

The relief in petroleum prices experienced in the mid-1980s delayed any commercialization of this process. At present, the plants required to perform this process are too expensive and the slightly lower raw material costs are insufficient to cover the costs of the increased capital expenditures. However, if petroleum prices again experience a sustained rise, this process may become one of the next generation of non-petroleum-based processes as it is nearly cost competitive with existing technology at this time.

VI. CONCLUSION

Vinyl acetate is expected to continue to grow as a significant derivative of acetic acid. The technology is well established, and new technology, namely the synthesis gas–based routes, will guarantee that vinyl acetate will continue to be an important commodity chemical well into the future, even if the petrochemical industry suffers severe setbacks due to petroleum shortages.

REFERENCES

1. Daniels, W., *Kirk-Othmer Encyclopedia of Chemical Technology*, 3rd ed., *23*, 817 (1978).
2. Leonard, E. C., *High Polym.*, *24*, 263 (1970).
3. Kripylo, P., Ulrich, D., Adler, R., and Jankowski, H., *Wiss. Z. Tech. Hochsch. "Carl Scorlemmer" Leuna-Mersburg*, *31*, 133 (1982).
4. Cincera, D. L., *Kirk-Othmer Encyclopedia of Chemical Technology*, 3rd ed., *23*, 848 (1978).
5. Lavin, E., and Snelgrove, J. A., *Kirk-Othmer Encyclopedia of Chemical Technology*, 3rd ed., *23*, 798 (1978).
6. For discussions of the Wacker process, see Collman, J. P., Hegedus, L. S., Norton, J. R., and Finke, R. G., *Principles and Applications of Organotransition Metal Chemistry*, University Science Books, Mill Valley, CA, 1987, pp. 412–415, 826.
7. *Hydrocarbon Process.*, *46* (4), 146 (1967).
8. Robinson, R. E., U.S. Patent 3,190,912 (1965).
9. *Eur. Chem. News*, *11* (272), 40 (1967).
10. Sennewald, K., and Vogt, W., U.S. Patent 3,637,819 (1972).

11. Fernholu, H., and Kauffelt, M., *Ger. Offen.* 2,745,174 (1979).
12. For a review of these processes, see Gauthier-Lafaye, J., and Perron, R., *Methanol et Carbonylation*, Rhone-Poulenc Recherches, Courbevoie, France, 1986 (English translation, *Methanol and Carbonylation*, Editions Technip, Paris, and Rhone-Poulenc Recherches, Courbevoie, France, 1987, p. 201).
13. Rizkalla, N., and Goliaszewski, A., *ACS Symp. Ser.*, *328*, 136 (1987).
14. Ehrler, J. L., and Juran, B., *Hydrocarbon Process., Int. Ed.*, *61*, 109 (1982).
15. Hoxley, H. F., Thomas, E. B., and Mills, W. G., U.S. Patent 2,425,389 (1947).
16. For a review of the conversion of methanol to acetaldehyde and ethanol, see Gauthier-Lafaye, J., and Perron, R., *Methanol et Carbonylation*, Rhone-Poulenc Recherches, Courbevoie, France, 1986 (English translation, *Methanol and Carbonylation*, Editions Technip, Paris, and Rhone-Poulenc Recherches, Courbevoie, France, 1987, p. 39).
17. Wegman, R. W., Busby, D. C., and Letts, J. B., *ACS Symp. Ser.*, *328*, 126 (1987).
18. Steinmetz, G. R., and Larkins, T. H., Jr., *Organometallics*, *2*, 1879 (1983) and references cited therein.
19. Steinmetz, G. R., *J. Mol. Catal.*, *26*, 145 (1984).
20. Larkins, T. H., Jr., and Steinmetz, G. R., U.S. Patent 4,389,532 (1983).
21. Gauthier-Lagaye, J., and Perron, R., U.S. Patent 4,306,091 (1981).
22. Lin, J. J., and Knifton, J. F., U.S. patent 4,433,176 (1984).
23. Lin, J. J., and Knifton, J. F., U.S. Patent 4,433,177 (1984).
24. Lin, J. J., and Knifton, J. F., U.S. Patent 4,433,178 (1984).
25. Fujimoto, K., Setoyama, T., and Tominaga, H., *Chem. Lett.*, 1811 (1983).
26. Wegman, R. W., and Busby, D. C., *J. Chem. Soc., Chem. Commun.*, 332 (1986).
27. Graff, J. L., and Romenelli, M. G., *J. Chem. Soc., Chem. Commun.*, 337 (1987).
28. (a) Rizkalla, N., and Winnick, C. N., GB Patent 1,538,782 (1979). (b) Porcelli, R. V., GB Patent 2,038,829 (1980). (c) Porcelli, R. V., U.S. Patent 4,302,611 (1981).
29. Vanhoye, D., Melloul, S., Castanet, Y., Mortreux, A., and Petit, F., *Angew. Chem., Int. Ed. Eng.*, *27*, 683 (1988).
30. Drent, E., European Patent Appl. 108,437 (1984).
31. Drent, E., U.S. Patent 4,429,150 (1984).
32. Isshiki, T., Kijima, Y., Ito, A., Miyauchi, Y., Kondo, T., and Watanabe, T., European Patent Appl. 48,174 (1982).
33. Rizkalla, N., and Minnick, C. N., GB Patent 1,538,782 (1979).
34. Mitsubishi Gas Chem. KK, JP 58/157,746 (1983).
35. Isshiki, T., Ito, A., Miyauchi, Y., Kondo, T., and Watanabe, T., European Patent Appl. 35,860 (1981).
36. Isshiki, T., Ito, A., Miyauchi, Y., Kondo, T., and Watanabe, T., European Patent Appl. 48,173 (1982).
37. Isshiki, T., Ito, A., Miyauchi, Y., Kondo, T., and Watanabe, T., European Patent Appl. 48,174 (1982).

38. Isshiki, T., Kilima, Y., Miyauchi, Y., and Yasunaga, T., U.S. Patent 4,659,865 (1987).
39. Rizkalla, N., U.S. Patent 4,323,697 (1982).
40. Cook, J., and Drury, D. J., European Patent Appl. 108,539 (1984).
41. Vogt, W., Glaser, H., and Koch, J., European Patent Appl. 81,152 (1983).
42. Kuebbeler, H. K., Erpenbach, H., Gehrmann, K., and Kohl, G., U.S. Patent 4,319,038 (1982).
43. (a) Paulik, F. E., and Schultz, R. G., European Patent Appl. 77,116 (1983).
    (b) Paulik, F. E., and Schultz, R. G., U.S. Patent 4,810,821 (1989).
44. Kudo, K., Mori, S., and Sugita, N., *Chem. Lett.*, 265 (1985).
45. Mitsubishi Gas Chem. KK, JP 57/64,644 (1982).
46. Mitsubishi Gas Chem. KK, JP 56/150,042 (1982).
47. Isshiki, T., Ito, A., Kijima, Y., and Watanabe, T., European Patent Appl. 34,062 (1981).
48. Shell Int. Res. Mij BV, European Patent Appl. 58,442 (1982).
49. Mitsubishi Chem. Ind. KK, JP 55/079,346 (1980).
50. Isshiki, T., Kijima, Y., Ito, A., Miyauchi, Y., Kondo, T., and Watanabe, T., European Patent Appl. 48,173 (1982).
51. Moy, D., GB Patent 2,075,508 (1981).
52. Daicel Chem. Inds. Ltd., JP 56/025,132 (1981).
53. Larkins, T., U.S. Patent 4,581,473 (1986).
54. Larkins, T., U.S. Patent 4,886,905 (1990).
55. Union Oil of Calif., U.S. Patent 3,579,566 (1971).
56. Larkins, T. H., U.S. Patent 4,374,265 (1983).
57. Suzuki, S., U.S. Patent 4,221,918 (1980).
58. Aikazayan, A. M., and Grigoryan, A. S., *Arm. Khim. Zh.*, *42*, 615 (1989).
59. Kent, A. G., GB Patent 2,129,430 (1984).
60. Larkins, T. H., U.S. Patent 4,337,351 (1982).
61. Rizkalla, N., U.S. Patent 4,618,705 (1986).
62. Rizkalla, N., U.S. Patent 4,328,362 (1982).
63. Jagers, E., and Koll, H. P., Ger. Offen. 3,407,092 (1985).
64. Moy, D., U.S. Patent 4,329,512 (1982).
65. Moy, D., U.S. Patent 4,356,328 (1982).
66. Rizkalla, N., GB Patent 2,079,753 (1982).

13

Cellulose Acetate

Lanny C. Treece and Griffin I. Johnson
Eastman Chemical Company, Kingsport, Tennessee

I. INTRODUCTION

Cellulose, consisting of long chains of anhydroglucose units with three reactive hydroxyl groups each, forms esters with organic acids. Cellulose esters from almost any organic acids can be prepared; however, owing to steric hindrance, industrial applications are limited to esters of acids containing four or fewer carbon atoms. Cellulose acetate is produced in the largest volume and has considerable commercial significance. Cellulose acetate propionate (CAP) and cellulose acetate butyrate (CAB) are produced in lower volumes but are also industrially important. Because of their instability, esters formed from formic acid are not produced commercially.

A. Cellulose Acetate

Direct esterification of cellulose with free acetic acid requires such high temperatures and catalyst concentrations that significant molecular weight degradation of the cellulose molecule occurs. Therefore, this method is not practiced commercially. All current industrial processes utilize acetic anhydride to effect the esterification step. Generally the triacetate is formed by reacting 3 mol of acetic anhydride per anhydroglucose unit with the liberation of 3 mol of acetic acid. The cellulose triacetate thus formed can

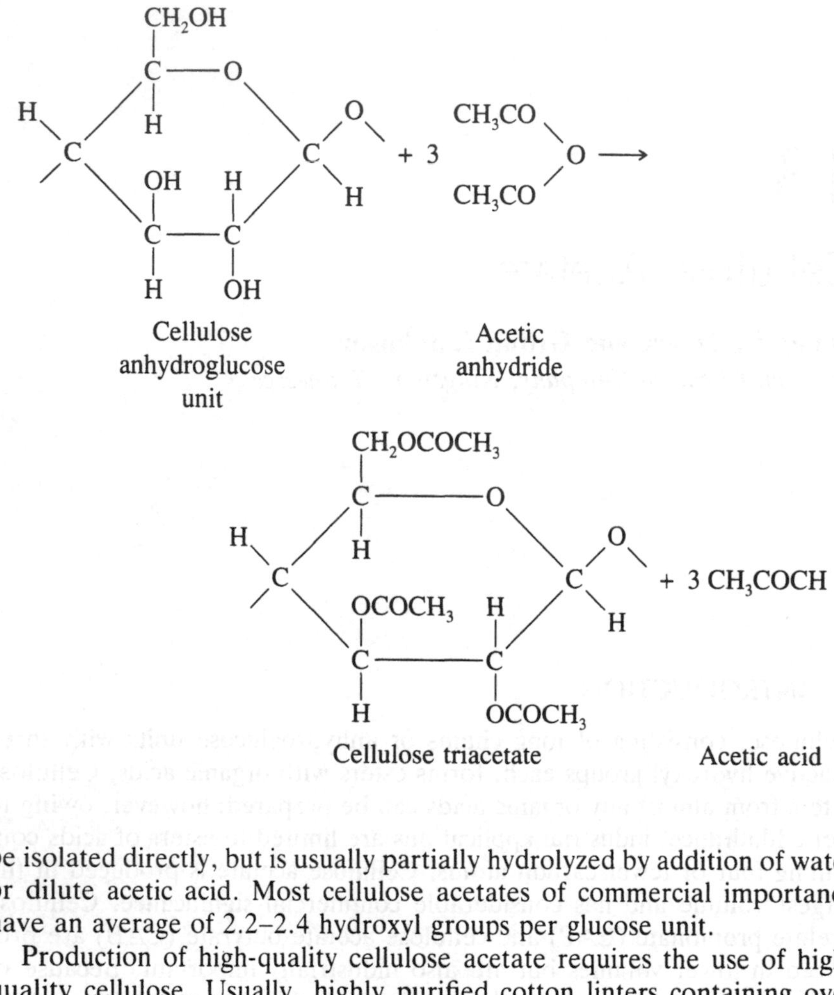

CH$_2$OH

Cellulose
anhydroglucose
unit

+ 3

Acetic
anhydride

→

CH$_2$OCOCH$_3$

Cellulose triacetate

+ 3 CH$_3$COCH

Acetic acid

be isolated directly, but is usually partially hydrolyzed by addition of water or dilute acetic acid. Most cellulose acetates of commercial importance have an average of 2.2–2.4 hydroxyl groups per glucose unit.

Production of high-quality cellulose acetate requires the use of high-quality cellulose. Usually, highly purified cotton linters containing over 99% alpha-cellulose or cellulose from wood pulp with an alpha-cellulose content of 90–97% is used. Cellulose acetate has a high molecular weight owing to the high degree of polymerization of the starting cellulose.

Manufacturers of cellulose acetate usually recover and convert the side product acetic acid to acetic anhydride on site. The concentration, as required for acetylation, is usually between 90 and 95%.

B. Mixed Cellulose Esters

Pure cellulose propionate and cellulose butyrate esters are difficult to produce and have not attained industrial importance. Only mixed esters of acetic acid and butyric acid or acetic acid and propionic acid are commercially important. The composition of mixed esters is characterized by the acetyl to propionyl (or butyryl) content and the free hydroxyl content. For both commercial grades of CAPs and CABs, the hydroxyl content ranges between 0.5 and 4.5%. For CABs, the acetyl content typically ranges from 1.5 to 30% and the butyryl content ranges from 15 to 54%. For CAPs, the acetyl content typically ranges from 3 to 8% and the propionyl content ranges from 55 to 62%. Many more grades of cellulose acetate butyrates are available commercially than cellulose acetate propionates.

The preparation of mixed esters is similar to that of cellulose acetate. Typically, the esterification mix consists of a mixture of acetic anhydride and butyric anhydride for CABs or acetic anhydride and propionic anhydride for CAPs. After esterification, a hydrolysis step is used to achieve the desired hydroxyl content.

II. HISTORICAL PERSPECTIVE

Cellulose acetate was first synthesized by Schutzenberger in 1865 [1]. The low-molecular-weight material was produced by heating cellulose and acetic acid under pressure. In 1879, Franchimont used sulfuric acid as a catalyst [1]. Sulfuric acid is still the most frequently used catalyst today. Industrial development and application of cellulose acetate was somewhat slowed by its limited solubility in less expensive solvents. In 1904, Miles and Eichengrun overcame this problem when they simultaneously succeeded in synthesizing an acetone-soluble acetate by partially hydrolyzing a triacetate [1].

During World War I, cellulose acetate replaced the highly flammable nitrocellulose used as a coating on early airplane wings and fuselage fabrics. Shortly after World War I, the manufacturing processes for cellulose acetate films, fibers, and moldings resins were developed. During the period 1920–1940, many patents and publications related to manufacture and use of cellulose esters were published. Today, advancement of cellulose ester manufacturing technology is generally regarded as complete, although efforts toward applying lower-cost technology, producing higher-quality products, and developing new applications continue.

III. SUPPLY AND DEMAND

The United States is the largest producer of cellulose acetate flake, with 55% of the world capacity. Western Europe has 16% and Japan has 13%. Worldwide capacity for production of cellulose acetate flake in 1988 was 832,000 metric tons/year. The annual U.S. production of cellulose acetate flake has grown from 183,000 metric tons in 1955, peaked at 409,000 in 1980, and decreased slightly to 378,000 by 1988. Eastman Chemical Co.

and Hoechst Celanease Corp. are the two major U.S. producers, with each having approximately 50% of domestic capacity. Other producers of cellulose acetate flake outside the United States are Rhone-Polene Chimie, France; Agfa-Gevaert nv and Tubize Plastics sa, Belgium; Rodia AG, Germany; Industrias del Acetato de Celulosa, Spain; Courtaulds Fibers Ltd. and Nelson Acetate Ltd., U.K.; Daicel Chemical Industries, Ltd., Acetati, Italy; and Teijin Acetate Ltd., Japan.

In 1987, Bayer AG (Germany) ceased the production of CAP and CAB esters. With this cessation, Eastman Chemicals Co. became the sole world producer of these materials. In 1988 the approximate U.S. production of CAB for moldings, sheeting, and extrusion was 14,000 metric tons, and of CAP, 7000 metric tons.

Depending on the particular grade, the 1988 year-end price for cellulose acetate flake was $2.80–3.52/kg. The 1988 year-end price for compounded CAB and CAP resins was $3.07/kg. In general, the prices for cellulose esters have increased steadily through the years, approximately following the rate of inflation.

The worldwide demand for cellulose acetate flake for use in plastics and textile fibers is expected to decline gradually, primarily as a result of competition from other plastics. Likewise, the demand for CAB and CAP resins is expected to decline for the same reason. Worldwide cellulose acetate flake demand for the cigarette filter tow market has increased steadily since 1980 and is expected to continue through the 1990s.

IV. PROPERTIES

A. Effect of Composition on Properties

In general, cellulose acetate esters are high melting, high strength, resistant to hydrocarbons and UV light, have low flammability and high clarity, but have limited solubility and compatibility. The mixed esters of propionate and butyrate have increased solubility, compatibility, and moisture resistance. Table 1 gives typical ester properties for three cellulose esters [2,3].

B. Degree of Substitution

Cellulose acetate is the most important cellulose ester because of its use in plastics, fibers, and coatings. Table 2 lists the acetyl content of cellulose acetate at various levels of degree of substitution (DS).

Three hydroxyl groups per anhydroglucose unit are available for esterification. Esterification of all three sites is only practiced with cellulose triester used in photographic film base. Hydrolyzing a portion of the acetyl groups results in free hydroxyl groups, which improves the solubility in

Table 1 Properties of Cellulose Esters

Properties	Acetate	Butyrate	Propionate
Acyl	39.7	13.0	2.5
Butyryl	—	37.0	—
Propionyl	2.5	—	46.0
Melting, °C	240–250	195–205	188–210
Density, g/cc	1.31	1.20	1.22
Dielectric, kV/cm	512–669	787–984	
Molecular Weight, Mn	60,000	70,000	75,000

more polar solvents such as alcohol and water systems. The solubility is important in spinning for fiber production and applications such as coatings. In esters for coating, the free hydroxyl groups provide the primary sites for reactions involving cross-linking agents and resins. In general, the lower the hydroxyl content, the greater the moisture resistance of the ester. For mixed ester, the normal hydroxyl content is 1.5–2.0%; however, some types have a hydroxyl content as low as 0.5% or as high as 4.3%. For cellulose acetate, the hydroxyl content is normally 3.5–4.0% (39.8–39.5% acetyl content). The solubility ranges of the hydrolyzed cellulose acetate in certain common solvents are listed in Table 3 [4,5].

C. Molecular Weight

Relative to most man-made polymers, cellulose esters can have a very high molecular weight or degree of polymerization (DP). This is a result of the starting material (cellulose) having a DP of 600–1500 and the ability to limit the rate of the degradation reaction step relative to the acetylation

Table 2 DS Versus Acetyl Content for Cellulose Acetate

DS	Acetyl (wt%)
0.5	11.7
1.0	21.1
1.5	28.7
2.0	35.0
2.5	40.3
3.0	44.8

Table 3 Cellulose Acetate Solubility

Acetyl content (%)	Solvent
44–43	Methylene chloride
44–38	Tetrachloroethane
44–14	Pyridine
44–30	Acetic acid
42–36	Acetone
32–24	2-Methoxyethanol
35–18	Acetone/water (1/1)
24–19	Water/methoxyethanol (1/1)
19–13	Water
Below 13	Insoluble in all above

reaction rate. The commercial esters have a molecular weight average as low as 15,000 and as high as 80,000 for some mixed esters of cellulose acetate propionate and cellulose acetate butyrate. The DP of the ester affects the strength of fiber esters and the toughness and mechanical strength of films and coatings. Increasing the DP of the ester has a small effect on solubility and generally increases the melting point. Low-DP esters permit use at high solids concentration in solvents. Normally the density and hardness are affected very little by the DP of the ester.

D. Mixed Esters—Effect of Increasing Acetyl Content

Pure esters of alkyl radicals such as butyryl are more soluble in nonpolar solvents like ketones, chlorinated hydrocarbons, and aromatic hydrocarbons. In addition, the resulting ester has greater flexibility, greater compatibility with other polymers and resins, and greater moisture resistance. However, without some acetyl radicals, the ester has a melting point, making it unusable in many applications. By esterifying with both butyryl and acetyl, the advantages of both can be combined. Cellulose acetate propionate has properties between those of cellulose acetate and cellulose acetate butyrate but without the odor problem of CAB esters.

V. MANUFACTURE OF CELLULOSE ACETATE

A. Basic Chemistry

The commercial method of preparing cellulose acetate is the esterification reaction with acetic anhydride in the presence of a mineral acid catalyst.

A high-purity cellulose is needed for esterification; therefore, either cotton linters or >95% cellulose wood pulps are used. Because of the price difference, wood pulps are normally used except where plastic ester clarity or stability requires the use of cotton linters. Sulfuric acid has remained the catalyst of choice for the commercial acetylation process. Other catalysts include perchloric acid, hydrochloric acid, and methanesulfonic acid. Sulfuric acid and perchloric acid are both exceptional catalysts under anhydrous conditions, but sulfuric acid does not present any difficult corrosion or safety problems compared to the other catalysts. Perchloric acid does not combine with the cellulose, which is an advantage that sulfuric acid does not have.

In general, some activation of the cellulose is necessary before a uniform esterification can be carried out. The function of the activation process is to increase the accessibility of the cellulose hydroxyl groups to the acetylation reactants. The rate of diffusion for the activating liquids is a function of time, temperature, and solvents. In the absence of acetic anhydride, sulfuric acid is reversibly adsorbed from acetic acid onto the cellulose [6].

The actual mechanism for acetylation is not completely understood. Cellulose esterification kinetics can be subdivided into two regions. The initial rate is rapid and follows first-order kinetics, whereas the final rate is slower. The initial rate is first order in anhydride, zero order in cellulose, and has first-order dependence on sulfuric acid. As mentioned previously, the acetylation of cellulose is promoted in the presence of numerous catalysts, such as sulfuric acid, perchloric acid, hydrochloric acid, and methanesulfonic acid. The mechanism of these catalysts is generally accepted to be proton activity (see mechanism I below). However, since it is known that sulfuric acid combines with the cellulose at the beginning of acetylation, a mechanism for acetylation with sulfuric acid could proceed by a cellulose sulfate intermediate, as shown in mechanism II below [6–9].

1. Mechanism I

$$R(OH)n + nH_2SO_4 + Ac_2O \longrightarrow R(OSO_2OH) + 2AcOH \tag{1}$$

$$R(OSO_2OH) + Ac_2O \Longleftrightarrow R(OSO_2O^-) + Ac_2O(H^+) \tag{2}$$

$$Ac_2O(H^+) + R(OH)_n \longrightarrow R(OH)_{n-1}OAc + AcOH \tag{3}$$

2. Mechanism II

$$R(OH)n + nH_2SO_4 + Ac_2O \longrightarrow R(OSO_2OH) + 2AcOH \tag{4}$$

$$R(OSO_2OH) + Ac_2O \Longleftrightarrow R(OSO_2OAc) + AcOH \tag{5}$$

$$R(OSO_2OAc) + R(OH)_n \longrightarrow R(OSO_2OH) + R(OH)_{n-1}OAc \tag{6}$$

B. Process Overview

Almost all organic cellulose esters are produced by the solution process whereby the product ester dissolves during reaction. In the production of cellulose acetate, cellulose and acetic acid are combined and mixed to activate the cellulose. Cooled acetic anhydride and sulfuric acid are added to the mixture. The acetylation reaction is actually slightly endothermic, but the reaction of acetic anhydride with the water by-product is highly exothermic. As the esterification is completed, the solution becomes clear, at which point the reaction is stopped by adding a water/acetic acid solution. To obtain a uniform hydroxyl substitution pattern, the ester must be completely acetylated and then back-hydrolyzed to a desired hydroxyl content. The ester is then precipitated, water-washed, stabilized, and dried.

C. Activation

Before the cellulose can be esterified, the fibers must be activated or swollen. The activated fiber is much more accessible to the catalyst and anhydride than unactivated fiber. Of the commercial processes, the best activation is obtained with water. However, the water must be extracted with acetic acid before acetylation to prevent the reaction of the activation water with acetic anhydride during acetylation. This extraction step increases acid recovery cost and capital cost. As alternatives, the cellulose fiber can be activated in acetic acid/water solutions or in acetic acid alone. Figure 1 is a flow diagram for a typical cellulose acetate manufacturing process.

Activation of the pulp is a function of the acetic acid/water ratio. When 0.5 part acetic acid is used, the ratio of acetic acid to water (including moisture in the pulp) is approximately 8:1, and good activation is obtained. When 3.5 parts of acetic acid is used, the ratio is approximately 60:1 and poor activation is obtained [10]. The degree of activation determines the rate of the following acetylation reaction and the degree of completeness achievable.

The cellulose acetate must have a considerably shorter chain length than native cellulose to be soluble in organic solvents and to give solutions with the desired viscosity for commercial applications. This chain length reduction is sometimes accomplished prior to esterification by addition of a small amount of catalyst; however, the molecular weight of the cellulose decreases rapidly, depending on the temperature and concentration of the catalyst. The added catalyst also aids in activation. The amount of catalyst is not as important as uniform distribution on the cellulose [3]. The effectiveness of various liquid systems for the activation of cellulose is shown in Table 4.

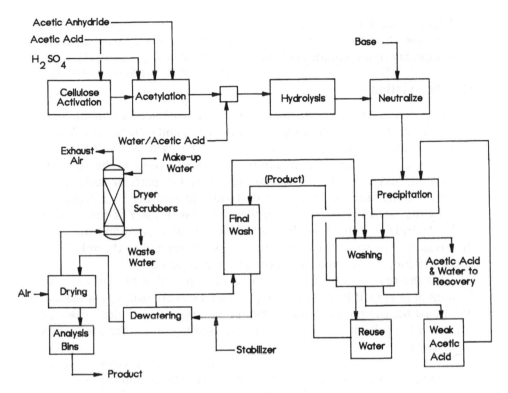

Figure 1 Cellulose acetate process flow diagram.

Table 4 Effectiveness of Activation

Activation liquid[a]	Esterification time[b]
10% NaOH	1.2
Water	1.0
75% acetic	0.85
Formic acid	1.1
Acetic acid	0.7
Propionic acid	>2.5
Butyric acid	>2.5

[a]Removed with acetic before acetylation.
[b]Relative to water.

Adding water to the catalyst treatment solution decreases the rate at which the cellulose chain length is reduced. This can be explained by the decrease in superacidity of the sulfuric acid in aqueous acetic acid solution [11].

D. Acetylation

The solution process is the one most widely used, with acetylation mixtures containing acetic anhydride, cellulose, acetic acid, and sulfuric acid. Methylene chloride, in addition to acetic acid, can be used as a solvent, and perchloric acid with or without sulfuric acid can be used as the catalyst. The normal usage for the acetic acid process is 4–5 parts of acetic acid, 2–4 parts anhydride, and 0.01–0.1 part sulfuric acid per part cellulose. Initially the rate of reaction is controlled by mass transfer of the catalyst and anhydride into the fiber. But as the fiber is acetylated and the ester dissolves into the solvent, the reaction rate is determined by the anhydride concentration, catalyst concentration, and temperature.

Since the reaction of water with the acetic anhydride liberates a large amount of heat (300 Btu/lb of ester), the reactants are usually cooled and additional heat removed through external cooling. If methylene chloride is used as the solvent, vaporization can be used to control the acetylation temperature.

The acetylation reaction time and temperature of the acetylation mixture determine the final molecular weight of the ester. The degradation reaction that occurs during the acetylation process has a larger activation energy than the acetylation reaction. This suggests that degradation reactions are more sensitive to temperature than acetylation reactions and that an increase in reaction temperature will increase the amount of degradation obtained for a given change in the degree of substitution of acetyl [12]. An increase in reaction temperature of 10°C doubles the rate of acetylation and more than doubles the rate of degradation [13]. The penalty of a lower reaction rate due to lower temperature must be balanced with the results of an incomplete reaction and/or lower molecular weight.

The choice of solvent, excess anhydride, and the degree of activation also affect the relative rates of these two competing reactions. As the cellulose is completely acetylated to the triester, solubility in the acetic acid decreases. Sulfonation by the sulfuric acid catalyst and and the use of solvents such as methylene chloride increase the solubility and prevent gel formation in the acetylation mixture.

E. Hydrolysis

If sulfuric acid catalyst is used in acetylation, it may also be used for hydrolysis at low temperatures (38–65°C) or neutralized for hydrolysis at high temperatures (>110°C). To obtain an ester stable to normal drying conditions, the sulfonate ester must be reduced to less than 200 ppm. The sulfonated ester is more quickly removed at low water concentration (<1%), but the degradation reaction (reduction of polymer chain length) of the ester at low water is also high. Therefore, the water concentration should be held at low concentrations for a few minutes to remove the sulfate esters before increasing for hydrolysis. This approach effectively minimizes ester degradation [7,14].

The rate of degradation during hydrolysis increases with increasing concentration of sulfuric acid, water, and temperature. Table 5 shows the relative degradation rates at different water levels compared to anhydrous conditions [5,7,14].

The viscosity of the final cellulose ester is usually controlled during esterification. The objective of the hydrolysis process is to obtain the specified hydroxyl value with the least amount of viscosity reduction in a reasonable time. Because of the cost associated with separating water from acetic acid and the fact that esterified esters are relatively intolerant to water, the water concentration is normally limited to a range from 5 to 15%. The hydrolysis time is fixed by the active catalyst concentration and ester/acetic acid solution temperature.

F. Precipitation

To isolate the ester, the reduced solubility of the ester in acetic acid/water solutions is used to effect precipitation. With an acetyl of 39.5%, the ester is insoluble in acetic acid/water solutions containing less than 55% acetic acid. The acetic acid/water solution can be added to the ester solution or the ester solution can be added to the acetic/water solution. If the final water concentration is too low, considerable ester is lost, and if it is too high, the acetic acid recovery cost is considerably higher. The optimum acid concentration is between 25 and 35% acetic acid.

Table 5 Degradation of Cellulose at Various Water Levels

Liquid	Degradation rate[a]
Acetic anhydride (5.0%)	1.0
Glacial acetic acid (0.08% H_2O)	0.08
Water (3.0%)	0.004
Water (6.0%)	0.001

[a]Relative to acetic anhydride.

Table 6 Equilibrium Moisture Versus
Relative Humidity

	Equilibrium moisture (%) at various levels of relative humidity			
Material	25	50	75	100
Cellulose	5.4	10.8	15.5	30.5
Acetate	0.6	2.0	3.8	7.8
Propionate	0.1	0.5	1.5	2.4
Butyrate	0.1	0.2	0.7	1.0

G. Washing and Stabilization

To meet the stability requirements of the ester, the salts and solvent must be removed, normally by washing with water. In addition, long-term stability can be improved by addition of a weak base to neutralize the sulfuric acid that may be produced by hydrolysis of any residual sulfonated ester. Autoclaving the ester at 115°C for 30 min can also effectively remove most of the combined sulfate ester and stabilize the ester without additives.

H. Drying

Before the ester is dried, various techniques such as pressing or centrifuging can be used to remove the free water. After the free water is removed, any of the common steam dryers can be used to reduce the water content to less than 5%. The equilibrium moisture content at four levels of relative humidity of cellulose acetate (acetyl content of 39.5%), compared to cellulose and other esters, is given in Table 6 [15].

VI. MANUFACTURE OF MIXED CELLULOSE ESTERS

In general, the mixed cellulose acetates used in the production of molding and coatings require higher-purity cellulose than that of cellulose acetates used in fiber applications. The manufacture of mixed esters also generally requires better activation of the cellulose relative to cellulose acetates. The composition of the mixed ester is determined by the composition of the reaction mixture. To obtain the desired mixed-ester composition, mixtures of the anhydrides in addition to mixtures of the acids are normally used during esterification. Even with better activation, the time for esterification of mixed esters is considerably longer (4–5 h) than that of acetylation of cellulose acetate, and esterification of mixed esters must be carried out at a lower temperature (<100°F).

VII. ANALYTICAL TEST METHODS

A. Degree of Polymerization

Solution viscosity in process solvents and at process concentrations is normally used to estimate the DP of the ester. However, the solution temperature, solids concentration, and solvent composition must be controlled very accurately for this estimate to be valid. For example, the ester viscosity is proportional to the solids concentration raised to the sixth power of 20% solutions of ester in acetic acid or acetone.

$$\text{Viscosity} \simeq A(\text{solids})^6 \tag{7}$$

Also, other composition variables, such as acetyl content, divalent salts, and hemicellulose content, can affect the solution viscosity.

Intrinsic viscosity (IV) provides a better estimate of the ester DP but requires more analytical preparation and equpiment than the concentrated solution viscosity test [16]. The viscosity of the ester is measured at low concentrations (0.5 wt%). The zero concentration viscosity is calculated to remove the viscosity effect of the solvent. However, this method is still slightly affected by the acetyl (or propyionyl/butyryl) content, as shown in Table 7 [16].

B. Thermal Stability

Stability is best tested by the requirements of the application. Temperature and moisture contribute to the hydrolysis of the residual sulfonated ester and result in ester degradation. Sometimes the thermal stability can be

Table 7 Intrinsic Viscosity Versus Percent Propionate

Propionate[a](%)	IV (in acetone)
51.7	2.31
48.9	2.41
48.5	2.42
47.6	2.47
46.7	2.55

[a]Cellulose propionate series all with the same DP.

tested by accelerating the hydrolysis at high temperature. The sample can then be observed for signs of ester degradation.

C. Degree of Substitution

For mixed esters, saponification followed by liquid chromatography is the method of choice for determining the degree of substitution. For cellulose acetate, there are a variety of solubility tests that depend on the ester tolerance for water or a nonpolar solvent, which can provide an estimate of degree of acetylation. However, the method of choice is saponification followed by titration. All these methods give an average acetyl content of the ester. The spread (or acetyl distribution) is an important composition property of the ester that is not estimated in these tests. The only test procedure in the literature that measures acetyl distribution is fractional precipitation and analysis of the fractions with the above tests.

VIII. END USES

A. Critical Properties as Related to Application

Cellulose acetate, CAP, and CAB are white amorphous materials that are commercially available as a powder or flake. They are nontoxic, tasteless, odorless, and less flammable than cellulose nitrates, resistant to weak acids, and stable to mineral and petroleum oils. Cellulose acetate ester plastics in general have excellent clarity, good mechanical properties, a wide range of colorability, are nonyellowing, have good toughness, are moisture resistant, have a good "hand" or "feel," and have good weatherability.

Important properties related to application are solution viscosity, degree of esterification, and, for mixed esters (CAPs and CABs), the ratio of acetyl to propionyl or butyryl content. Viscosity is a primary indicator of the DP, which greatly influences its mechanical properties and processability. The degree of esterification primarily determines solubility and compatibility with other resins, varnishes, plasticizers, and solvents. In general, the presence of the propionyl or butyryl groups improves the flexibility, moisture resistance, and compatibility with other resins and solvents.

A wide range of cellulose acetate ester compositions with a corresponding wide range of applications can be (and are) manufactured commercially. This capability has made cellulose esters one of the most versatile families of polymers in use today, with new formulations to meet the needs of specific applications continually being found.

B. Cellulose Acetate

Cigarette filter tow is the largest single end use for cellulose acetate ester, with approximately two-thirds of domestic production used for this purpose.

The textile fibers market represented approximately 20% of cellulose acetate consumed in 1988; however, this market is declining.

Cellulose triacetate is extensively used in the manufacture of amateur and professional photographic and movie film because of its excellent handling and processing characteristics. At one time, cellulose acetates dominated the X-ray film, microfilm, and graphic art film markets, but in recent years more and more of these markets have converted to polyethylene terephthalate (PET) resins. Although declining in use, cellulose acetate still has limited use for high-quality display packaging and decorative signs because of its excellent clarity.

Special castings of cellulose acetate films serve as separation membranes used in reverse-osmosis purification of brackish water, purification of blood in dialysis units, air separation units, and a wide range of other biotechnology applications.

Cellulose acetate is molded and extruded into various consumer products, such as toothbrush handles, combs, brushes, eyeglass frames, appliance housing, tool handles, toys, steering wheels, and other miscellaneous items.

Cellulose acetates with a low viscosity are used in lacquers and protective coating for various substrates such as paper, glass, metal, leather, and wood products. Very thin cast films are used to provide glossy finishes for postcards, books, and other graphic art items. The largest use of extruded thin film is in pressure-sensitive tape, the most familiar being 3M Company's Scotch Brand Magic Transparent Tape. Cellulose acetate is also used as an adhesive for photographic film and as a heat-sensitive adhesive for textiles.

C. Mixed Cellulose Esters

CAP and CAB esters, like cellulose acetates, have numerous applications as films, sheetings, moldings, and coatings. As thermoplastics, they can be injection-molded, extruded, or dissolved and cast into films. Molding and extrusion applications are similar to those for acetates. Mixed esters are more compatible with various plasticizers and synthetic resins than cellulose acetate. For example, cellulose acetate butyrate is compatible with polyesters, acrylics, vinyl, and alkyd resins, depending on the butyryl content and degree of hydrolysis.

As coatings, mixed esters are formulated to meet a wide variety of needs. Cellulose acetate butyrates with a high butyryl content and low viscosity are soluble in lacquer solvents widely used in protective and decorative coating applied to furniture and automobiles. Low-viscosity cellulose propionate butyrate esters containing 3–5% butyryl, 40–50% propionyl, and 2–3% hydroxyl groups are compatible with oil-modified alkyd resins used in wood furniture coating. CABs are also used in "wet-on-wet" automotive coatings where the CAB is used as the pigmented basecoat before application of a clear topcoat [17]. CAPs are used in printing inks because of their low odor, high surface hardness, and high tolerance for alcohol solvents and substantial quantities of water.

REFERENCES

1. *Ullmans's Encyclopedia of Industrial Chemistry*, Eicher, Theo., VCH Publishers, N.Y., N.Y. Vol. A5, 5th Edition, 1986, p. 438.
2. Internal publication of Eastman Kodak.
3. Malm, C. J., Barkey, K. T., May, D. C., and Lefferts, E. B., *Ind. Eng. Chem.*, *44*, 2904 (1952).
4. Malm, C. J., Barkey, K. T., Salo, Martin and May, D. C., *Ind. Eng. Chem.*, *49*, 79 (1957).
5. Malm, C. J., Fordyce, C. R., and Tanner, H. A., *Ind. Eng. Chem.*, *34*, 430 (1942).
6. Ricter, G. A., Herdle, L. E., and Gage, I. L., *Ind. Eng. Chem.*, *45*, 2773, (1953).
7. Malm, C. J., Tanghe, L. J., and Laird, B. C., *Ind. Eng. Chem.*, *38*, 77 (1946).
8. Rosenthal, A. J., *Pure Appl. Chem.*, *14*(304), 535–546 (1967).
9. Araki, Tsunao, *Text Res. J.*, *22*, 630 (1952).
10. Manning J. H. and Fuller, I. H., *Tappi*, *52*(4), 612 (1969).
11. Chevalet, P. A., British Patent 264,181 (Jan. 11, 1926).
12. Hiller, L. A., *J. Polym. Sci.*, *10*(4), 385–423 (1953).
13. Malm, C. J., Barkey, K. T., Schmitt, J. T., and May, D. C., *Ind. Eng. Chem.*, *49*(4), 763 (1957).
14. Malm, C. J., Glegg, R. E., Salzer, J. T., Ingerick, D. F., and Tanghe, L. J., *I&EC Process Des. Dev.*, *5*, 81 (1966).
15. Malm, C. J., Mench, J. W., Kendall, D. L., and Hiatt, G. D., *Ind. Eng. Chem.*, *43*, 688 (1951).
16. Genung, L. B., *Anal. Chem.*, *36*(9), 1819 (1964).
17. Bogan, R. T., and Brewer, R. J., *Encyclopedia of Polymer Science and Engineering*, Vol. 3, 2nd ed., John Wiley & Sons Inc., N.Y., N.Y., 1985, pp. 158–188.
18. Rosenthal, A. J., *J. Polym. Sci.*, *51*, 111–122 (1961).

14

Alcohol Acetates

J. Stewart Witzeman and Victor H. Agreda
Eastman Chemical Company, Kingsport, Tennessee

I. INTRODUCTION

The lower acetate esters constitute a significant proportion of the total market for acetic anhydride derivatives. Because of the ability of these esters to solubilize polymers such as cellulosics, nitrocellulosics, acrylics, and polyesters, these materials are useful as solvents in coatings, adhesives, cosmetics, and film base applications. These materials have also found use as process solvents in the specialty chemical, pharmaceutical, and agrichemical industries. These esters are also used as carrier solvents in agrichemical formulations and in extractive procedures, such as coffee decaffeination

Physical properties such as evaporation rate, boiling point, and "solvent power" (loosely defined as the ability of the solvent to reduce the viscosity of a polymer solution) often determine which ester is used in a given application. Evaporation rate is often a key factor in coatings and adhesive applications, and thus these acetate esters are often blended with each other or other types of solvents (such as ketones or alcohols) to give a formulation with the desired combination of solvent properties and evaporation rate.

Table 1 [1] compares the physical properties such as boiling point, evaporation rate, vapor pressure, and resistivity of these materials. Data in this table indicate that a wide range in boiling point and evaporation rate can

Table 1 Physical Properties of Various Acetate Esters

	Methyl	Ethyl	Propyl	iso-Propyl	Butyl	iso-Butyl
Molecular weight	74.1	88.1	102.1	102.1	116.2	116.2
Evap. rate[a]	5.3	4.1	2.3	3.0	1.0	1.4
Specific gravity	0.94	0.90	0.89	0.87	0.88	0.87
Solubility in water[b]	29.6	7.4	2.3	2.9	0.7	0.7
Water in	7.4	3.3	2.6	1.8	1.6	1.6
Vapor P[c]	171.3	86	23	47.5	10	12.5
Boiling range	55.8–58.2	75.5–78.0	99–103	85–90	122–128	112–119
Flash point[d]	−13	−4	13	2	27	20

[a]BuOAc = 1.
[b]wt%, at 20°C.
[c]At 20°C, in mmHg.
[d]Tag closed cup, in °C.
Source: Data from Ref. 1.

be obtained merely by selection of the ester. In addition to the evaporation rate, the propensity of the material to diffuse out of a polymer is also important. Branching of the ester component inhibits the rate of diffusion of the solvent out of the polymer, thus making esters such as iso-butyl acetate slower to diffuse from a polymer matrix than the linear *n*-butyl acetate. Another property that is becoming increasingly important owing to the ever-increasing usage of electrostatic spraying operations in coating applications is the resistivity of the solvent. Resistivity values for these esters are generally greater than 20 megohms while the resistivity of typical ketone solvents is generally less than 0.3 megohms. The much higher resistivity for the ester solvents indicates that these materials are more readily sprayed electrostatically and can also be used to help adjust the resistivity of a given formulation.

A more difficult property to quantify, which is well known to formulators, is the solvent power of the material. One way to express solvent power is by comparing the reduction in viscosities obtained when a given polymer is dissolved in various solvents. The viscosity of various polymer solutions is given in Table 2 [1]. It can be seen that although the acetate esters are effective in reducing the viscosity of many polymeric materials, they are typically not as effective as ketones. The evaporation characteristics, low toxicity, resistivity properties, and low odor of acetate esters, however, often compensate for the slightly higher resulting solution viscosities [2]. By blending these materials with other solvents, it is often possible to overcome these slightly higher viscosity levels.

Table 2 Comparison of Solvent Activity of Various Materials

	Solution viscosity at 25°C (cP)				
	Nitrocellulose[a]	Nitrocellulose[a]	Acrylic resin[b]	Epoxy resin[c]	FPC[d]
Solvent	10 wt%	20 wt%	20 wt%	40 wt%	20 wt%
Acetone	15	25	73	63	14
MEK	20	38	28	80	16
Methylene chloride		Insol.	58	543	
Methyl isobutyl ketone	30		64		24
Methyl acetate		58	53	130	
Ethyl acetate[e]	27	73	85	170	26
Propyl acetate	30		57	28	
iso-Propyl acetate	38		59		29
Butyl acetate	46		77		33
iso-Butyl acetate	49		83		38

[a]RS 1/2-sec NC from Hercules, Inc.
[b]Elvacite 2010 Resin from Du Pont.
[c]Epon 1007F from Shell.
[d]Polyvinyl chloride/polyvinyl acetate polymer formerly available from Firestone.
[e]99% assay material; similar values are noted for 85% assay material
Source: Data from Ref. 1.

II. USES/APPLICATIONS OF PARTICULAR ACETATE ESTERS

A. Methyl Acetate

The boiling point of methyl acetate is such that it is often used as a component in systems where rapid dry-to-touch times are needed [3]. This ester is often blended with less volatile solvents, which impart flow and leveling properties to the formulation. Methyl acetate has also found some use as a process solvent and as a reagent for transesterifcation reactions (see below). Large quantities of methyl acetate are used in the preparation of acetic anhydride (see Chapter 9).

B. Ethyl Acetate

Ethyl acetate is commonly used as a solvent for coatings, adhesives, inks, cosmetics, and film base. It is also used as a synthesis solvent for the preparation of synthetic organic chemicals for pharmaceutical and agrichemical applications. This ester is also used in the processing of wood pulp, in the decaffeination of coffee, and in synthetic fruit essences. In addition to coffee decaffeination, this material is a useful coextractant (usually in conjunction with ethers) for many naturally occurring materials such as camphor, fats, oils, antibiotics, and several resins and gums. The market proportion of this material is approximately 50% coatings applications, 31% inks, and 14% process solvent based, with the remainder being miscellaneous applications.

C. Propyl Acetates

1.   n-Propyl Acetate

n-Propyl acetate is used as a solvent for formulations that use cellulose esters and ethers, to solublize various waxes, and in nitrocellulose-based lacquers. This ester also has found applications in insecticide formulations and as a food preservative. The main use of this material, however, is as a solvent in the printing inks industry.

2.   iso-Propyl Acetate

iso-Propyl acetate is used in similar applications as the n-propyl derivative. Because of the higher evaporation rate of this ester relative to the n-propyl analog, it can be blended with the n-propyl acetate to adjust the drying speed of the ink and the viscosity of the formulation. iso-Propyl acetate tends to swell the rubber rollers used in flexographic inks and thus is used mainly in applications that utilize metal rollers, such as gravure printing applications. The lower dissolving power of iso-propyl acetate relative to ethyl acetate makes this material suitable for multicolor printing because these solubility characteristics enable clear color resolution without "bleeding." This ester is also useful as a solvent for formulations that use natural gums such as kauri gum, Manila gum, and pontianak dammar. This ester is also used in coatings, particularly systems based on cellulosics and polyurethanes [4a].

D. Butyl Acetates

1.   n-Butyl Acetate

n-Butyl acetate is used primarily in coatings, where its solvent power and lower relative volatility make it useful for adjustment of evaporation rate and viscosity. This material is particularly useful as a solvent for acrylic polymers, vinyl resins, and nitrocellulose lacquers. This ester is also used as a reaction medium for adhesives, as a solvent for leather dressings, as a process solvent in various applications, and in cosmetic formulations. As with iso-propyl acetate, natural gums are also readily solubilized with this material. n-Butyl acetate has also found a great deal of use in the printing ink industry. The current worldwide market for this material is 55% as a solvent for lacquers, 10% for adhesives and pharmaceutical solvents, and 35% exported [4b].

2. iso-Butyl Acetate

iso-Butyl acetate has similar uses as the unbranched derivative, but is generally thought to be less effective as a process solvent. As with the propyl acetates, this acetate is often blended with other solvents to adjust the evaporation rate and viscosity. This acetate has recently been used as a replacement solvent for MIBK and toluene owing to the requirements of Rule 66, which restricts the levels of branched ketones and aromatic solvents used in coating formulations.

3. Other Butyl Acetates

sec-Butyl and tertiary-butyl acetates account for only a small part of the total butyl acetate market. These materials are primarily used in laboratory and specialty chemical applications.

III. PREPARATION OF ACETATE ESTERS

There are a variety of methods for the preparation of acetate esters in the patent and open chemical literature. These methods include reaction of the acid with the alcohol (esterification), addition of acid to olefins, carbonylation of acetic acid or anhydride, reductive methods involving the carboxylic acid or anhydride, and the Tischenko reaction. Although these reactions seem quite varied, the number of methods used in industrial practice is much more limited.

A. Esterification

The acid-catalyzed reaction of a carboxylic acid with an alcohol to produce the corresponding ester and water [Eq. (1)] is known as the esterification reaction [5]. This reaction is an equilibrium process with equilibrium constant for the process given by Eq. (2). The equilibrium constant, in turn, reflects [Eq. (3)] the overall free energy of the process ($\Delta G°$), which is defined by the thermodynamic stabilities of the various reagents and products. The equilibrium nature of the reaction means that to effect a high conversion from this process, it is necessary to conduct the reaction under conditions that either use excess reagents or remove one or more of the reaction products. The mechanistic possibilities of the esterification reaction are given in Eq. (4) and (5). In general, the reaction of most acetate esters follows an $A_{Ac}2$ reaction pathway [Eq. (5)]. In cases where either the R group or the reaction medium is capable of stabilizing the charge associated with the acylium ion 1 in Eq. (4), a unimolecular $A_{Ac}1$ process may occur. The symmetrical nature of the acylium ion has been demonstrated by ^{18}O labeling experiments, which have shown complete scrambling of the oxygen label. The esterification reaction is an acid-catalyzed process whereas the reverse hydrolysis reaction can be catalyzed by either acid or base.

$$ROH + CH_3CO_2H \rightleftharpoons CH_3COR + H_2O \tag{1}$$

$$K_{eq} = \frac{[CH_3CO_2R][H_2O]}{[CH_3CO_2H][ROH]} \tag{2}$$

$$\Delta G^\circ = -RT \ln(K_{eq}) \tag{3}$$

This process can be conducted in the gas phase, in condensed phases, or, in the case of many industrial processes, in a combination of both phases. A related reaction is the transesterification reaction [Eq. (6)], which involves the exchange of alcohol components in an ester. The transesterification reaction follows a similar equilibrium pathway.

1.   Role of the Catalyst

A strong acid catalyst is necessary in this process to achieve an appreciable rate of reaction. Typically, laboratory procedures as well as industrial practice often involve the use of mineral acids such as sulfuric and hydrochloric acids [6]. Other strong acids, such as organic sulfonic acids [7], perfluorosulfonic acid resins [8], and mixtures of sulfuric and para-toluene sulfonic acid (tosic acid) [9], have been used. A sulfonated microporous polystyrene has also been studied in some detail [10]. The use of phosphoric acid adsorbed onto silica in the gas phase esterification of butyl acetate has also been described [11].

The literature also contains many examples of the use of Lewis acids as esterification catalysts. These catalysts are often incorporated on various polymer supports to make use and recovery of the catalyst more efficient. One of the more commonly used Lewis acids is titanium tetrachloride. This material has been used in conjunction with a variety of polymer binders, including a styrene-*N*-vinylpyrrolidone-divinylbenzene copolymer [12], poly(styryl)bipyridine [13], a styrene-*N*-vinylcarbazole-divinylbenzene copolymer [14], and a styrene-acrylonitrile-divinylbenzene copolymer [15]. In all the above cases, divinylbenzene is used to impart chemical resistance to the polymeric support by cross linking. This titanium catalyst has also been absorbed on weakly basic polystyryl anionic exchange resins. Other similar catalytic systems that have been described in the literature include polystyrene-divinylbenzene–bound $TeCl_4$ [16] and BBr_3 [17] reagents. Other Lewis acids that have been employed in various esterification applications include silica/alumina blends [18], magnesium lignosulfonates, $AlPO_4$ supported on either silica or alumina [19], ZrO_2 [20], $AlCl_3$, and a fluoranated Al_2O_3 [21]. Other solid phase catalysts, such as cation exchange resins [22] and oxidized carbon [23], have also been used for esterification processes.

In addition to protic and Lewis acids, highly active "superacids" (highly ionizing strong acids) have also found utility in the preparation of various acetate esters by esterification. Some workers have claimed that a solid superacid known as HRB is useful in the preparation of a wide variety of esters from the corresponding alcohol and carboxylic acid [24]. Similarly, quantitative yields of ethyl acetate at 96% conversion [25] are claimed from ethanol and acetic acid using a packed column of the superacid 12-tungstosilicic acid on activated carbon at 150°C. A solid superacid prepared by calcining $Zr(OH)_2$ and NH_4SO_4 at 500–750°C is also claimed to provide good yields of methyl, ethyl, propyl, butyl, and iso-butyl acetates from the corresponding alcohols and acetic acid [26].

2. Role of the Alcohol

Because of the previously mentioned equilibrium and thermodynamic constraints of the esterification reaction, various structural effects of both the reactants and the products will be reflected in both the rate and the position of the equilibrium. In the case of simple primary alcohols and the corresponding esters, these effects are subtle. For more highly substituted secondary and tertiary alcohols, these structural effects can adversely effect both the rate and the position of the equilibrium. For tertiary alcohols,

Table 3   Rate of Esterification (k_e) and Alkaline Hydrolysis (k_{OH}) for Various Acetic Acid Esters

R (in ROAc)	$k_e \times 10^{4a}$ ($m^{-1}sec^{-1}$)	$k_{OH} \times 10^{3b}$ ($m^{-1}sec^{-1}$)
iPr	0.289	7.07
nBu	0.496	23.0
iBu	0.096	18.2
nPr	0.500	27.0
Et	1.02	46.6
Me	1.47	108
sBu		3.27
tBu		0.27

[a]Rate of esterification of acetic acid and corresponding alcohol at 25°C in water.
[b]Rate of hydrolysis of corresponding acetate ester at 24.7°C in 70% acetone-water.

this can be further complicated by competing dehydration of the alcohol under the acidic reaction conditions. This often necessitates other synthetic approaches to these materials.

The effect of alcohol on the rate of esterification (k_e) and alkaline hydrolysis (k_{OH}) for various acetate esters is given in Table 3 [27]. The table indicates that both processes are sensitive to steric effects, with substitution at the position adjacent to the ester carbonyl particularly influencing the rate of the processes.

B. Addition of Acetic Acid to Olefins

A second method for the preparation of acetic acid esters is by addition of the carboxylic acid to the corresponding olefin [Eq. (7)]. This process is particularly useful for the preparation of highly hindered esters such as *t*-butyl and *sec*-butyl acetates since the steric bulk associated with the alcohol component in these esters makes the equilibrium for the esterification reactions of these materials much less favorable. This procedure has been used widely in both industrial and laboratory settings. This route is used commercially by Rhone-Poulenc to prepare iso-propyl acetate from acetic acid and propylene [28].

(7)

Common laboratory methods for the preparation of tertiary esters typically involve reaction of isobutylene with a carboxylic acid using a strong protic or Lewis acid catalyst, such as sulfuric acid, nitric acid, or BF_3-OEt_2 [29]. These acids presumably function by forming an intermediate carbonium ion, which is subsequently trapped by the carboxylic acid. These methods are effective for the preparation of tertiary esters from the incipient highly stabilized, tertiary carbonium ion.

Industrial methods for this reaction generally involve the use of a catalyst that is capable of both absorbing the reagents and subsequently activating the olefin, and thus various supported catalysts are used (see below). The mechanism of this reaction presumably involves addition of the acid to the adsorbed olefin, which has been either highly polarized by the solid support or fully protonated to produce the corresponding carbonium ion. In most instances, the reaction involves use of the excess olefin, heat, and sometimes pressure. Kinetic studies on these systems have shown that sorption is often the key kinetic step in the process [30].

The need for adsorption of the reactants as well as effective dispersion of the catalytic species often necessitates the use of supports for the catalytic species. One of the commonly used supports used for this reaction is silica (SiO_2). This material has been impregnated with inorganic acids such as sulfuric [30] and phosphoric [31] acids, as well as with strong organic acids such as halopolysulfonic acids [32] and aromatic disulfonic acids (either alone or in combination with sulfuric acid) [33]. A 1961 U.S. patent on the preparation of *t*-butyl acetate describes the use of a vanadium oxide/potasium sulfate mixture supported on silica. This catalyst is effective for catalysis of reactions involving substituted olefins, such as isobutylene, but is inactive toward unsubstituted olefins. Thus it is possible to selectively react carboxylic acids with isobutylene in petroleum cracking streams that also contain paraffins and various 1- and 2-butenes [34].

Various treated clays, such as montmorillonite, bentonite, and vermiculite, have also been used to catalyze this reaction [35]. Use of these systems enables the preparation of acetates derived from less substituted secondary olefins. Treatment of these clays with small amounts of water extends the catalytic lifetime of these systems whereas ion exchange with protic acids [36] or metals such as Al^{3+} or Cr^{3+} improves both the activity and selectivity of the catalyst system [37]. When long-chain linear olefins are used in this reaction, complicated isomeric mixtures are often noted owing to Wagneer-Meerwin shifts of the cationic center in the olefin. Thus reaction of 1-hexene with acetic acid in the presence of a Cr^{+3}-exchanged montmorillonite produces a 70:30 mixture of the 2- and 3-hexyl acetates [37b].

The use of zeolites to catalyze this process has also been described [38]. The zeolite HZSM-12 has a very different selectivity than other catalytic systems. For example, reaction of acetic acid with *trans* 4-octene using HSZM-12 zeolite gives the 2-, 3-, and 4-octyl acetates in 61, 24, and 15% yields, respectively, whereas catalysts such as BF_3-OEt, amorphous SiO_2/AlO_3, or a large-pore-size zeolite such as REY produce predominantly the 4-octyl acetate in 73–83% conversion.

Polymer-bound catalysts have also been used to effect this reaction. For example, sulfonated fluoropolymers have been used to prepare ethyl acetate from ethylene and acetic acid [39]. Porous ion exchange resins have also been used to prepare both acetate and fatty acid esters [40].

This process has been used to prepare diacetates by reaction of the olefin with oxygen and acetic acid over a SnO/I_2 catalyst; both the di- and mono-acetates [41] are obtained from this reaction.

C. Methods Involving Carbonylation and/or Hydrogenation

Methods are also known that involve the catalyzed addition of carbon monoxide and hydrogen to either acetic acid or acetic anhydride. These processes are related to the processes discussed in Chapters 4 and 9. Thus methyl, ethyl, and propyl acetates have been prepared by the reaction of synthesis gas (CO and H_2) with acetic acid at 375°C and 600 psig using mixed oxides comprised of Ru, Ni, and alkali metals as catalysts, such as $RuNiNa_{0.1}O_{10}$. This catalyst can also contain oxides of metals such as Co, Cd, and Zn [42]. Similarly, mixtures of ethyl acetate, ethanol, and diethyl ether have been prepared from acetic anhydride, carbon monoxide, and hydrogen at 220°C and pressures of 3000–4600 psig using a homogeneous catalyst system prepared from a ruthenium compound such as $RuCl_3$ or $Ru_3(CO)_{12}$ halide and a Lewis acid such as tin, boron, or iron trihalides [43]. Similar preparations of ethyl acetate that use ruthenium trichloride, either alone [44] or in combination with HCl and phosphine oxides such as Ph_3PO, have also been described [45]. The latter two reactions also produce small amounts of ethylidene diacetate (1,1-diacetoxy ethane). A mixture of methyl and ethyl acetates has been prepared from acetic acid, carbon monoxide, and hydrogen using a 3:1 complex of $NaRe(CO)_2$ and $Ru_3(CO)_{12}$ [46]. A similar process involves the same reagents and a catalyst comprised of $Ru_3(CO)_{12}$, $Mn_2(CO)_{10}$, and HI [47].

Another variation of this reaction involves the hydrogenation of acetic acid to produce mixtures of ethanol and ethyl acetate. These processes are generally used to produce ethanol, with the ethyl acetate by-product being sold as a solvent. The proportion of alcohol to ester can usually be controlled by the operating conditions and/or the extent of conversion, with lower conversions favoring ester formation. Thus hydrogenation of acetic acid at 10.9 bar and 249°C on a graphite- or silica- supported catalyst comprised of molybdenum or tungsten, used in conjunction with a group VIII metal such as Pd, Rh, or Ru and (optionally) potassium as a modifying metal, gives a 2.4:1 mixture of ethanol to ethyl acetate [48]. A similar process that involves hydrogenation of either acetic acid or acetic anhydride with the aid of group VIII noble metals used in combination with metals capable of alloying with these metals (i.e., blends of Ru and/or Re alloyed with gold or cooper) is claimed to provide ethanol (54%) and ethyl acetate (18%) at 55.4% conversion [49]. This reaction also produces methane (17%) and ethane (8%). Exclusion of the alloying agent increases the relative porportions of methane and ethane formed. Use of Re on silica gel is claimed to give a high yield of ethyl acetate (86%) at low conversions (24%) of acetic acid [50], whereas Ru on a similar support gives nearly equimolar ratios of ethanol and ethyl acetate [51]. Use of a CuO catalyst on a $ZnAl_2O_4$ support in the hydrogenation of acetic acid is claimed to provide ethanol in 99% yield even at complete conversion of the acetic acid [52]. This work notes that lower temperatures and pressures favor ester formation. The conditions used in this process can vary from 175 to 300°C and 50 to 5000 psig, with 275°C/500 psig being optimal for ethanol conversion.

Use of homogeneous catalysts such as $(Bu_3P)_4Ru_4H_4(CO)_8$ in the hydrogenation of acetic acid has also been studied [53]. This homogeneous catalyst exchanges with acetic acid to form various acetate complexes during the course of the reaction and eventually evolves toward the most thermodynamically stable and least catalytically active species. A method for obtaining high selectivities of ethyl acetate over ethanol in the hydrogenation is claimed in an unexamined Japanese patent [54]. This method involves hydrogenation of a carboxylic acid using a mixture of ruthenium catalysts and amines such as alpha-picoline.

D. Tischenko Reaction

The reaction of two aldehydes under the influence of a metal alkoxide catalyst such as aluminum isopropoxide was first observed by Tischenko in 1906 [55]. This reaction offers an alternative route to ethyl acetate via the reaction of 2 mol acetaldehyde. The synthetic and mechanistic consequences of this reaction have been studied in some detail, and the reaction has been exploited in both industrial and laboratory applications. A large number of other esters are potentially also accessible from this process. Catalysts that have been used to effect this reaction include aluminum alkoxides, boric acid [56a], magnesium or calcium aluminum alkoxides

[56b], lithium tungsten (IV) oxide [56c], and hydridoruthenium complexes. Boric acid has been shown to be an effective catalyst for the Tischenko reaction of aldehydes such as formaldehyde or isobutyraldehyde, which do not undergo a competing aldol reaction. The magnesium and calcium aluminum alkoxides can give glycol esters as the major products. These materials presumably arise from aldol condensation of the two aldehydes to give the hydroxyaldehyde intermediate, which subsequently undergoes a Tischenko reaction to give the glycol ester, as shown in Eq. (8).

$$
\underset{R'}{\overset{O}{\parallel}}\!\!{}^{}H + \underset{RCH_2}{\overset{O}{\parallel}}\!\!{}^{}H \xrightarrow{(\text{aldol})} \underset{\underset{R}{R'}}{\overset{OH}{\parallel}}\!\!{\overset{O}{\parallel}}\!\!{}^{}H + \underset{R''}{\overset{O}{\parallel}}\!\!{}^{}H
$$

(2)

$$
\xrightarrow{(\text{Tischenko})} \underset{\underset{R}{R'}}{\overset{OH}{|}}\!\!\overset{O}{\parallel}\!\!{}^{}_{O}\ R''
$$

(8)

The mechanism of this reaction, as proposed by Ogata et al. [57a], is given in Eq. (9)–(13). The salient features of this process are the rapid initial complexation of the catalyst with the aldehyde [step 1, Eq. 9] followed by a shift of one of the metal-bound alkoxide moieties to the electron deficient carbonyl [step 2, Eq. 10]; reaction of a second mole of aldehyde with the metallic center [step 3, Eq. (11)]; and a hydride transfer to the metal-complexed aldehyde followed by dissociation of the metal-ester complex [step 4, Eq. (12)]. Evidence for the alkoxide migration from the metal center includes the rapid initial formation of isopropenyl acetate from the aluminum isopropoxide–catalyzed reaction of acetaldehyde and the significant induction periods noted in the aluminum *tert*-butoxide–catalyzed reaction of benzaldehyde [57b]. The decomposition of the intermediate metal-ester complex is thought to occur via a cyclic transition state [3 in Eq. (13)]. The second step [Eq. (10)] of this reaction is believed to be rate-determining when the reactants are aliphatic aldehydes such as acetaldehyde [57a]. When benzaldehyde is used, however, the second-order kinetic behavior of the aldehyde is more consistent with step 3 [Eq. (11)] being rate-determining.

$$
\underset{R}{\overset{O}{\parallel}}\!\!{}^{}H + Al(OR')_3 \rightleftharpoons \underset{R}{\overset{O^{\nearrow}\!\!{}^{Al(OR')_3}}{\parallel}}\!\!{}^{}H
$$

(9)

$$
\underset{\substack{R \quad H}}{\overset{\overset{\displaystyle \bar{Al}(OR')_3}{\displaystyle |}}{\overset{\displaystyle O}{\|}}} \rightleftharpoons \underset{\substack{OR'}}{\overset{\overset{\displaystyle Al(OR')_2}{\displaystyle |}}{\overset{\displaystyle O}{|}}} R-\overset{H}{\underset{}{|}} \tag{10}
$$

$$
\underset{\substack{OR'}}{R-\overset{\overset{\displaystyle Al(OR')_2}{\displaystyle O}}{\underset{H}{|}}} + \underset{\substack{R \quad H}}{\overset{\displaystyle O}{\|}} \rightleftharpoons \underset{\substack{OR'}}{R-\overset{\overset{\displaystyle \overset{+}{O}CHR}{\displaystyle |}}{\overset{\overset{\displaystyle Al(OR')_2}{\displaystyle |}}{\overset{\displaystyle O}{|}}}}\overset{H}{\underset{}{|}} \tag{11}
$$

$$
\underset{\substack{OR'}}{R-\overset{\overset{\displaystyle \overset{+}{O}CHR}{\displaystyle |}}{\overset{\overset{\displaystyle Al(OR')_2}{\displaystyle |}}{\overset{\displaystyle O}{|}}}}\overset{H}{\underset{}{|}} \longrightarrow \underset{\substack{R \quad OR'}}{\overset{\displaystyle O}{\|}} + Al(OR')_2(OCH_2R) \tag{12}
$$

$$
\underset{\substack{R'O \quad H}}{Ar-\overset{\displaystyle O}{\underset{}{|}}} \cdot \overset{}{\underset{\substack{Ar'}}{\overset{\displaystyle Al}{\overset{O \quad O}{}}}} \longrightarrow \underset{\substack{Ar \quad OR'}}{\overset{\displaystyle O}{\|}} + -Al(OCH_2Ar') \tag{13}
$$

$$(\underline{3})$$

    Studies on the mixed Tischenko reaction of aldehydes with varying electron demands indicate that the more electron-deficient species prefers to be the acid component [57b] [Eq. (14)]. Thus reaction of equimolar mixtures of benzaldehyde and *p*-chlorobenzaldehyde gives a slight (ca. 12%) excess of benzyl *p*-chlorobenzoate. This preference is even more pronounced in the reaction of benzaldehyde and *p*-anisaldehyde, which produces benzyl benzoate and benzyl *p*-methoxybenzoate in a 1.70:1 ratio at 30% conversion. Only traces of *p*-methoxy benzyl benzoate are noted, and none of the product of self-reaction of anisaldehyde is noted in either this reaction or control reactions of anisaldehyde alone. This preference has been rationalized by invoking intermediate 3 in which the electron-rich aromatic species Ar stabilizes the incipient charge due to the hydride migration while the electron-deficient Ar' is well suited to accept the migrating group. Collapse of this intermediate gives an aluminum complex in which the electron-deficient aldehyde is complexed to the metal. This sets up the more electron-poor species to participate as the alcohol component in a subsequent reaction.

$$
\underset{Ar}{\overset{O}{\|}}\underset{H}{\quad} + \underset{Ar'}{\overset{O}{\|}}\underset{H}{\quad} \longrightarrow
$$

$$
\underset{Ar}{\overset{O}{\|}}\underset{OAr}{\quad} + \underset{Ar}{\overset{O}{\|}}\underset{OAr'}{\quad} + \underset{Ar'}{\overset{O}{\|}}\underset{OAr}{\quad} + \underset{Ar'}{\overset{O}{\|}}\underset{OAr'}{\quad} \qquad (14)
$$

Other studies [56b] have shown that the relative rate of reaction of aliphatic aldehydes is much greater than that of the aromatic analogs. Mixed reaction of aliphatic aldehydes such as acetaldehyde or butyraldehyde with benzaldehyde gives (in addition to the corresponding aliphatic esters) benzyl acetate and benzyl butyrate as the predominant mixed products [58]. Studies on the mixed Tischenko reaction of various aldehydes using $RuH_2(PPh_3)$ as the catalyst have led to the reactivity order of propanylaldehyde > acetaldehyde > butyraldehyde > iso-butyryladehyde >> benzaldehyde. The reasons underlying this reactivity order are presumably much more subtle than for the substituted aromatic aldehydes.

IV.   INDUSTRIAL METHODS OF MANUFACTURE

The most important industrial methods for the manufacture of acetate esters are based on the acid-catalyzed reaction of acetic acid with alcohols. The reaction of acetic anhydride with alcohols and the Tischenko reaction are also used industrially. Eastman Chemical Company (a division of Eastman Kodak Company), Hoechst-Celanese Corporation, and Union Carbide Corporation are among the largest producers of a variety of acetate esters. Many other companies produce acetate esters in several countries. Most of them produce smaller quantities and/or fewer types of acetate esters. Yearly U.S. production and sales volumes of acetate esters are available [59], as are estimates of U.S. consumption [60]. Additional general information concerning the chemistry and production methods of acetate esters can be found in several sources [61–63]. This section reviews esterification technologies used industrially for the large-scale production of acetate esters.

A.   Esterification of Acetic Acid with Alcohols

As discussed previously, the acid-catalyzed, liquid phase reaction of acetic acid with an alcohol is an equilibrium-limited, reversible reaction. The

Table 4 Equilibrium Constants for the Esterification of Acetic Acid with Selected Alcohols [61]

Alcohol	K_{eq}
Methanol	5.2
Ethanol	4.0
Propanol	4.1
Butanol	4.2
Isopropanol	2.4
2-Butanol	2.1
3-Pentanol	2.0
t-Butanol	0.005

liquid equilibrium constant (K_{eq}) given by Eq. (1) varies depending on the alcohol. Equilibrium constants for the liquid phase esterification of acetic acid with various alcohols are shown in Table 4. The heat of reaction is generally small [61]. Thus, thermodynamic considerations determine their equilibrium constants to be relatively independent of temperature.

The uncatalyzed esterification of acetic acid with alcohols is slow at temperatures ranging from ambient to the normal boiling point of the reacting mixtures. Industrial production of acetate esters requires the use of catalysts. The kinetics of the esterification reaction between acetic acid and alcohols have been extensively studied. Mechanisms, rate equations, and rate constants are given in detail by several workers [64–70]. Generally, the acid-catalyzed reaction of acetic acid with alcohols is proportional to the acid or hydrogen ion concentration, the alcohol concentration, and the acetic acid concentration. The most common mineral acid catalyst used in these applications is probably sulfuric acid. It is more effective and less expensive than phosphoric acid and not as corrosive to metals as hydrochloric acid. Relatively low concentrations of sulfuric acid are used to avoid dehydration of alcohols to ethers and olefins and to avoid the formation of color bodies. Typically, concentrations ranging from 0.1 to 3% by weight of reacting mixture are employed. Sulfonic acids, in particular the isomers of toluenesulfonic acid, are also used because of their selectivity and reasonably good catalytic activity. In general, sulfonic acids are more expensive and need to be used in higher concentrations than sulfuric acid, but they generally cause fewer side reactions than sulfuric acid. Another catalyst type used industrially is the ion exchange resin. Resins usually employed are sulfonic acid cation exchangers in the hydrogen form. These catalysts are used either in fixed beds or in back-mixed reactors with mechanical separation devices for recovery and recycle of the resin beads. The effect of temperature on the reaction rate of acetic acid with alcohols generally follows the Arrhenius equation, with activation energies generally between 10 and 20 Kcal/mol.

Reactions of acetic acid with alcohols are generally characterized by high yields and few side reactions. However, since the esterification of acetic acid with an alcohol is limited by equilibrium, high liquid phase reaction conversions of one reactant can be achieved either by having a very large excess of the other reactant or by removal of the ester and/or water from the reacting mixture. If the reacting mixture is allowed to flash, the vapor composition (y) is related to the liquid composition (x) by the activity coefficients (γ) and vapor pressures (p°) of the components in the mixture (assuming no catalyst, and therefore no reaction in the vapor phase) by the equation:

$$\frac{y_{\text{ester}} y_{\text{H}_2\text{O}}}{y_{\text{HOAc}} y_{\text{alcohol}}} = \frac{[\gamma p^\circ x]_{\text{ester}} [\gamma p^\circ x]_{\text{H}_2\text{O}}}{[\gamma p^\circ x]_{\text{HOAc}} [\gamma p^\circ x]_{\text{alcohol}}} \tag{15}$$

Depending on the vapor pressures and activity coefficients of the components in a given system, at vapor-liquid equilibrium, the ratio of the products to reactants in the vapor phase may be higher than in the liquid phase, as is the case with methyl acetate because of the higher volatility of methyl acetate relative to the other components. In other systems this may be achieved indirectly by the formation of low-boiling azeotropes between products, a product and a reactant, or a product and an inert azeotroping agent added for that purpose. Toluene is frequently used to form a binary low-boiling azeotrope with water. The net result is that the overall extent of reaction achieved between the vapor and liquid phases is greater than the extent achievable in the liquid phase alone. Furthermore, because these reactions are reversible, the rate of reaction in the liquid phase is typically increased by removing the product(s) preferentially to the other components in the mixture. The actual process configuration and operating conditions depend on the alcohol reacted with acetic acid. If the alcohol is not appreciably soluble in water, and it forms a minimum boiling azeotrope with water, the by-product water may be removed as the alcohol-water azeotrope vapor. The vapor can then be condensed and decanted, and the alcohol layer recycled to the reaction mixture. As discussed before, an inert azeotroping liquid can also be used should the alcohol have low volatility or be soluble in water. Some low-boiling acetate esters form binary azeotropes with water and with the alcohols from which they are reacted and also form low-boiling ternary azeotropes with the alcohol and water. In these cases, the ester, water, and alcohol vapor are condensed and decanted, and the aqueous and organic layers are fractionated further before the alcohol is recycled. Vapor-liquid relations have been determined for the esterification of acetic acid with methanol [71], ethanol [72], and butanol [73]. An analysis of the principles underlaying the influence of equilibrium chemical reactions on vapor liquid phase diagrams, applicable to the formation of acetate esters, is available [74]. A generalized algorithm for simultaneous chemical and phase equilibrium calculation is given by Xiao et al. [75].

B. Esterification of Acetic Anhydride with Alcohols

Unlike the esterification of acetic acid, the esterification of acetic anhydride with alcohols is not reversible. It can go to completion without removal of the acetate ester or acetic acid produced. Acetic anhydride is significantly more reactive than acetic acid. Although the reaction is catalyzed by acids and bases, it is often sufficient to heat a mixture of acetic anhydride and alcohol for complete reaction to occur in a brief period of time. The same acid catalysts as those used for the esterification of acetic acid can be used because reaction of the by-product acetic acid is governed by the principles discussed in the preceding section. In practice, many uncatalyzed systems are operated with sufficient excess acetic anhydride such that no by-product water is present in the final product.

C. Industrial Processes for the Commercial Production and Purification of Acetate Esters via Esterification

Many industrial acetic acid esterification processes have been developed based on the chemistry and physical principles described in the previous sections. Numerous patents have been issued for different schemes for the production of acetate esters [76–80]. This section provides generalized descriptions of processes for the esterification of acetic acid with an alcohol to produce high-volume, low-boiling acetate esters. They are intended only as guides for the design of specific processes. Generalized descriptions of process flowsheets for the production of low-volatility (high-boiling) esters, use of azeotroping agents, and use of anhydride instead of acid are given by Simons [61]. A systematic procedure for the selection of a reactor type (reactor column, batch reactor with attached column, or series of back-mixed reactors with attached columns) for a given esterification, based on the equilibrium constant and relative volatilities of the components, is discussed by Mayer and Worz [81]. Selected esterification process flow diagrams and equipment are reviewed by Keyes [82].

In general, low-boiling acetate esters are refined by distillation and require no neutralization of the catalyst. High-boiling acetate esters can be produced in like manner, but costly and energy-intensive high-vacuum distillation may be required. More commonly, these processes involve the removal of excess reactant(s) by distillation, neutralization followed by drying, filtration to remove salts, steam treatment for deodorization, decolorization by activated carbon, and final filtration.

Plant size and mode of operation are dependent on the volumes and types of products produced. Continuous processes are generally used for large-volume acetate esters. These processes can be used for the production of more than one acetate ester, or they can be specifically designed as single product lines. Batch processes are usually employed when several different products are made on the same equipment in relatively small volumes. Plant materials of construction range from stainless steels to specialty alloys depending on the specific application and section of the plant. Equipment handling boiling strong acids typically requires high-nickel alloys, and reboiler tubes in contact with boiling mineral acids, such as sulfuric acid, may require zirconium-based alloys.

A batch process flow diagram for the production of volatile (low-boiling acetate esters is shown in Figure 1. Acetic acid, alcohol, and catalyst are

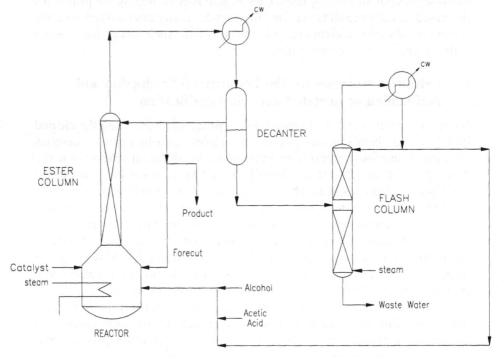

Figure 1   Batch esterification system for low-boiling alcohol acetates.

fed to the base or reboiler of a batch distillation column, which serves as the batch, or fed-batch flashing reactor. The distillation column can be trayed or packed; the required number of stages depends on the acetate esters being manufactured and the reflux flow ranges to be used. The composition of the reaction charge depends on the acetate being produced, but will usually contain excess alcohol since it will likely be part of a minimum-boiling azeotrope. Furthermore, the acetic acid and the acetate ester can have close boiling points, which would require further distillation to separate them if the acetic acid is not completely reacted. This may be particularly important in the manufacture of commercial solvents that may require very low acetic acid concentrations in the product. The process is started by heating the reactor such that the column is put on total reflux at an adequate rate. In time, liquid from the condenser will form organic and aqueous phases in the decanter. The organic phase is returned to the column as reflux while the aqueous phase is removed and fed to a flash column where organics are steam distilled for recycle. The by-product water is underflowed along with the condensed steam. At this point, the top column temperature corresponds to the acetate-alcohol-water azeotropic mixture.

As the aqueous phase is continually removed, the reaction mixture may be adjusted by addition of more alcohol. As conversion approaches completion, water formation ceases, and the column temperatures increase. Takeoff of organic material proceeds until the top column temperature reaches the boiling point of the acetate ester. The material taken off between the water and the ester (forecut) can be used in subsequent production runs. The product (heart cut) is removed while the column is operated at a sufficiently high reflux ratio to ensure adequate fractionation until the reactor volume is lowered sufficiently. The reactor is then cooled, the forecut material, fresh acetic acid, alcohol, and additional catalyst (if needed) are charged, and the cycle is repeated. This procedure is repeated until production is changed to another alcohol acetate or until sludge buildup requires reactor cleanout. As discussed previously, in some cases it may be desirable to use an inert azetroping agent for water removal. The process would then be operated with water removal being achieved by the added azeotropic agent. If acetic acid is more volatile than the acetate ester, excess acetic acid can be used to complete conversion of the alcohol. The excess acetic acid can then be removed by distillation, after completion of the reaction, followed by distillation of the acetate ester. An optimization and control strategy for batch rectification accompanied by chemical reaction, as is the case during the batch reaction of acetic acid and alcohols, is described by Egly et al. [83].

A continuous process scheme for the continuous esterification of acetic acid, to form low-boiling esters using a back-mixed flashing reactor, is shown in Figure 2. In this process scheme, acetic acid and an alcohol are continuously fed to the esterification reactor. The distillate is decanted, and the aqueous layer, containing some alcohol and acetate ester, is continuously decanted and fed to a flash column where it is steam-stripped. The alcohol and acetate are the stripping column's distillate and are returned to the reactor. The underflow is the by-product water and steam condensate. The acetate ester is in the primary component of the organic layer in the decanter. This layer usually also contains some alcohol, water, and sometimes traces of acetic acid. A portion of the organic phase is refluxed back to the reactor's column, and the rest is fed to a low boiler column. The low boiler column distillate, containing alcohol, water, and some acetate ester, is returned to the reactor. The dry, alcohol-free acetate ester underflows the low boiler column and is fed to a high boiler column where final refining of the acetate ester takes place.

A special case, with interesting processing challenges, is the manufacture of high-purity methyl acetate. The schemes discussed above do not work

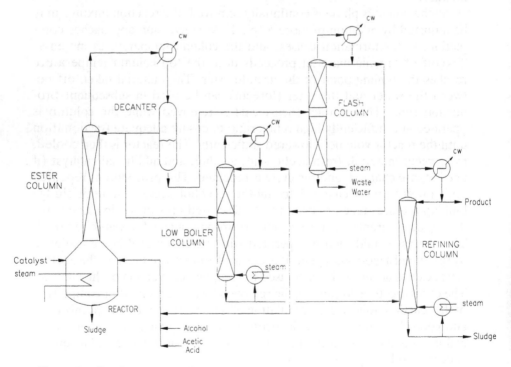

Figure 2 Continuous esterification system for low-boiling alcohol acetates.

well in the case of methyl acetate because the aqueous phase envelope in the ternary system of methyl acetate, methanol, and water is very small [84]. This limitation, coupled with the reaction equilibrium limitations and the formation of methyl acetate–methanol and methyl acetate–water minimum-boiling azeotropes, makes the manufacture of high-purity methyl acetate quite difficult. Conventional processes use schemes with multiple reactors where a large excess of one of the reactants (typically acetic acid) is used to achieve high conversion of the other reactant. Some schemes use a series of vacuum and atmospheric distillation columns to change the composition of the methyl acetate–water azeotrope [85]. The refined methyl acetate is separated from the unconverted reactants, and the methyl acetate–methanol azeotrope is recycled to the reactors. Other schemes use several atmospheric distillation columns and a column with an extractive agent, such as ethylene glycol monomethyl ether [86], to act as an entrainer to separate the methyl acetate from the methanol. Other workers [80] have resorted to the use of entrainers, such as toluene and methyl isobutyl ketone, to improve the purity of the methyl acetate in the distillation systems following the reactor. In addition to refining the methyl acetate, all these processes also have to deal with the problem of fractionating the by-product water and impurities that may be present in the methanol and acetic acid feedstocks. Such impurities may be, or may form, intermediate-boiling compounds that will contaminate the product, accumulate in the process recycle streams, and require additional distillation steps for their removal. Unreacted acetic acid can be separated easily from methyl acetate and methanol in a distillation step in which acetic acid and water are underflowed. Refining the acetic acid for recycle can be done via standard distillation, which is very energy intensive, or by well-known [87], but still energy- and capital-intensive, procedures that involve extraction, azeotropic distillation, and decanting equipment.

A good procedure for the production of high-purity methyl acetate is reactive distillation. However, early attempts [76] to produce refined methyl acetate utilizing procedures applicable to alcohol acetates in general did not recognize the difficulties inherent in the manufacture of high-purity methyl acetate caused by the presence of azeotropes and the very small liquid aqueous phase in the ternary system envelope. More recent work [88] finds that enrichment beyond the methyl acetate–methanol azeotrope can occur in a suitably configured, packed reactive distillation column, but concludes that a second column is required to fractionate methyl acetate from the methyl acetate–methanol azeotrope.

A process that overcomes these problems and provides an energy-efficient method for the production of high-purity methyl acetate [89] is shown in Figure 3. The key technology in this process is the use of trayed coun-

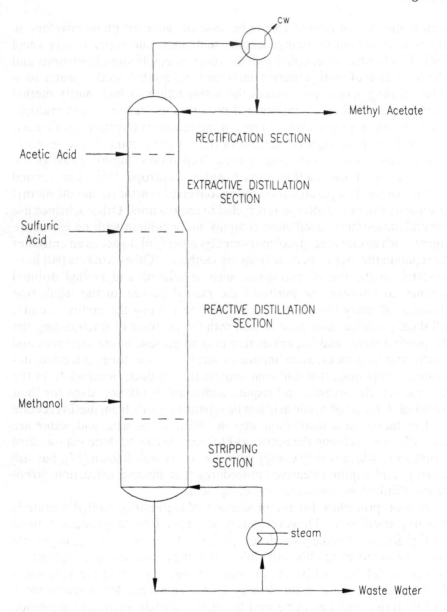

Figure 3 Countercurrent reactive distillation process for the manufacture of methyl acetate.

tercurrent reactive distillation combined with extractive distillation, in one unit operation. In the reactor column, most of the reaction occurs in the middle or reaction section, in a series of countercurrent flashing stages catalyzed with sulfuric acid. Above this section, water (and, to a lesser extent, methanol) is extracted with acetic acid in the extractive distillation section below the acetic acid feed. Acetic acid and methyl acetate are separated above the acetic acid feed, in the rectification section. In the lowest column section, methanol is stripped from by-product water. The countercurrent flow of reactants permits the attainment of very high conversions by creating zones in which one reactant exists in great excess, relative to the other reactant, even though the overall feeds to the reactor column may be stoichiometrically balanced. Thus, refined methyl acetate is the overhead product, and water is the bottom product. Special issues, concerning holdup requirements, appropriate control schemes, and heat integration, need to be addressed in the design of a reactor column of this type. A commercial application of this technology, where the acetic acid used is recycled acid containing several impurities that are removed via a vapor side-draw and impurity removal columns, is described by Agreda et al. [90]. Additional process improvements permit the manufacture of ultra-high-purity methyl acetate via reactive distillation of methanol with acetic acid and a small amount of acetic anhydride [91].

Simultaneous reaction and distillation is an approach that can provide significant advantages in energy usage and process intensification in the manufacture of acetate esters. However, the design and operation of reactive distillation processes can pose significant challenges. Principles and methods have been published concerning the design [92] and modeling [93–99] of continuous reactions in distillation equipment. Literature is also available concerning the conditions for steady state [100,101], the prediction of multiple steady states in reaction-rectification processes [102], the optimization of reactive distillation processes [103], and the synthesis of reactive distillation systems [104]. The design of continuous reactive distillation columns may be aided significantly by the development of a new mass and heat transfer rate-based approach for modeling staged separations [105,106], including reactive distillation systems.

REFERENCES

1. Data from Eastman publications M-145E, M-144E, M-140C, M-1422D, M-139D, and M-266, available from Eastman Chemical Co., Kingsport, TN by request.
2. *Chem. Market. Rep.*, p. 11 (March 31, 1986).
3. *Chem. Market. Rep.*, p. 46 (Jan. 9, 1989).
4. (a) *Chem. Market. Rep.*, p. 973 (November 1980). (b) *Chem. Market. Rep.*, p. 58, (Oct. 26, 1987).

5. Euranto, E. K., in *The Chemistry of Carboxylic Acids and Esters* (S. Patai, ed.), Wiley Interscience, New York, 1969, Chapter 11.
6. See, for example, *Organic Synthesis*, Col. Vol. *1*, 237, 241, 254, 451 (1941); Col. Vol. *2*, 261, 264, 276, 292, 365, 414 (1943); Col. Vol. *3*, 381 (1955); Col. Vol. *4*, 329, 532, 635 (1963); Col. Vol. *5*, 762 (1973). For a recent review of esterification reactions, see Haslim, E., *Tetrahedron, 36*, 2409 (1980).
7. Yeomans, B., European Patent Appl. EP 60719 A1 (Sept. 22, 1982) (*Chem. Abstr., 98*(5), 34271c).
8. Xu, Y. Zhang, G., Ma, Z., and Zhou, G. *Huaxue Tongbao*, (*5*), 14–15 (1983) (Chem. Abstr., *99*(23), 194404x).
9. Jpn. Kokai Tokkyo Koho JP 55/38399 [80/38399] (March 17, 1980) (*Chem. Abstr., 93*(13), 132086n).
10. Gomzi, Z., and Zrncevic, S., *Croat. Chem. Acta, 53*(1), 25–32 (1980) (*Chem. Abstr., 93*(13), 131701x).
11. Thakar, H. P., Puranik, S. A., and Kher, M. G., *Fluid. Ses Appl., C.-R. Congr. Int.*, 1973, 615–627 (H. Angelino, J. P. Couderc, and H. Gilbert, eds.), Toulouse, Fr. (*Chem. Abstr., 87*(3), 22303z).
12. Ran, R., Huang, J., and Shen, J., *Gaofenzi Xuebao, 6*, 476–480 (1987) (*Chem. Abstr. 110*(11), 94630v).
13. Pei, W., Liu, X., and Ran, R., *Beijing Daxue Xuebao, Ziran Kexueban, 24*(2), 143–149 (1988) (*Chem. Abstr. 110*(7), 56683g).
14. Ran, R., Huang, J., Jia, X., and Shen, J., *Cuihua Xuebao, 8*(4), 440–444 (1987) (*Chem. Abstr., 110*(1), 7168j).
15. Ran, R., Huang, J., and Shen, J., *Yingyong Huaxue, 4*(5), 49–54 (1987) (*Chem. Abstr., 109*(7), 54437k).
16. Ran, R., Wu, X., Jia, X., and Pei, W., *Gaofenzi Xuebao, 1*, 67–71 (1988) (*Chem. Abstr., 110*(1), 7170d).
17. Ran, R., Pei, W., Jia, X., and Wu, X., *Youji Huaxue, 4*, 268–272 (1987) (*Chem. Abstr., 108*(17), 150013u).
18. Bilbao Elorriage, J., Gonzalez Marcos, J. A., Gonzalez-Velasco, J. R., and Arandes Esteban, J. M., *Afinidad, 40*(387), 459–463 (1983) (*Chem. Abstr., 100*(15), 120210y).
19. Vara, E., Saura Calixto, F., and Marinas, J. M., *Afinidad, 37*(368), 327–331 (1980) (*Chem. Abstr., 94*(7), 46729p).
20. *Kirk-Othmer Encylcopedia of Chemical Technology*, Vol. 9, p. 300, John Wiley & Sons, New York, 1978.
21. Gurevich, V. R., Bogod, I. A., Fatali-Zade, Ch. A., Malyutin, N. R., and Lapshev, A. I., U.S.S.R. Patent SU 411891 (Jan. 25, 1974) (*Chem. Abstr., 81*(11), 63169e).
22. Vesely, V., Nyvlt, V., Plisek, J., Glosser, Z., and Hojer, J., Czech. Patent CS 165254 (Oct. 15, 1976) (*Chem. Abstr., 86*(25), 189219t).
23. Strazhesko, D. N., Tovbina, Z. M., and Stavitskaya, S. S., *Ukr. Khim. Zh.* (Russ. ed.), *40*(4), 354–359 (1974) (*Chem. Abstr., 81*(3), 13062u).
24. Fu, X., He, M., and Zeng, G., *Xinan Shifan Daxue Xuebao, Ziran Kexueban, 2*, 91–94 (General Organic Chemistry) (1988) (*Chem. Abstr., 110*(19), 172339y).

25. Jpn. Kokai Tokkyo Koho JP 57/130954 A2 (82/130954) (Aug. 13, 1982), 4 pp. (*Chem. Abstr.*, *98*(1), 4314b).

26. Hino, M., and Arata, K., *Chem. Lett.*, *12*, 1671–1672(1981) (*Chem. Abstr.*, *96*(7), 51780u).

27. (a) Bhide, B. V., and Sudborough, J. J., *J. Indian Inst. Sci.*, *A8*, 89 (1925). (b) Jones, R. W. A., and Thomas, J. D. R., *J. Chem. Soc.* (*B*), 661 (1966).

28. *Eur. Chem. News*, p. 6 (March 17, 1986).

29. *Organ. Syn.*, Col. Vol. *4*, 261, 417 (1963).

30. (a) Dettmer, M., and Renken, A., *Chem.-Ing.-Tech.*, *55*(2), 146–147 (1983) (*Chem. Abstr.*, *98*(19), 159940b). (b) Leupold, E. I., and Renken, A., *Chem.-Ing.-Tech.*, *49*(8), 667 (1977) (*Chem. Abstr.*, *87*(21), 167509f). (c) Leupold, E. I., Arpe, H. J., Renken, A., and Schlosser, E. G., Ger. Offen. DE 2545845 (April 28, 1977) (*Chem. Abstr.*, *87*(5), 38924p).

31. Dockner, T., and Platz, R., Ger. Offen. DE 2511978 (Sept. 30, 1976) (*Chem. Abstr.*, *86*(3), 16308t).

32. Jpn. Kokai Tokkyo Koho JP 58/183640 A2 (83/183640) (Oct. 26, 1983) (*Chem. Abstr.*, *100*(15), 120529j).

33. (a) Jpn. Kokai Tokkyo Koho JP 57/183743 A2 (82/183743) (Nov. 12, 1982) (*Chem. Abstr.*, *98*(17):142972r). (b) Jpn. Kokai Tokkoyo Koho JP 57/176926 A2 (82/176926) (Oct. 30, 1982) (*Chem. Abstr.*, *98*(13), 106827p).

34. Kerr, E. R., and Throckmorton, M. C., U.S. Patent 3,014,066 (Dec. 19, 1961).

35. Gregory, R., European Patent Appl. EP 73141 A2 (March 2, 1983) (*Chem. Abstr.*, *98*(24), 205105h).

36. Ballantine, J. A., Purnell, J. H., Westlake, D. J., Thomas, J. M., and Gregory, R., European Patent Appl. EP 31687 (July 8, 1981) (*Chem. Abstr.*, *95*(19), 168572y).

37. (a) Ballantine, J. A., Purnell, J. H., and Thomas, J. M., European Patent Appl. EP 31252 (July 1, 1981) (*Chem. Abstr.*, *95*(19), 168569c). (b) Ballantine, J. A., Davies, M., Purnell, H., Rayanakorn, M., Thomas, J. M., and Williams, K. J., *J. Chem. Soc., Chem. Commun.*, *1*, 8–9 (1981) (*Chem. Abstr.*, *95*(1), 6457n).

38. Young, L. B., U.S. Patent 4365084 (Dec. 21, 1982) (*Chem. Abstr.*, *98*(11), 88841g).

39. Gruffax, M., and Micaelli, O., European Patent Appl. EP 5680 (Nov. 28, 1979) (*Chem. Abstr.*, *92*(23), 197944q).

40. (a) Takamiya, N., Jpn. Kokai JP 49/100016 (74/100016) (Sept. 20, 1974) (*Chem. Abstr.*, *82*(23) 155383u). (b) Murakami, Y., Hattori, T., and Uchida, H., *Kogyo Kagaku Zasshi*, *72*(9), 1945–1948 (1969) (*Chem. Abstr.*, *72*(7), 31173z).

41. Neri, C., and Esposito, A., Ger. Offen. DE 2700538 (July 21, 1977), CL C07C69/12 (Jan. 9, 1976) (*Chem. Abstr.*, *87*(17), 133911r).

42. Pesa, F. A., and Graham, A. M., U.S. Patent 4510320 (April 9, 1985) (*Chem. Abstr.*, *102*(25), 220463t).

43. McGinnis, J. L., U.S. Patent 4480115 (Oct. 30, 1984), 4 pp. (*Chem. Abstr.*, *102*(5), 45494e).

44. Kent, A. G., U.K. Patent Appl. GB 2129430 A1 (May 16, 1984) (*Chem. Abstr.*, *101*(15), 130246z).

45. Drent, E. European Patent Appl. EP 117575 A1 (Sept. 5, 1984) (*Chem. Abstr.*, *102*(3), 24121r).

46. Jpn. Kokai Tokkyo Koho JP 58/144347 A2 (83/144347) (Aug. 27, 1983) (*Chem. Abstr.*, *100*(1), 5877j).

47. Dombek, B. D., European Patent Appl. EP 64287 A1 (Nov. 10, 1982) (*Chem. Abstr.*, *98*(23), 197612d).

48. Kitson, M., and Williams, P. S., European Patent Appl. EP 198682 A2 (Oct. 22, 1986) (*Chem. Abstr.*, *111*(1), 6918n).

49. Kitson, M., and Williams, P. S., European Patent Appl. EP 285420 A1 (Oct. 5, 1988) (*Chem. Abstr.*, *110*(26), 233609k).

50. Saito, Y., and Takahashi, O., Jpn. Kokai Tokkyo Koho JP 63/179845 A2 (88/179845) (July 23, 1988) (*Chem. Abstr.*, *110*(3), 23351u).

51. Caillod, J., and Sauvion, G., N. European Patent Appl. EP 192587 A1 (Aug. 27, 1986) (*Chem. Abstr.*, *105*(19), 171850t).

52. Moy, D., European Patent Appl. EP 175558 A1 (March 26, 1986) (*Chem. Abstr.*, *104*(24), 209171s).

53. Matteoli, U., Menchi, G., Bianchi, M., Frediani, P., and Piacenti, F., *Gazz. Chim. Ital.*, *115*(11–12, Pt. A), 603–606 (1985) (*Chem. Abstr.*, *104*(25), 224426x).

54. Jpn. Kokai Tokkyo Koho JP 58/192850 A2 (83/192850) (Nov. 10, 1983), 7 pp. (*Chem. Abstr.*, *100*(13), 102772d).

55. Tischenko, V. E., *J. Russ. Phys. Chem. Soc.*, *38*, 355, 482 (1908).

56. (a) Staff, P. R., *J. Org. Chem.*, *38*, 1433 (1973). (b) Childs, W. C., and Adkins, H., *J. Am. Chem. Soc.*, *45*, 3013 (1923); *47*, 798 (1925). (c) Villacorta, G. M., and San Filippo, J., Jr., *J. Org. Chem.*, *48*, 1151 (1983). (d) Ito, T., Horino, H., Koshiro, Y., and Yammamoto, A., *Bull. Chem. Soc. Jpn.*, *55*, 504 (1982).

57. (a) Ogata, Y., Kawasaki, A., and Kishi, I., *Tetrahedron*, *23*, 825 (1967). (b) Ogata, Y., and Kawasaki, A., *Tetrahedron*, *25*, 929 (1969).

58. Lins, I., and Day, A. R., *J. Am. Chem. Soc.*, *74*, 5133 (1952).

59. *Synthetic Organic Chemicals, U.S. Production and Sales*, U.S. International Trade Commission.

60. *Chemical Economics Handbook*, SRI International.

61. Simons, R. M., in *Encyclopedia of Chemical Processing and Design*, Vol. 19, (J. J. McKetta and W. A. Cunningham, eds.), Marcel Dekker, New York, 1983, p. 381.

62. Zey, E. G., in *Kirk-Othmer Encyclopedia of Chemical Technology*, Vol. 9, 3rd ed., Wiley, New York, 1980, p. 291.

63. Elam, E. U., in *Kirk-Othmer Encyclopedia of Chemical Technology*, Vol. 9, 3rd ed., Wiley, New York, 1980, p. 311.

64. Berthelot, M., and Pean de Saint-Gilles, L., *Ann. Chim. Phys.*, *65*, 385 (1862); *66*, 5 (1862); *68*, 225 (1863).

65. Rolfe, A. C., and Hinshelwood, C. M., *Trans. Faraday Soc.*, *30*, 935 (1934).
66. Williamson, A. T., and Hinshelwood, C. M., *Trans. Faraday Soc.*, *30*, 1145 (1934).
67. Smith, H. A., *J. Am. Chem. Soc.*, *61*, 256 (1939).
68. Fairclough, R. A., and Hinshelwood, C. N., *Trans. Faraday Soc.*, *30*, 593 (1939).
69. Smith, H. A., and Reichardt, C. H., *J. Am. Chem. Soc.*, *63*, 605 (1941).
70. Sniegoski, P. J., *J. Org. Chem.*, *41*(11), 2058 (1976).
71. Hirata, M., Komatsu, H., and Misaki, Y., *Kagaku Kogaku* (Chemical Engineering, Japan), *31*, 1184 (1967).
72. Hirata, M., and Komatsu, H., *Kagaku Kogaku* (Chemical Engineering, Japan), *30*, 989 (1966).
73. Hirata, M., and Komatsu, H., *Kagaku Kogaku* (Chemical Engineering, Japan), *30*, 129 (1966).
74. Barbosa, D., and Doherty, M. F., *Chem. Eng. Sci.*, *43*(3), 529 (1988).
75. Xiao, W., Zhu, K., Yuan, W., and Chien, H. H., *A.I.Ch.E. J.*, *35*(11), 1813 (1989).
76. Backhaus, A. A., U.S. Patent 1,400,849 (December 1921).
77. Backhaus, A. A., U.S. Patent 1,425,624 and 1,425,625 (August 1922).
78. Backhaus, A. A., U.S. Patent 1,454,462 and 1,454,463 (May 1923).
79. McKeon, T. J., Pyle, C., 3rd, and Van Ness, R. T., U.S. Patent 2,208,769 (July 1940).
80. Yeomans, B., European Patent Appl. 060717 and 0060719 (September 1982).
81. Mayer, H. H., and Worz, O., *Ger. Chem. Eng.*, *3* 252 (1980).
82. Keyes, D. B., *Ind. Eng. Chem.*, *24*(10), 1096 (1932).
83. Egly, H., Ruby, V., and Seid, B., *Ger. Chem. Eng.*, *6*, 220 (1983).
84. Crawford, A. G., Edwards, G., and Lindsay, D. S., *J. Chem. Soc.*, Part I, 1054 (1949).
85. Harrison, J. M., U.S. Patent 2,704,271 (March 1955).
86. Kumerle, K., German Patent 1,070,165 (December 1959).
87. Othmer, D. F., *Kirk-Othmer Encyclopedia of Chemical Technology*, Vol. 3, 3rd ed., Wiley, New York, 1978, p. 352.
88. Sawistoski, H., and Pilavakis, P. A., *Chem. Eng. Sci.*, *43*, 355 (1988).
89. Agreda, V. H., and Partin, L. R., U.S. Patent 4,435,595 (March 1984).
90. Agreda, V. H., Partin, L. R., and Heise, W. H., *Chem. Eng. Prog.*, *86*(2), 40 (1990).
91. Agreda, V. H., and Lilly, R. D., U.S. Patent 4,939,294 (July 1990).
92. Balashov, M. I., and Serafinov, L. A., *Teoreticheskie Osnovy Khimicheskoi Tekhnol.*, *14*(4), 515 (1980).
93. Belck, L. H., *A.I.Ch.E. J.*, *1*(4), 467 (1955).
94. Huneck, J., Foldes, P., and Sawinsky, J., *Int. Chem. Eng.*, *19*(2), 248 (1979).
95. Holland, C. D., *Fundamentals of Multicomponent Distillation*, McGraw-Hill, New York (1981).
96. Chang, Y. A., and Seader, J. D. *Comput. Chem. Eng.*, *12*(12), 1243 (1988).
97. Bogacki, M. B., Alejski, K., and Szymanowski, J., *Comput. Chem. Eng.*, *13*(9), 1081 (1989).

98. Barbosa, D., and Doherty, M. F., *Chem. Eng. Sci.*, *43*(7), 1523 (1988).
99. Barbosa, D., and Doherty, M. F., *Chem. Eng. Sci.*, *43*(9), 2377 (1988).
100. Pisarenko, Y. A., Epifanova, O. A., and Serafimov, L. A., *Teoreticheskie Osnovy Khimicheskoi Tekhnol.*, *21*(4), 466 (1987).
101. Pisarenko, Y. A., Epifanova, O. A., and Serafimov, L. A., *Teoreticheskie Osnovy Khimicheskoi Tekhnol.*, *22*(1), 38 (1988).
102. Timofeev, V. S., Solokhin, A. V., and Kalerin, E. A., *Teoreticheskie Osnovy Khimicheskoi Tekhnol.*, *22*(6), 729 (1988).
103. Duprat, F., Gassend, R., and Gau, G., *Comput. Chem. Eng.*, *12*(11), 1141 (1988).
104. Barbosa, D., and Doherty, M. F., *Chem. Eng. Sci.*, *43*(3), 541(1988).
105. Seader, J. D., *Chem. Eng. Prog.*, *85*(10), 41 (1989).
106. Sivasubramanian, M. S., and Boston, J. F., in *Computer Applications in Chemical Engineering* (H. Th. Bussemaker, and P. D. Iedema, eds.), Elsevier, New York, 1990, p. 331.

15

Halogenated Derivatives

Paul R. Worsham
Eastman Chemical Company, Kingsport, Tennessee

I. INTRODUCTION

Halogenated acetic acid derivatives are most commonly used as intermediates for synthesis of pharmaceuticals and other industrial chemicals, but halogen substitution gives acetic acid unique properties that have allowed this class of derivatives to be exploited also as reaction catalysts, specialty solvents, and biocides. The halogenated derivatives can be divided into two classes—the acetyl halides and the alpha-halogenated compounds. Acetyl halides typically are used as acetylating reagents in organic synthesis. Alpha-halogenated derivatives, however, have been utilized in a wide variety of applications, including uses as reactive intermediates, catalysts, solvents, and pesticides.

Replacement of hydrogen by halogen at the alpha position results in an increase in the acidity of the acetic acid molecule. The acidity increases with the addition of each subsequent halogen atom. Within the mono-, di-, and trihalogenated series, the acidity generally increases with increasing electronegativity of the halogen atom, i.e., in the order $F > Cl > Br > I$. The reported acidities of acetic acid derivatives are shown in Table 1. Other physical properties are reported in Table 2.

Acetic acid is a relatively nontoxic material, but all alpha-halogenated derivatives should be regarded as hazardous because of their high acidity and potential for toxicity to plants and animals. The biological activity of

Table 1 Acidity of Halogenated Acetic Acid Derivatives

Derivative	Ka	pKa	Ref.
Acetic acid	1.8×10^{-5}	4.75	90
Fluoroacetic acid	2.6×10^{-3}	2.59	91
	2.2×10^{-3}	2.66	92
Difluoroacetic acid	5.7×10^{-2}	1.24	92
	5.7×10^{-2}	1.24	93
	4.6×10^{-2}	1.34	94
Trifluoroacetic acid	5.9×10^{-1}	0.23	92
	3.2×10^{-1}	0.50	94
Chloroacetic acid	1.4×10^{-3}	2.85	1
	1.4×10^{-3}	2.85	90
Dichloroacetic acid	5.1×10^{-2}	1.29	1
	3.3×10^{-2}	1.48	90
	4.4×10^{-2}	1.36	94
Trichloroacetic acid	2.2×10^{-1}	0.67	1
	2.0×10^{-1}	0.7	90
	3.0×10^{-1}	0.52	94
Bromoacetic acid	1.3×10^{-3}	2.90	91
	2.0×10^{-3}	2.69	90
Dibromoacetic acid	4.1×10^{-2}	1.39	95
Tribromoacetic acid	1.4	−0.147	95
Iodoacetic acid	7.1×10^{-4}	3.15	1
	7.5×10^{-4}	3.12	90

the halogenated derivatives has made them useful as bactericides, fungicides, herbicides, insecticides, and rodenticides. But this high degree of biological activity indicates a strong potential for high human toxicity, and these derivatives should be handled with great care.

The discussion of halogenated acetic acid derivatives in this chapter presents a brief description of the common halogenated derivatives, including methods of preparation, significant uses, and unique properties and reactions. No attempt has been made to be comprehensive because excellent reviews have been published that present further detail on this important class of acetic acid–based compounds [1,2].

II. ALPHA-HALOGENATED DERIVATIVES

A. Alpha-fluorinated Derivatives

Fluoroacetic acid occurs naturally as a constituent of "Gifblaar," a South African poisonous plant [3]. It is an extremely potent toxin, having an

Table 2 Physical Properties of Halogenated Acetic Acid Derivatives

Derivatives	Melting point (°C)	Boiling point (°C)	Density (g/cm³)
Haloacids			
Fluoroacetic	35.3 [96]	168.3 [97]	1.369 at 36°C [96]
Difluoroacetic	−0.4 [93]	134.2 [93]	1.530 at 20/4°C [98]
Trifluoroacetic	−15.4 [99]	72.4 [100]	1.489 at 20/4°C [100]
Chloroacetic			
Alpha	61.3	189	1.398 at 65/65°C [101]
Beta	56.2		
Gamma	52.5		
Dichloroacetic	5–6	194	1.563 at 20/4°C [3]
Trichloroacetic	58	196–197	1.629 at 61/4°C [3]
Bromoacetic	50	208	1.927 at 50/4°C [102]
Dibromoacetic	48	195–197 [101] at 250 torr	
Tribromoacetic	131	235 dec	
Iodoacetic	83		
Diiodoacetic	110		
Triiodoacetic	150		
Haloacetyl halides			
Fluoroacetyl fluoride		50.5–51	
Fluoroacetyl chloride		71.5–73	
Fluoroacetyl bromide		95–96	
Difluoroacetyl chloride		25	
Trifluoroacetyl chloride		−14	
Trifluoroacetyl bromide		2–3	
Chloroacetyl chloride		108–110	
Chloroacetyl bromide		127	
Dichloroacetyl chloride		107–108	1.532 at 16/4°C [3]
Trichloroacetyl chloride		118	
Trichloroacetyl bromide		143	
Bromoacetyl chloride		134	
Bromoacetyl bromide		150	
Dibromoacetyl bromide		194	
Tribromoacetyl bromide		210–215	
Iodoacetyl chloride		49–52 at 15 mm	

Source: Data from Ref. 13, unless otherwise indicated.

estimated oral lethal dose of 2–5 mg/kg for the sodium salt. Fluoroacetic acid, its salts, and amide derivatives have been used mainly as rodenticides.

Fluoroacetic acid has been prepared from methyl iodoacetate by reaction with silver fluoride and from methyl chloroacetate by reaction with potassium fluoride followed by ester hydrolysis [3]. However, the reaction of some chlorocarboxylic acids with KF in glacial acetic acid has been shown to give acetoxy rather than fluoro derivatives [4]. This unanticipated reaction was attributed to the strength of hydrogen bonding between fluoride and acetic acid solvent. Strong hydrogen bonding was said to enhance the nucleophilicity of hydroxyl O atom of acetic acid while reducing that of the fluoride ion.

Trifluoroacetic acid has become a common reagent in organic chemistry laboratories because of its usefulness as a solvent and acid catalyst for many chemical transformations. One of the most impressive catalysis applications is the sequential cyclization reaction shown in Eq. (1). This chemistry was designed to mimic biological synthesis of the steroid nucleus and proceeds best using trifluoroacetic acid catalyst in methylene chloride at −78°C [5]. As a solvent this acid has found application in the platinum-catalyzed hydrogenation of ketones [6] and pteridines [7].

$$CF_3CO_2H, \quad CH_2Cl_2, \quad -78°\ C \xrightarrow{\quad LiAlH_4 \quad}$$

$$(1)$$

B. Alpha-chlorinated Derivatives

Chloroacetic acid is the principal representative of this category and the highest-volume chemical of all alpha-halogenated derivatives. Chloroacetic acid is prepared industrially by chlorination of acetic acid or by the hydrolysis of trichloroethylene with sulfuric acid. Also, the acid can be made by the hydrolysis of chloroacetyl chloride, which is available from carbonylation of methylene chloride, oxidation of vinylidene chloride, or addition of chlorine to ketene [1]. Another approach involves reaction of formaldehyde with formic acid and HCl [8].

The major use of chloroacetic acid is as an intermediate for the synthesis of other chemicals. Generally, these reactions involve nucleophilic substitution at the alpha carbon atom of a chloroacetic acid derivative, resulting in the introduction of a carboxymethyl group into the nucleophile. Some typical representatives of this class of reactions are shown in Figure 1.

Treatment of chloroacetate with hydroxide gives glycolic acid salts and treatment with ammonia gives the amino acid glycine. Reaction with cyanide affords cyanoacetate, an important chemical for the production of coumarin and malonic acid derivatives [1]. The reaction of the potassium salt of chloroacetic acid with potassium nitrite affords a nitro acid, which on thermal decomposition gives nitromethane [9].

Some important commercial and industrial chemicals are produced by substitution reactions of chloroacetic acid. A large volume of chloroacetic acid is used in the production of carboxymethylcellulose and other carboxymethylated polysaccharides. Also, the reaction of chloroacetic acid with 2,4-dichlorophenol and 2,4,5-trichlorophenol gives the widely used herbicides 2,4-D and 2,4,5-T.

The Darzens reaction is a synthetic procedure that involves condensation of aldehydes or ketones with alpha-haloacetic acid esters to form alpha,beta-epoxy esters (glycidic esters). The reaction requires a strong nonnucleophilic base. Usually sodium ethoxide, sodium *t*-butoxide, or sodamide is used, and the reaction has been reported to work best with alpha-chloro esters, although bromo esters have been used successfully

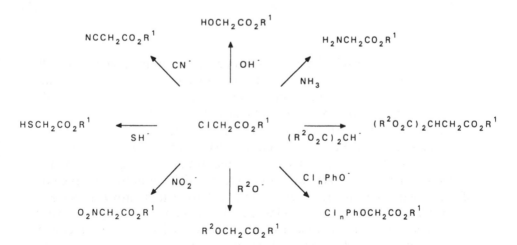

Figure 1 Substitution reactions of chloroacetic acid esters.

[10]. The mechanism probably involves an enolate anion of the alpha-haloacetate ester. As shown in Eq. (2), this anion attacks the carbonyl carbon of the aldehyde or ketone, followed by nucleophilic displacement of the halogen. This final step can be viewed as an internal nucleophilic substitution reaction.

The Darzens reaction has been useful in synthesis since hydrolysis and decarboxylation of the resulting glycidic ester produces a carbonyl compound with one carbon atom more than the original starting material [10]. Both higher aldehydes and ketones can be produced using this methodology.

(2)

Dichloroacetic acid is similar to chloroacetic acid in the reactions it will undergo. It has been prepared by the reaction of chloral hydrate [$Cl_3CCH(OH)_2$] with cyanide and calcium carbonate, followed by hydrolysis, by the chlorination of chloroacetic acid, by the reduction of trichloroacetic acid, and by the hydrolysis of pentachloroethane.

Dichloroacetic acid can be used in a variation of the Darzens reaction. This condensation occurs, as shown in the example in Eq. (3), when a dichloroacetate ester reacts with a ketone or aldehyde in the presence of dilute magnesium amalgam [10]. Alpha-chloro beta-hydroxy esters are formed in almost quantitative yields from ketones and are easily converted into glycidic esters by treatment with sodium ethoxide. The intermediate alpha-chloro beta-hydroxy ester also can be dehydrated with phosphorous pentoxide to an alpha-chloroacrylate in high yield.

$$CI_2CHCO_2R^1 \quad + \quad \underset{R^2 \overset{O}{\diagup} R^3}{\overset{O}{\diagdown}} \quad \xrightarrow[Mg \cdot Hg]{Et_2O} \quad R^2 \overset{CICHCO_2R^1}{\underset{R^3}{\diagup}} OH$$

$$P_2O_5 \diagup \qquad\qquad NaOEt \Big| -CI^-$$

$$\underset{R^3}{\overset{R^2}{\diagdown}} C{=}CCICO_2R^1 \qquad\qquad R^2 \overset{CHCO_2R^1}{\underset{R^3}{\diagup}} O \qquad (3)$$

The toxicity of dichloroacetic acid is substantially lower than that of chloroacetic acid. The oral LD_{50} in rats has been reported as 2.82 g/kg [11] and 4.48 g/kg [12] for the dichlorocompound compared to 76 mg/kg for the monochloro derivative [11]. Dichloroacetic acid has virucidal, fungicidal, and bactericidal activity [12]. In fact, the dichloroacetyl group appears in nature in the antibiotic chloromycetin [13]. The acid or inositol-nicotinate ester derivative has been suggested for medical use in treatment of lactic acidosis, diabetic mellitus, hyperlipidemia [11,14] and renal ischemia [15].

Exhaustive chlorination of acetic acid affords trichloroacetic acid. Chlorination is the route practiced by industry, but this derivative can also be prepared by oxidation of chloral hydrate with nitric acid [16] and oxidation of tetrachloroethylene [17]. Mild hydrogenation over palladium on carbon converts both dichloro- and trichloroacetic acid to chloroacetic acid [18]. Trichloroacetic acid is somewhat unstable in dilute aqueous solutions, decomposing into chloroform, hydrochloric acid, carbon dioxide, and carbon monoxide [19]. Also, trichloroacetic acid decomposes when heated in the presence of bases, affording chloroform and carbon dioxide or carbonates.

The formation of carbene species in trichloroacetic acid decomposition is suggested by the observation that decomposition of sodium trichloroacetate in 1,2-dimethoxyethane in the presence of carbene acceptors gives the product expected from addition of dichlorocarbene. Thus, as shown in Eq. (4), decomposition in the presence of cyclohexene gives 7,7-dichloronorcarane in 65% yield [20]. Dichlorocarbene can also be generated by a much harsher alternative procedure that involves the reaction of chloroform with strong base.

$$Cl_3CCO_2Na \xrightarrow[\text{(CH}_3\text{OCH}_2)_2]{\Delta} Cl_2C: + NaCl + CO_2 \xrightarrow{\hspace{2cm}} \quad (4)$$

The oral LD_{50} in rats for trichloroacetic acid is approximately 5 g/kg [21]. Although not as toxic as some other halogenated acetic acid derivatives, this compound is quite corrosive to skin. Trichloroacetic acid or its salts have been used as herbicides, as aids for microscopy [19], and in medicine as an astringent and antiseptic [1]. Dilute trichloroacetic acid is frequently used to precipitate protein from aqueous solutions in protein purification procedures.

C. Alpha-brominated Derivatives

The preparation, reactions, and uses of the brominated acetic acid derivatives parallel those of the chloroacetic acids. Bromoacetic acid has been prepared by bromination of acetic acid, from reaction of chloroacetic acid with HBr [22–24], or from reaction of glycolic acid and HBr [25].

Halogenation of acetic acid with either chlorine or bromine, but not fluorine or iodine, can be catalyzed by phosphorus in a process known as the Hell-Volhard-Zelinsky reaction. This reaction is illustrated in Eq. (5).

$$CH_3CO_2H \xrightarrow{Br_2 \, , \, P} CH_3COBr \xrightarrow{Br_2 \, , \, P} BrCH_2COBr$$

$$BrCH_2COBr + CH_3CO_2H \xrightarrow{\hspace{1.5cm}} BrCH_2CO_2H + CH_3COBr$$

$$(5)$$

In contrast to most halogenation reactions, which involve radical reactions and often give poor selectivity, the Hell-Volhard-Zelinsky reaction occurs by an acid-catalyzed process through an acid halide intermediate and is specific for preparation of alpha-halogenated derivatives. The required acid halide is produced from a phosphorous trihalide, which in turn is produced from the reaction of the phosphorous catalyst with halogen. The phosphorous trihalide is not involved in the alpha halogenation since the reaction of an acid with chlorine in the presence of phosphorous tribromide gives chlorination and not bromination. The Hell-Volhard-Zelinsky reaction is suitable for the preparation of a wide variety of higher alpha-halogenated acid derivatives in addition to its usefulness for preparation of halogenated acetates [2].

Other reagents that have been used to promote or catalyze the bromination reaction are acetic anhydride/pyridine [26,27] and dry hydrogen chloride [28]. The bromination reaction will occur without catalysts at sufficiently high temperature [29]. Bromoacetic acid has also been prepared by oxidation of bromine-containing compounds. Oxidation of ethylene bromide with fuming nitric acid [30] and air oxidation of bromoacetylene [31] both give bromoacetic acid.

The bromo derivative is more reactive than the chloro analog and is susceptible to hydrolysis to glycolic acid under mild conditions. Bromoacetic acid esters are synthetically useful as starting materials in the Reformatsky reaction for preparation of beta-hydroxyacids or alpha,beta-unsaturated acids. This synthesis involves treatment of an aldehyde or ketone with zinc and a bromoacetic acid ester as shown in Eq. (6).

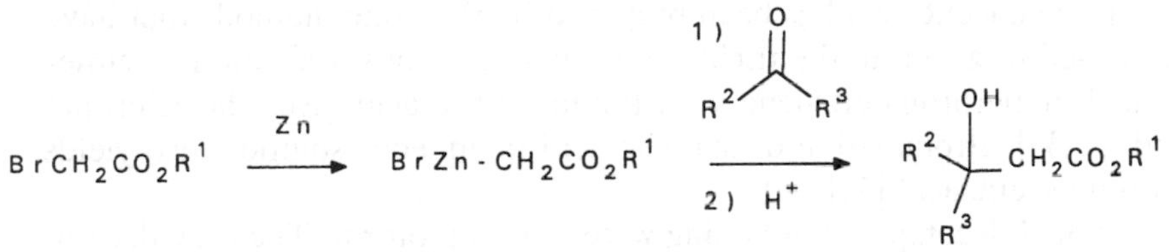

$$BrCH_2CO_2R^1 \xrightarrow{Zn} BrZn\text{-}CH_2CO_2R^1 \xrightarrow[2)\ H^+]{1)\ \underset{R^2 \diagdown \diagup R^3}{\overset{O}{\parallel}}} R^2\underset{R^3}{\overset{OH}{\diagup\diagdown}}CH_2CO_2R^1 \quad (6)$$

Crystallography reveals the dimeric structure shown in Figure 2 for the solid reaction product from treatment of *t*-butyl bromoacetate with zinc [32]. The reaction can be considered analogous to the Grignard reaction conducted with organic halides and magnesium, but the Grignard reagent is too reactive to be prepared from haloesters. Thus, the Reformatsky reaction provides an important extension of the scope of Grignard-type chemistry and has been widely used in synthesis [33].

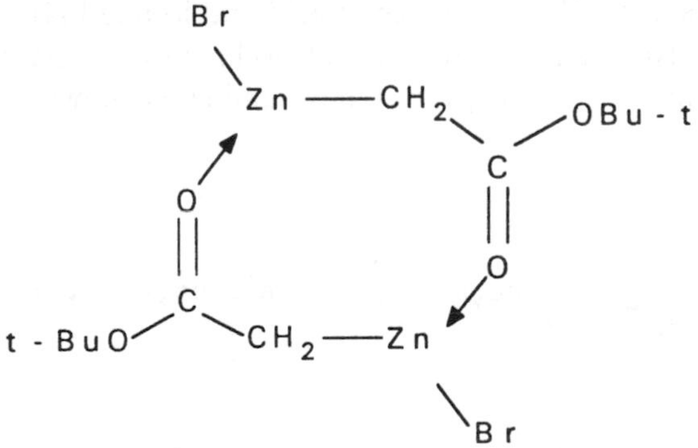

Figure 2 Dimeric structure of a Reformatsky intermediate.

Dibromoacetic acid has been prepared by addition of bromine to acetic acid in the presence of sulfur [34] and by oxidation of tribromoethylene with peracetic acid [35].

Dibromoacetate esters can be alkylated with organoboranes. As shown in Eq. (7), an organoborane is generated from diborane and an unsaturated hydrocarbon such as cyclohexene. Treatment of the resulting organoborane with ethyl dibromoacetate followed by t-butoxide in t-butanol affords the alkylated alpha-bromoester [36].

$$\text{cyclohexene} \xrightarrow{B_2H_6} (C_6H_{11})_3\text{-B} \xrightarrow[\text{2) KOBu-t}]{\text{1) Br}_2\text{CHCO}_2\text{Et}} \text{cyclohexyl-CHCO}_2\text{Et} \quad (7)$$
$$\text{Br}$$

Tribromoacetic acid has been prepared by the same methods that have been used to generate the trichloro derivatives. Thus, oxidation of either bromal or perbromoethylene with fuming nitric acid gives the tribromo acid [37,38]. Bromination of malonic acid in aqueous solution also yields tribromoacetic acid [39].

The acid decomposes in boiling water to bromoform. The easy decomposition of the acid is a property that has allowed this derivative to be used as a polymerization catalyst and brominating agent [40].

D. Alpha-iodinated Derivatives

Iodoacetic acid generally is prepared by treating the chloro or bromo derivative with sodium or potassium iodide in acetone [41]. The acid also has been prepared by iodination of acetic anhydride in the presence of sulfuric or nitric acids [42]. Recently a method has been developed for direct iodination of acetic acid using an iodine-copper salt [43]. Iodoacetic acid reacts with hydrogen iodide at 85°C to form acetic acid and iodine [44].

Alpha-iodo ester derivatives can be prepared by iodination of ester enolates prepared using lithium N-isopropylcyclohexylamide as shown in Eq. (8) [45].

$$R^2CH_2CO_2R^1 + LiN\text{(cyclohexyl)(Pr-i)} \xrightarrow[-80°C]{THF} R^2CHCO_2R^1 \xrightarrow{I_2} R^2CHCO_2R^1 + LiI \quad (8)$$
$$\qquad\qquad Li \qquad\qquad\qquad I$$

Iodoacetates are frequently used in biochemistry to selectively derivatize thiol groups in proteins. This reaction inactivates certain enzymic proteins and often serves as an analytical tool in structural or mechanistic studies [46,47]. Iodoacetic acid has been reported as an ingredient in a herbicidal formulation and is claimed to act in synergism with other components to provide weed control and desiccation of potatoes, rape, and seed legumes [48].

Diiodoacetic acid has been prepared from diiodomaleic acid and water [49] and from malonic acid and iodic acid in water [50]. Triiodoacetic acid has also been prepared from malonic acid and iodic acid in water. Triiodoacetic acid is unstable at elevated temperature and decomposes to iodine, iodoform, and carbon dioxide [51].

III. ACETYL HALIDES

Another class of halogenated acetic acid derivatives are the acetyl halides, in which the acidic hydrogen has been replaced by halogen. These compounds represent one of the most powerful classes of acylating reagents in organic synthesis. Acetyl halides prepared from acetic acid or anhydride are common reagents for preparation of acetates, acetamides, and many other acetyl derivatives. The most well-known and widely used of the acetyl halides is acetyl chloride, but acetyl fluoride, bromide, and iodide also are easily prepared.

A. Acetyl Fluoride

Acetyl fluoride has been prepared by the reaction of acetic anhydride with fluorosulfonic acid [52] or hydrogen fluoride [53]. In each case an acid by-product is formed. Addition of hydrofluoric acid to ketene gives acetyl fluoride without acid by-product formation [54].

Acetyl fluoride also can be prepared by fluoride ion exchange between acetyl chloride and hydrogen fluoride [53] or tetraethyl-ammonium fluoride in acetic acid. Kinetic and equilibrium studies of the latter reaction have shown that it involves rapid formation of acetic anhydride and hydrogen fluoride, which react slowly to give acetyl fluoride [55].

B. Acetyl Chloride

Preparation of acetyl chloride is easily accomplished by reaction of acetic acid or anhydride with a variety of inorganic and organic chlorides. One of the principal reagents for this reaction is thionyl chloride. The reaction of thionyl chloride with carboxylic acids is frequently catalyzed by organic nitrogen-containing bases or inorganic compounds. Among the frequently used catalysts are pyridine, N,N-dimethylformamide, hexamethylphosphoramide, and iodine. Also, alkali metal chlorides have been shown to enhance the rate of reaction of thionyl chloride with halogenated acetic acid derivatives. Trichloroacetic acid reacts very slowly in refluxing thionyl chloride, remaining essentially unreacted after 12 h [56]. However, the addition of 2 g of potassium chloride per mole of thionyl chloride permitted the acid chloride to be obtained in 87% yield after 7 h at reflux [57]. A similar rate enhancement is obtained by using DMF as the catalyst [59]. Addition of catalyst affords only a small rate increase in the conversion of monochloroacetic acid to the acid chloride [57].

Kinetic measurements on the reaction of thionyl chloride with carboxylic acids indicate the reaction is slower with strong acids and sterically hindered acids. The reaction was shown to be second order overall and first order in each reactant. Table 3 shows the relative rates of reaction of a series of halogenated acetic acid derivatives [59].

Another common reagent for conversion of carboxylic acids to acid chlorides, especially on the industrial scale, is phosgene. Production of highly pure acetyl chloride has been claimed in the reaction of acetic acid and phosgene when carried out in an aromatic hydrocarbon solvent using aliphatic amide and metal catalysts [60]. Other traditional reagents useful for preparation of acetyl chloride are PCl_3, PCl_5, $POCl_3$, and oxalyl chloride.

A new manufacturing process for acetyl chloride from PCl_3 and acetic acid produces two phases, which after separation and distillation affords acetyl chloride with high yield and selectivity. The distillation step is carried out with a countercurrent flow of HCl gas [61].

Sulfur monochloride, S_2Cl_2, will react with carboxylic acids in the presence of $Fe(OAc)_3$ or other iron catalyst to form acid chlorides [62]. Also, acetyl chloride can be obtained by direct reaction of 2 mol of acetic acid with 1 mol of sulfur and 2 mol of chlorine over ferric acetate catalyst [63].

Bromine chloride, BrCl, when added in an equimolar quantity to boiling acetic acid gave an 82% yield of acetyl chloride, which was recovered by distillation from the reaction mixture. After cooling and hydrolysis of the distillation residue, a 95% yield of bromoacetic acid was obtained [64].

A recently developed method for synthesis of acetyl chloride involves carbonylation of methyl chloride [65–68]. Early patents describe a process

Table 3 Relative Reaction Rates of Halogenated Acetic Acids with Thionyl Chloride

CH_3CO_2H	ICH_2CO_2H	$BrCH_2CO_2H$	$ClCH_2CO_2H$	Cl_3CCO_2H
1.0	0.30	0.25	0.14	No reaction

in which methyl chloride reacts with carbon monoxide in a heptane solvent under pressure and at elevated temperature. The reaction employs a catalyst mixture that contains rhodium and chromium chlorides, methyl iodide, triphenyl phosphine oxide, and methyltriphenylphosphonium iodide [65,66]. An attractive feature of the process is that the catalyst system forms a separate phase, which can be separated from the product and reused.

An alternative process that avoids organophosphorous compounds has also been patented [67]. The reaction mixture in this case includes acetic acid and lithium acetate in addition to rhodium and chromium chlorides and methyl iodide. The carbonylation is performed with a gas mixture containing carbon monoxide and hydrogen in a 10:1 mole ratio. Acetyl chloride yields of between 47% and 56% have been obtained using these carbonylation methods. Cobalt also has been reported as a catalyst for the carbonylation of methyl chloride [68]. This catalyst system includes acetic acid, methyl iodide, and triphenylphosphine.

C. Acetyl Bromide

In analogy to the reactions of inorganic chlorides, brominating reagents such as thionyl bromide [69] and bromine/phosphorous tribromide [70] can be used for preparation of acetyl bromide. A synthesis of carbon 14–labeled acetyl bromide was accomplished with a 94% yield using bromine exchange between [^{14}C] acetic acid and benzoyl bromide [71]. Acetyl bromide can also be prepared by the exchange of halogen between acetyl chloride and hydrogen bromide [22].

Acetyl bromide could be used in many of the synthetic reactions that employ acetyl chloride, but acetyl chloride is more common because of the wide availability of the required chlorination reagents. Acetyl bromide has found use in a process for recovery of cobalt and manganese from spent catalysts used in air oxidation of alkyl aromatics [72]. Acetyl bromide in acetic acid solubilizes cobalt or manganese oxalate by-products that precipitate from oxidation reaction mixtures. In this way the catalyst metals can be returned to the reaction. Acetyl bromide has also proved useful as an analytical reagent for the determination of lignin in herbaceous plants [73].

D. Acetyl Iodide

Halogen exchange is useful for preparation of acetyl iodide as well as acetyl bromide and fluoride. Thus acetyl iodide can be made by reaction of acetyl chloride with hydrogen iodide [74] and with sodium iodide [75,76]. Acetyl iodide has also been prepared by the reaction of trimethylsilyl iodide with acetyl chloride [77]. Diiodosilane (SiH_2I_2) is a new reagent for the conversion of acids, anhydrides, esters, and acyl chlorides to acyl iodides [78]. The reactions are significantly accelerated in the presence of iodine. Anhydrides react in the presence of iodine with one equivalent of diiodosilane to give two equivalents of acyl iodide.

Acetyl iodide can be produced by carbonylation of methyl iodide over a rhodium catalyst [79]. This chemistry is similar to the carbonylation of methyl chloride to acetyl chloride, but in this case the acetyl iodide product is typically an intermediate in a carbonylation process to produce acetic acid or anhydride. The involvement of acetyl iodide in the carbonylation of methyl acetate and methanol is discussed more fully in Chapter 9.

E. Haloacetyl Halides

Haloacetyl halides are an important class of halogenated acetic acid derivatives that find wide use in the synthesis of herbicides. Chloroacetyl chloride is used in large volume for synthesis of the preemergence herbicides alachlor and butachlor. Pesticide requirements for this intermediate have been estimated to be approximately 100 million pounds annually. Smaller, but still substantial, volumes of chloroacetyl chloride go into pharmaceutical manufacture and the production of chloroacetophenone, a tear gas.

Alachlor is produced from methanol, formaldehyde, chloroacetyl chloride, and 2,6-diethylaniline. Butachlor is produced by the same synthesis with substitution of butanol for methanol. The structure of alachlor and butachlor is shown in Figure 3.

Alachlor is used mainly for corn, soybeans, cotton, peanuts, and sorghum. Butachlor is used almost exclusively for rice. Both herbicides

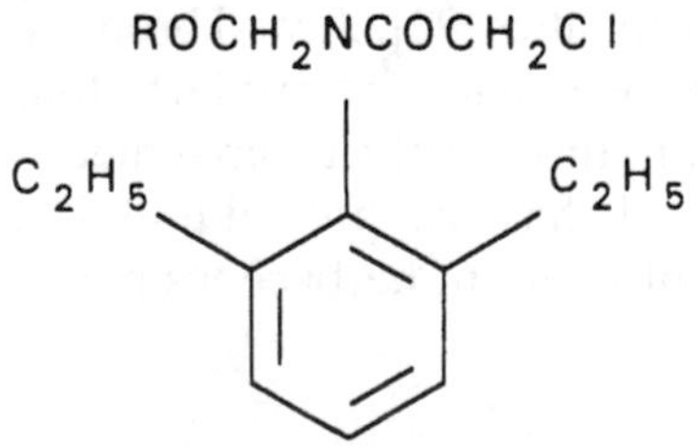

Alachlor R=CH_3
Butachlor R=$CH_3CH_2CH_2CH_2$

Figure 3 Structure of alachlor and butachlor.

were developed by Monsanto and are sold under the trade names Lasso (alachlor) and Machete (butachlor).

Several patented routes to chloroacetyl chloride are shown in Eq. (9)–(14). As indicated, chloroacetyl chloride can be produced from chloroacetic acid by phosgenation [80] or reaction with thionyl chloride [81]. A catalyst such as dimethylformamide or a substituted urea normally is used to promote the reactions. Alternate routes to this compound are chlorination of acetyl chloride [82] and acetic acid [83]. Although direct chlorination of acetic acid appears attractive, the synthesis has low yields. Two of the more exotic syntheses of this compound are carbonylation of methylene chloride [84] and chlorination of ketene [85]. Preparation of chloroacetyl chloride from chloroacetic acid appears to be the industrially preferred route because of the low cost and wide availability of this starting material.

Other halogenated acetyl chlorides have been made by iodination or bromination of acetyl chloride [86]. Bromination has been accomplished with N-bromosuccinimide in thionyl chloride. Iodination has been performed with iodine in thionyl chloride. Thionyl chloride is a necessary ingredient for successful iodination because it scavenges hydrogen iodide from the reaction mixture by reacting to form iodine and sulfur.

Bromoacetyl bromide has been prepared by heating acetic acid with white phosphorus and bromine at 20–70°C to prepare acetyl bromide and then adding a supplemental amount of bromine with further heating to 100°C to give bromoacetyl bromide in 94% yield [87].

The kinetics for the bromination of acetyl bromide and acetyl chloride have been reported by Watson [88]. The rate constants determined in this work indicate that the bromination of acetyl bromide proceeds at about three times the rate of bromination of acetyl chloride. Other kinetic experiments on bromination of acetyl bromide in nitrobenzene over a range of temperatures provided thermodynamic data, including an energy of activation of 20.3 kcal/mol [89]. The iodinated and brominated acetyl chlorides and acetyl bromides are not of commercial significance.

Synthetic Routes to Chloroacetyl Chloride

1. Phosgenation of chloroacetic acid:

$$ClCH_2COOH + COCl_2 \xrightarrow{DMF} ClCH_2COCl + HCl + CO_2 \qquad (9)$$

2. Reaction of thionyl chloride and chloroacetic acid:

$$ClCH_2COOH + SOCl_2 \longrightarrow ClCH_2COCl + HCl + SO_2 \qquad (10)$$

3. Chlorination of acetyl chloride:

$$CH_3COCl + Cl_2 \xrightarrow{H_2SO_4} ClCH_2COCl + HCl \qquad (11)$$

4. Chlorination of acetic acid:

$$4CH_3COOH + 7Cl_2 + S_2Cl_2 \xrightarrow{FeCl_3} 4ClCH_2COCl + 8HCl + 2SO_2$$

(12)

5. Carbonylation of methylene chloride:

$$CH_2Cl_2 + CO \xrightarrow[77\ atm]{PdCl_2} ClCH_2COCl$$

(13)

6. Chlorination of ketene:

$$CH_2{=}C{=}O + Cl_2 \longrightarrow ClCH_2COCl$$

(14)

REFERENCES

1. Freiter, E. R., in *Kirk-Othmer Encyclopedia of Chemical Technology*, Vol. 1, 3rd ed., Wiley Interscience, New York, 1978, pp. 171–178.
2. Cox, A., in *Comprehensive Organic Chemistry*, Vol. 2 (I. O. Sutherland, ed.), Pergamon Press, Oxford, 1979, pp. 719–737.
3. *The Merck Index*, 11th ed. (S. Budavari, ed.), Merck and Co., Rahway, N.J., 1989.
4. Clark, J. H., and Emsley, J., *J. Chem. Soc., Dalton Trans.*, *20*, 2129 (1975).
5. Johnson, W. S., *Acct. Chem. Res.*, *1*, 1 (1968).
6. Peterson, P. E., and Casey, C., *J. Org. Chem.*, *29*, 2325 (1964).
7. Bobst, A., and Viscontini, M., *Helv. Chim. Acta*, *49*, 875 (1966).
8. Kaplan, L., *J. Org. Chem.*, *50*, 5376 (1985).
9. Whitmore, F. C., and Whitmore, M. G., *Org. Syn., Coll.*, *1*, 401 (1941).
10. Newman, M. S., and Magerlein, B. J., *Org. Reactions*, *5*, 413 (1949).
11. *The Merck Index*, 11th ed. (S. Budavari, ed.), Merck and Co., Rahway, N.J., 1989, pp. 326, 481.
12. Momotari, Y., Jpn. Kokai 74,109,525 (Oct. 18, 1974) (*Chem. Abstr.*, *82*, 165885y).
13. Johnson, A., Dalgliesh, C., and Walker, J., in *Chemistry of Carbon Compounds*, Vol. 1 Part A (E. H. Rodd, ed.), Elsevier, Amsterdam, 1951, p. 620.
14. Stacpoole, P., and Bodor, N., U.S. Patent 4,801,597 (1989).
15. Otsuka Pharm., Japanese Patent 1,242,523 (1989).
16. Parkes, G. D., and Hollingshead, R. G. W., *Chem. & Ind. (Lond.)*, 222, issue no. 8, February 20, (1954) (*Chem. Abstr.*, *48*, 6377d).
17. Yasnitskii, B., Dolberg, E., and Kovelenka, G., *Metody Poluch. Khim. Reakt. Prep.*, *21*, 106 (1970) (*Chem. Abstr.*, *76*, 85321x).
18. van Messel, G., Neth. Patent 109,768 (Oct. 15, 1964) (*Chem. Abstr.*, *62*, 7643e).
19. *The Merck Index*, 11th ed. (S. Budavari, ed.), Merck and Co., Rahway, N.J., 1989, p. 1515.

20. Wagner, W. M., *Proc. Chem. Soc.*, 229, August, (1959).
21. Bailey, G., and White, J., *Residue Rev.*, *10*, 97 (1965).
22. Lake, D. E., and Asadorian, A. A., U.S. Patent 2,553,518 (1951) (*Chem. Abstr.*, *46*, 2561).
23. Asadorian, A. A., and Burk, G. A., U.S. Patent 3,130,222 (1964).
24. Jenker, H., and Karsten, R., Ger. Offen. 2,151,565 (April 19, 1973) (*Chem. Abstr.*, *79*, 18121f).
25. Johnston, J. D., U.S. Patent 2,876,255 (1959) (*Chem. Abstr.*, *53*, 12157g).
26. Natelson, S., and Gottfried, G., *Org. Syn.*, *Coll. Vol. 3*, 381 (1955).
27. Gidez, L. I., and Karnovsky, M. L., *J. Am. Chem. Soc.*, *74*, 2413 (1952).
28. Hell, C., and Muhlhauser, O., *Ber. Deutsch. Chem. Ges.*, *11*, 241 (1878).
29. Michael, A., *Am. Chem. J.*, *5*, 202 (1883).
30. Kachler, J., *Monatschr. Chem.*, *2*, 559 (1881).
31. Gloeckner, W., *Justus Liebigs Ann. Chem. Suppl.*, *7*, 115 (1870).
32. Dekker, J., Boersma, J., and van der Kerk, G., *J. Chem. Soc. Chem. Commun.*, 553, issue no. 10, (1983).
33. Hauser, C. R., and Breslow, D. S., *Org. Syn.*, *Coll. Vol. 3*, 408 (1955).
34. Genvresse, M. P., *Bull. Soc. Chim. Fr.*, *7*, 365 (1892).
35. Shakkhnazaryan, G. M., *Arm. Khim. Zh.*, *27*, 177 (1974) (*Chem. Abstr.*, *81*, 49222k).
36. Brown, H. C., and Rogic, M. M., *J. Am. Chem. Soc.*, *91*, 2146 (1969).
37. Shaffer, L., *Ber. Deutsch. Chem. Ges.*, *4*, 366 (1871).
38. Neff, J. V., *Justus Liebigs Ann. Chem.*, *308*, 264 (1899).
39. Petrieff, W., *Ber. Deutsch. Chem. Ges.*, *8*, 730 (1875).
40. Szczepek, W. J., *Polish J. Chem.*, *55*, 709 (1981).
41. Prochazka, E., Czech. Patent 152,947 (April 15, 1974) (*Chem. Abstr.*, *81*, 135465y).
42. Novikou, A. N., and Siyanko, P. I., U.S.S.R. Patent 213,014 (March 12, 1968) (*Chem. Abstr.*, *69*, 51528k).
43. Horiuchi, C. A., and Satoh, J. Y., *Chem. Lett.*, 1509, issue no. 9, September, (1984).
44. Ichikawa, K., and Miura, E., *J. Chem. Soc. Jpn.*, *74*, 798 (1953).
45. Rathke, M. W., and Lindert, A., *Tetrahedron Lett.*, 3995, issue no. 43, October, (1971).
46. Goswami, A., and Rosenberg, I., *Biochem. Int.*, *19*, 361 (1989).
47. Batra, P., Sasa, K., Ueki, T., and Takeda, K., *Int. J. Biochem.*, *21*, 857 (1989).
48. Bergmann, H., Burth, E., Kochmann, W., Kramer, W., Lyr, H., Radzuhn, B., Siering, G., and Steinke, W., German (DDR) Patent 221,058 (1985).
49. Clarke, L., and Bolton, E. K., *J. Am. Chem. Soc.*, *36*, 1899 (1914).
50. Fairclough, R. A., *J. Chem. Soc.*, 1186 (1938).
51. Cobb, R. L., *J. Org. Chem.*, *23*, 1368 (1958).
52. Heyboer, J., and Staverman, A. J., *Rec. Trav. Chim.*, *69*, 787 (1950).
53. Olah, G. A., and Kuhn, J. S., *J. Org. Chem.*, *26*, 237 (1961).
54. Chick, F., and Wilsmore, N. T. M., *J. Chem. Soc.*, *24*, 77 (1908).

55. Emsley, J., Gold, V., Hibbert, F., and Szeto, A., *J. Chem. Soc., Perkin Trans.*, *2*, 923 (1988).
56. Gerrard, W., and Thrush, A. M., *J. Chem. Soc.*, 2117 (1953).
57. Ansell, M. F., in *The Chemistry of Acyl Halides* (S. Patai, ed.), Interscience, London, 1972, p. 38.
58. Bosshard, H.M., Moray, R., Schmid, M., and Zollinger, H., *Helv. Chim. Acta*, *62*, 1633 (1959).
59. Beg, M. A., and Singh, H. N., *Z. Phys. Chem.*, *237*, 129 (1968).
60. Mitsui Toatsu Chem., Japanese Patent 56,103,134 (1981).
61. Damjan, J., Benczik, J., Kolonics, Z., Pelyva, J., Laborczy, R., Szabolcs, J., Soptei, C., Barcza, I., and Kayos, C., British Patent 2,213,144 (1989).
62. Takada, Y., Matsuda, T., and Inoue, G., Japanese Patent 6,812,123 (1968).
63. Matsuda, T., Yokota, K., and Takata, Y., *Hokkaido Daigaku Kogakubu Kenkyu Hokoku*, 87–90, issue no. 95, (1979) (*Chem. Abstr.*, *92*, 76059w).
64. Swietoslawski, J., and Ratajczak, A., *Org. Prep. Proced. Int.*, *11*, 253 (1979) (*Chem. Abstr.*, *92*, 58196j).
65. Erpenbach, H., Gehrmann, K., Lork, W., and Prinz, P., German Patent 3,016,900 (1981) (*Chem. Abstr.*, *96*, 19680a).
66. Erpenbach, H., Gehrmann, K., Lork, W., and Prinz, P., U.S. Patent 4,352,761 (1982).
67. Erpenbach, H., Gehrmann, K., Lork, W., and Prinz, P., U.S. Patent 4,543,217 (1985).
68. Kent, A. G., British Patent Application 2,133,792 (1984) (*Chem. Abstr.*, *102*, 24120q).
69. Saraf, S. D., and Zakai, M., *Synthesis*, 612, issue no. 10, October, (1973).
70. Chemische Fabrik Kalk G.m.b.H., French Patent 1,556,480 (1969) [*Chem. Abstr.*, *72*, 42844v (1970)].
71. Heidelberger, C., and Hurlbert, R., *J. Am. Chem. Soc.*, *72*, 4704 (1950).
72. Feld, M., and Zoche, G., U.S. Patent 4,490,297 (1984).
73. Iiyama, K., and Wallis, A., *J. Sci. Food Agric.*, *51*, 145 (1990).
74. Gustus, E., and Stevens, P., *J. Am. Chem. Soc.*, *55*, 374 (1933).
75. Blum, J., Rosenman, H., and Bergman, E., *J. Org. Chem.*, *33*, 1928 (1968).
76. Hoffmann, H., and Haase, K., *Synthesis*, 715, issue no. 9, September, (1981).
77. Schmidt, A., Russ, M., and Grosse, D., *Synthesis*, 216, issue no. 3, March, (1981).
78. Keinan, E., and Sahai, M., *J. Org. Chem.*, *55*, 3922 (1990).
79. Hewlett, C., U.S. Patent 4,698,187 (1987).
80. Hertel, O., British Patent 1,361,018 (1974).
81. Heydkamp, W., German Patent 2,943,432 (1981).
82. Bressel, U., U.S. Patent 3,880,923 (1975).
83. Imperial Chemical Industries, British Patent 1,125,772 (1968).
84. Mador, I. L., U.S. Patent 3,454,632 (1969).
85. Gash, V. W., U.S. Patent 3,812,183 (1974).
86. Harpp, D. N., Bao, L. Q., Black, C. J., Gleason, J. G., and Smith, R. A., *J. Org. Chem.*, *40*, 3420 (1975).

87. Chemische Fabrik Kalk G.m.b.H., French Patent 1,556,481 (1969) [*Chem. Abstr.*, *72*, 42845w (1970)].
88. Watson, H. B., *J. Chem. Soc.*, 1137 (1928).
89. Cicero, C., and Matthews, D., *J. Phys. Chem.*, *68*, 469 (1964).
90. *Handbook of Chemistry and Physics*, 71st ed. (D. R. Lide, ed.), CRC Press, Boca Raton, FL, 1990, pp. 8–35.
91. Ives, D. J. G., and Pryor, J. H., *J. Chem. Soc.*, 2108, 2110 (1955).
92. Henne, A. L., and Fox, C. J., *J. Am. Chem. Soc.*, *73*, 2323 (1951).
93. Swarts, F., *Chem. Zentralbl.*, *II*, 709 (1903).
94. Kunz, J., and Farrer, J., *J. Am. Chem. Soc.*, *91*, 6057 (1969).
95. *Langes Handbook of Chemistry*, 13th ed. (J. A. Dean, ed.), McGraw-Hill, New York, 1985, pp. 5–58.
96. Jasper, J. J., and Grodzka, P. G., *J. Am. Chem. Soc.*, *76*, 1453 (1954).
97. Jasper, J. J., and Miller, G. B., *J. Phys. Chem.*, *59*, 441 (1955).
98. Yarovenko, N. N., Raksha, M. A., Shemanina, V. N., and Vasileva, A. S., *Zhur. Obshch. Khim.*, 2246, 2248 (1957).
99. Lundin, R. E., Harris, F. E., and Nash, L. K., *J. Am. Chem. Soc.*, *74*, 4654 (1952).
100. Kauck, E. A., and Diesslin, A. R., *Ind. Eng. Chem.*, *43*, 2332 (1951) (*Chem. Abstr.*, *46*, 4478).
101. Perkin, W. H., *J. Chem. Soc.*, *65*, 421 (1894).
102. Sumarokova, T. N., and Khakhlova, N. V., *Zhur. Obshch. Khim.*, *26*, 2690–2693 (1956) (*Chem. Abstr.*, *51*, 6302).

16

Nitrogen Derivatives

Frank Cooke

Eastman Chemical Company, Kingsport, Tennessee

I. INTRODUCTION

In this chapter three major nitrogen derivatives of acetic acid will be discussed:

Acetamide
N,N-Dimethylacetamide
Acetonitrile

These major industrial chemicals can be considered derivatives of acetic acid as they all can be prepared directly from it and retain the acetyl group as part of the final molecule. However, their industrial production may not involve the use of acetic acid.

Acetic acid has been used in the preparation of a number of other nitrogen-containing compounds, in particular, heterocyclic systems where acetic acid is acting as a condensant. Several examples of this and its use as an acylating and alkylating agent will be given.

II. ACETAMIDE

Acetamide (CAS Registry Number 60-35-5) has been used in its molten state as a solvent and is excellent in this regard for many organic and inorganic compounds. It has been used as a cosolvent in aqueous systems;

a plasticizer for leather, cloth, films, and coatings; a stabilizer for peroxides; a flux ingredient in solder; an antacid because of its amphoteric nature; a starting material in the manufacture of methylamine; and as a denaturing agent for alcohol. It is triboluminescent.

Suppliers of acetamide include Whittaker Corp., Henley Manufacturing, Chemical Dynamics Corp., Hoechst Celanese Corp., and Wall Chemicals Corp.

A. Physical Properties

Molecular formula C_2H_5NO; molecular weight 59.07; melting point 82°C [1]; boiling point 221.2°C [2]; d_4^{80} 0.9985 [1]; n_D^{13} 1.4079 [3]; dielectric constant 65.3 (at 80°C) [4].

Acetamide is compatible with ketones, esters, organic acids, alcohols, water, ammonium salts, and chloroform. It is practically insoluble in ether and a number of saturated hydrocarbons but soluble in hot benzene. No odor is associated with acetamide when in a pure state. However, on standing it develops a characteristic mouselike odor, the origin of which has not been determined.

B. Preparations

1. From Acetic Acid

A number of papers and patents describe the preparation of acetamide from an acetic acid–type precursor. These references fall into two main categories: those that use an acetate and those that use acetic acid or its ammonium salt.

Ethyl or methyl acetate may be treated with aqueous ammonia [5] according to the following reaction, to give almost quantitative yields of the desired product.

$$CH_3COOC_2H_5 + NH_3 \xrightarrow{H_2O} CH_3CONH_2 + C_2H_5OH \qquad (1)$$

Acetamide can be prepared by heating ammonium acetate and collecting the product as the distillate [6,7]. This method is often used for small-scale captive needs.

$$CH_3COONH_4 \xrightarrow{\Delta} CH_3CONH_2 + H_2O \qquad (2)$$

An adaptation of the above involves passing ammonia through acetic acid and continuously removing the water to drive the reaction to completion [8].

Acetamide can also be prepared using other amides or ureas, at around 220°C, according to the following [9].

$$R^1CO_2H + R^2CONH_2 \longrightarrow R^2CO_2H + R^1CONH_2 \tag{3}$$

$$R^1CO_2H + H_2NCONH_2 \longrightarrow R^1CONH_2 + CO_2 + NH_3 \tag{4}$$

Many accounts of the kinetics of the formation of acetamide by the distillation of ammonium acetate have been written from as early as 1884 [10,11].

2. From Materials Other than Acetic Acid

Over the past century a plethora of variations for the hydrolysis of nitriles has been employed to prepare amides. This constitutes a major route to acetamide for large-scale needs using by-product acetonitrile from the Sohio acrylonitrile process. A great number of these variations, in particular those in patents, deal with the hydrolysis of acrylonitrile to acrylamide, a monomer of much importance. A number of catalytic systems for this hydrolysis have employed copper, in a variety of oxidation states, and copper activated with other metals, in particular chromium [12–16]. Other transition metals and their complexes have been used, for example, Pt, Pd, Ni [17–21], Rh [22], and Mn [23–25]. Two patents describe the use of microorganisms to effect the hydrolysis [26,27]. Other methods include the use of NH_3 or alcohol promoters [28], ammonium or tetraalkylammonium salts [29,30], zeolites [31], solid cation exchange resins [32], basic alcoholic media [33,34], KF on alumina [35], and via the conversion of acetonitrile as a ligand on an osmium cluster surface [36].

$$CH_3CN \longrightarrow CH_3CONH_2 \longrightarrow CH_3CO_2H \tag{5}$$

Aldoximes can be rearranged to the corresponding amide by using one of a number of metal catalysts at 50–250°C [37], cyanohydrin catalysts [38], or silica gel at pH 6.5–7.0 in anhydrous xylene [39].

$$\tag{6}$$

Ethylamine can be treated with ozone to give a number of products including acetamide [40].

$$C_2H_5NH_2 + O_3 \longrightarrow CH_3CONH_2 \qquad (7)$$

An intriguing patent application from Hitachi describes the fixation of gaseous nitrogen at atmospheric pressure according to the following reaction [41].

$$HCHO + KOH + N_2 \xrightarrow[20°C]{CuO} CH_3CONH_2 \qquad (8)$$

According to an ICI patent, acetamide has been made by contacting O_2 with a liquid phase containing a saturated ketone, a Cu(II) compound, and NH_3 [42]. In a somewhat similar fashion, 1,3-diketones have been treated with chloramine to yield amides and dichloroketones [43].

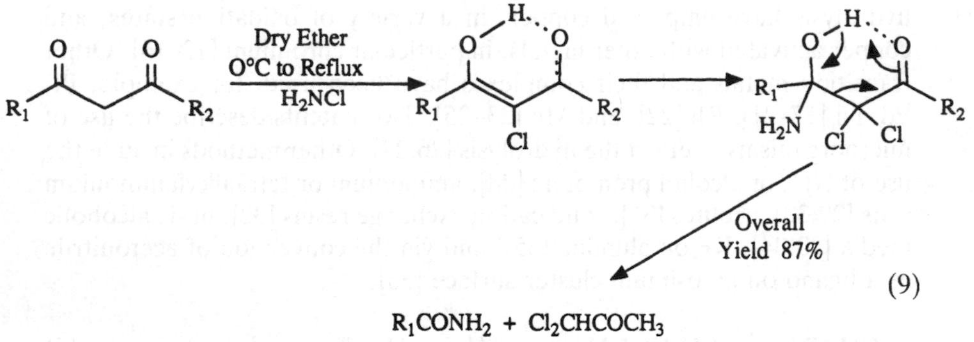

$$R_1CONH_2 + Cl_2CHCOCH_3 \qquad (9)$$

Acetamide is made in minor amounts by the carbonylation of amines using noble metal catalysts [44]. The reaction of ketene and ammonia has been reported to give acetamide [45].

$$H_2C{=}C{=}O + NH_3 \longrightarrow CH_3CONH_2 \qquad (10)$$

C. Acetamide Reactions

Acetamide has been used to prepare oxazoles [46] and thiazoles via conversion to the intermediate thioacetamide [47].

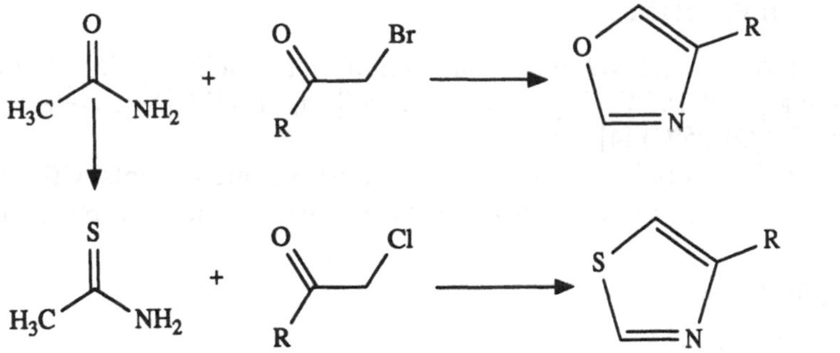

(11)

Acetamide can be used as an amine acylating agent [48].

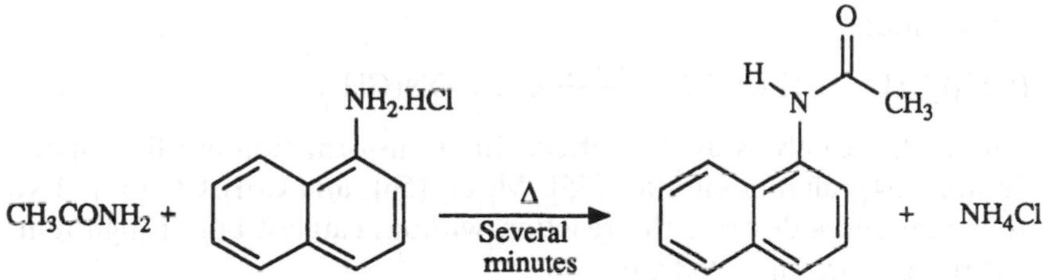

(12)

Work done at Texaco Development Corporation describes the preparation of *N*-acyl-∝-amino acids [49].

$$(CH_2O)_x + CH_3CONH_2 + CO/H_2 \xrightarrow[\substack{120^\circ C \\ 2\ h}]{Co(CO)_8 - Ph_2SO} \begin{matrix} COOH \\ | \\ CH_2 \\ | \\ NHCOCH_3 \end{matrix} + \begin{matrix} NHCOCH_3 \\ | \\ CH_2 \\ | \\ NHCOCH_3 \end{matrix}$$

$$\downarrow \text{Hydrolysis}$$

Glycine

(13)

III. *N,N*-DIMETHYLACETAMIDE

N,N-Dimethylacetamide (DMAC) (CAS Registry Number 127-19-5) is a dipolar aprotic solvent used in many organic reactions and industrial applications. It has been used as a solvent for electrolysis reactions, resins, polymers, complexes, solvates, crystallization, and purification.

Major suppliers of DMAC include Ashland Chemical Co., BASF, Du Pont, and Monsanto.

A. Physical Properties

Molecular formula C_4H_9NO; molecular weight 87.12; melting point $-20°C$ [50]; boiling point 165°C [51]; d_4^{25} 0.9372 [52]; n_D^{20} 1.4373 [53]; dielectric constant 37.8 (at 25°C) [4].

DMAC is fully miscible with water and most organic solvents with the exception of some saturated hydrocarbons. DMAC is mildly hygroscopic.

B. Preparations

1. From Acetic Acid

A number of references describe the reaction of acetic acid with dimethylamine or trimethylamine to yield DMAC. This constitutes the major industrial route.

$$(CH_3)_2NH + CH_3COOH \xrightarrow{140-300°C} CH_3CON(CH_3)_2 \tag{14}$$

Among the catalysts used to effect this transformation are fluorinated γ-alumina [54], aluminosilicates [55], MoO_3 [56], and CoI_2/CO [57]. Two further references describe the reaction without catalyst but at high temperatures and pressures [58,59].

Also, acetic acid has been converted using N,N-dimethylsulfamide [60].

$$CH_3COOH + (CH_3)_2NSO_2NH_2 \xrightarrow[3\ h]{100°C} CH_3CON(CH_3)_2 \tag{15}$$

Various alkyl acetates have been reacted with dimethylamine to yield DMAC. Typically, an alcohol is employed as the solvent and is recycled with the slight excess of ester used [61].

$$CH_3CO_2R + (CH_3)_2NH \xrightarrow{50-150°C} CH_3CON(CH_3)_2 + ROH \tag{16}$$

A number of patents describe this reaction wherein a basic catalyst system is employed, for example, sodium methoxide [62], hydroxide in methanol [63], or alkali metal hydroxides in glycols [64]. Two other patents describe the reaction being promoted by either cation exchange resin [65] or salts of cobalt [66]. Two reviews, written in 1980, give details of the history and reaction conditions for laboratory and commercial preparations of DMAC [67,68].

2.   From Materials Other than Acetic Acid

DMAC has been prepared from trimethylamine and CO using cobalt octacarbonyl [69,70]. A similar reaction has been reported using Fe, Co, Ni, or Hg and I_2/H_2, although the major products were of the formamide type [71].

$$CO + (CH_3)_3N \xrightarrow{Co_2(CO)_8} CH_3CON(CH_3)_2 \tag{17}$$

Cobalt, in the form of $CoCl_2$, has been used to convert acetonitrile, in the presence of methanol, to DMAC [72]. Also, HCl has been used as the catalyst [73].

$$CH_3CN \xrightarrow[380°C]{CoCl_2/CH_3OH} CH_3CON(CH_3)_2 \tag{18}$$

Acetonitrile has been reacted with primary, secondary, and tertiary amines in the presence of water, with or without catalyst, at 360°C to give DMAC [74].

$$CH_3CN + H_2O + R^1R^2R^3N \longrightarrow CH_3CON(CH_3)_2 \tag{19}$$

Amines have been used in conjunction with ethanol and O_2 and a Pt or Pd catalyst [75]. Similarly, acetaldehyde can be used as the starting material [76].

$$\begin{matrix} C_2H_5OH \\ \text{or} \\ CH_3CHO \end{matrix} + O_2 + (CH_3)_2NH \xrightarrow{50°C} CH_3CON(CH_3)_2 \tag{20}$$

Acetamide, in the form of ammonium acetate, has been alkylated with methanol using ammonium halides as catalysts [77].

$$NH_4OAc + CH_3OH \longrightarrow CH_3CON(CH_3)_2 \tag{21}$$

Highly pure DMAC has been reported as the product of the reaction of ketene and dimethylamine [78].

$$H_2C{=}C{=}O + (CH_3)_2NH \xrightarrow[\text{High yields}]{150-160°C} CH_3CON(CH_3)_2 \tag{22}$$

C. DMAC Reactions

DMAC is widely used for its properties as a solvent and little has been done with it chemically. Rathke has described the α-lithiation of DMAC and shown the resultant lithiated species to be quite stable. The α-lithio material can be alkylated readily [79]. The corresponding sodium enolate has been generated and reacted with RX at 65°C in benzene, suggesting that this enolate is quite stable also.

$$CH_3CON(CH_3)_2 + \overset{+}{Li}\,\bar{N}(Pr_2) \xrightarrow[0° \text{ or } -78°C]{RX} RCH_2CON(CH_3)_2 \tag{23}$$

DMAC can be made to undergo a self-condensation using $POCl_3$ to give the corresponding acetoacetamide [80].

$$2CH_3CON(CH_3)_2 + POCl_3 \xrightarrow[C_6H_6]{80°C} \underset{H_3C}{\overset{CH_3}{\underset{\|\ \|}{\overset{|}{N}}}}\!\!\!CH_3 + (CH_3)_2NH \tag{24}$$

Amides are somewhat resistant to reduction, but this can be effected to give methylamine and ethanol or N,N-dimethylethylamine, depending on the choice of conditions, catalyst, and reducing agent.

$$CH_3CON(CH_3)_2 \longrightarrow C_2H_5N(CH_3)_2$$
$$\longrightarrow (CH_3)_2NH + EtOH \tag{25}$$

Amidines can be prepared by the reaction of an isocyanate [81].

$$\tag{26}$$

IV. ACETONITRILE

Acetonitrile (CAS Registry Number 75-05-8) has been used extensively as a water-miscible, polar aprotic reaction solvent, and its low boiling point aids in recovery of this industrial solvent. It has been used as a solvent for polymers and gases, in photolytic processes, and for the separation of butadiene from C4 streams. It has been used for extraction and refining of copper and as a stabilizer for chlorinated solvents.

The major suppliers are Du Pont and B. P. America in the United States and a number of others from Europe and Japan, including Rhone-Poulenc, Degussa, Mitsubishi, and Asahi.

A. Physical Properties

Molecular formula C_2H_3N; molecular weight 41.06; melting point $-45°C$ [82]; boiling point 80°C [83]; d_4^{20} 0.7868 [84]; n_D^{20} 1.3442 [85]; dielectric constant 38.8 (at 20°C) [86].

Under neutral conditions acetonitrile is compatible and miscible with water and most organic solvents.

B. Preparations

1. From Acetic Acid

Acetonitrile is readily prepared by the high-temperature reaction of acetic acid and ammonia according to the following:

$$CH_3COOH + NH_3 \longrightarrow CH_3COO^- \ NH_4^+ \tag{27}$$

$$\downarrow$$

$$CH_3CONH_2 + H_2O$$

$$\downarrow$$

$$CH_3CN + H_2O$$

This reaction is usually carried out at around 400–450°C using catalysts such as phosphoric acid on alumina [87], metal oxides such as bauxite and ferric oxide [88], silica gel [89], and alumina [90]. Japanese acid clay has been used as a catalyst in the preparation of acetonitrile from ethyl acetate and ammonia at 350–400°C [91,92].

2. From Materials Other than Acetic Acid

The major source of acetonitrile is not via a specific synthetic route, but as a by-product of the Sohio process for the manufacture of acrylonitrile, which involves reacting ammonia with propene. HCN is liberated as the other major by-product.

$$NH_3 + H_2C\overset{\displaystyle\diagup\!\!\diagdown}{}CH_3 \xrightarrow[\substack{5-30 \text{ psi} \\ 400-500°C}]{\text{Fluidized bed}} H_2C\overset{\displaystyle\diagup\!\!\diagdown}{}CN + CH_3CN + HCN \tag{28}$$

Preparations of acetonitrile from acetic acid proceed via formation of acetamide and subsequent dehydration. Acetamide has been dehydrated by a number of high-temperature techniques using a variety of catalysts including ammoxidation catalysts such as bismuth phosphomolybdate [93]. Chemical means employing reagents such as phosgene [94] and pyrosulfo compounds [95] have also been used.

Acetaldehyde and ammonia have been reacted in the presence of a chlorinated oxidant such as chlorine dioxide or chloramine at 0–40°C [96] or with a vanadium/oxygen/silicon catalyst at 350–500°C [97].

$$CH_3CHO + NH_3 + ClO_2 \longrightarrow CH_3CN \tag{29}$$

Ammonia reacts with formaldehyde and methanol over an aluminosilicate zeolite to give moderate yields of acetonitrile [98]. Paraldehyde and ammonia generate the desired product at 400–500°C over a ZnO on pumice catalyst [99].

$$HCHO + CH_3OH + NH_3 \longrightarrow CH_3CN \tag{30}$$

Copper catalysts [100] have been used for the reaction of ethanol with ammonia to give acetonitrile.

$$EtOH + NH_3 \xrightarrow{300-350°C} CH_3CN \tag{31}$$

Activated carbon, silica gel, Al_2O_3, and ThO_2 have been employed as catalysts for the reaction of monohydric and polyhydric primary, secondary, and tertiary alcohols in the vapor phase with HCN [101].

$$ROH + HCN \xrightarrow{Vapor\ phase} CH_3CN \tag{32}$$

Monsanto Company has been actively pursuing the preparation of acetonitrile from $CO/NH_3/H_2$ at elevated temperatures using a variety of catalysts [102–104]. The major by-products from this reaction are HCN and CO_2.

$$CO + H_2 + NH_3 \xrightarrow[\sim 500\ psi]{\sim 500°C} CH_3CN + HCN + CO_2 \tag{33}$$

The ammonolysis of a number of alkanes, including butane, propane, and ethane, has been employed to prepare acetonitrile. Most of this work has been done by Russian authors using a variety of transition metal catalysts, metal oxides, and zeolites with temperatures in the range 350–500°C [105–110].

$$NH_3 + Alkyl-CH_3 + air \longrightarrow CH_3CN \tag{34}$$

Similarly, alkenes have been used in the same type of reaction. Alumina [111] and iron nitride [112] have been shown to catalyze this process at temperatures similar to those above. Group VIII noble metal catalysts enable the use of much less severe conditions. Near-atmospheric pressure and temperatures in the range 90–175°C promote the reaction [113].

Nitrogen in the form of sodium nitrite has been reacted with alkenes at 105–115°C to give nitriles including acetonitrile [114].

Acetylene has been reacted with ammonia to prepare acetonitrile. Much of this work was done in the 1920s and early 1930s. Temperatures of 350–400°C and metal oxide catalysts such as ZnO, FeO, and bauxite were used [115–119].

$$HC\equiv CH + NH_3 \longrightarrow CH_3CN \tag{35}$$

Other methods of preparing acetonitrile include methylation of KCN with dimethyl sulfate [120], dehydration of monoethanolamine in the gas phase at 340–440°C over γ-alumina activated with Cr_2O_3 [121], and conversion of methylamine, dimethylamine, and trimethylamine to CH_3CN by heating with a transition metal catalyst, e.g., Mo, in the presence of H_2 [122].

$$KCN + (CH_3)_2SO_4 \xrightarrow{85°C} CH_3CN \tag{36}$$

$$HOCH_2CH_2NH_2 \xrightarrow{\Delta} CH_3CN \tag{37}$$

$$R^1R^2R^3N + H_2 \xrightarrow{500°C} CH_3CN \tag{38}$$

C. Acetonitrile Reactions

Acetonitrile has been used to synthesize a variety of pyridines and other nitrogen-containing heterocycles, some of which are shown below [123–127].

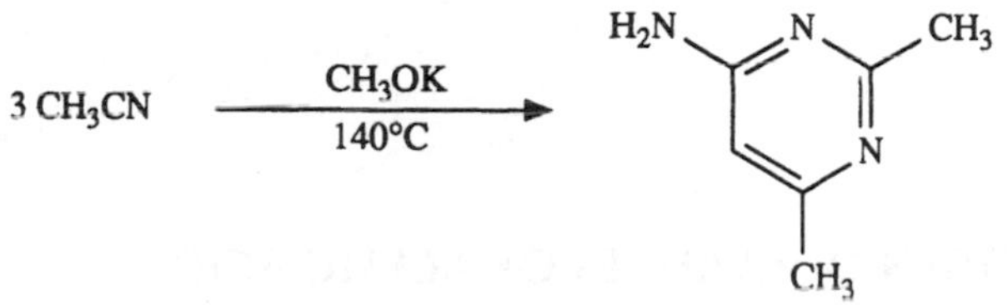

$$3\ CH_3CN \xrightarrow[140°C]{CH_3OK} \tag{39}$$

$$CH_3CN + 2\ R\text{-}C\equiv CH \xrightarrow[\substack{Co \\ catalysts}]{Various} \text{Isomers of trialkyl pyridines} \tag{40}$$

74% 2-methylpyridine for acetylene as reagent

$$\xrightarrow[150°C]{Cobaltocene} \text{60\% 2-methylpyridine for acetylene as reagent} \tag{41}$$

$$\xrightarrow[\substack{\geq 6\ atmos \\ 140°C}]{\substack{Co \\ catalysts}} \tag{42}$$

95% using acetylene

$$CH_3CN + NBu_4BF_4 + \xrightarrow[Co\ electrode]{e^-} \tag{43}$$

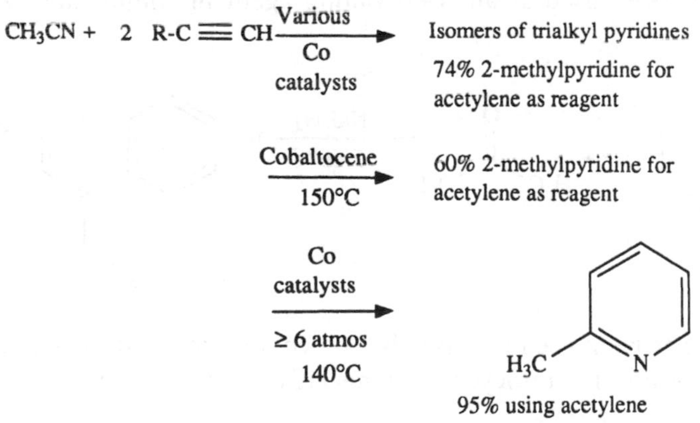

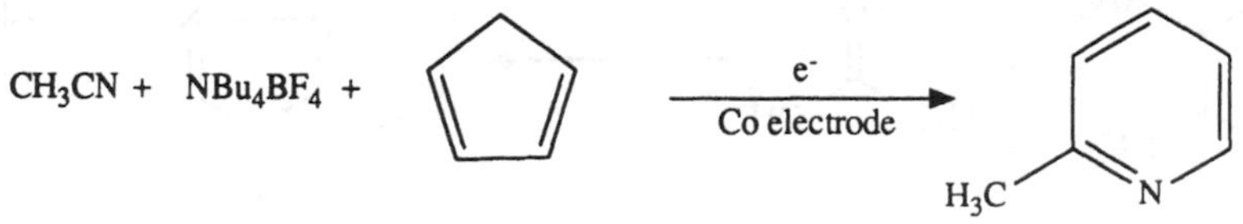

Some Russian workers have reacted acetonitrile with $NH_3/CO/H_2$ to give acrylonitrile [128].

$$CH_3CN + NH_3 + CO + H_2 \longrightarrow H_2C \diagup\diagdown CN \qquad (44)$$

Chemists at Lonza have converted acetonitrile into malononitrile in high-purity and moderate yields [129].

$$CH_3CN + ClCN \longrightarrow \overset{\displaystyle CN}{\underset{\displaystyle CN}{CH_2}} \qquad (45)$$

By-products $\longrightarrow$ HCN + Cl_2
of Sohio
acrylonitrile
process

V. OTHER NITROGEN DERIVATIVES OF ACETIC ACID

Acetic acid has been used as an *N*-ethylating agent in conjunction with $NaBH_4$ [130].

$$\qquad (46)$$

Acetic acid has been used extensively as a condensant and one-carbon unit for ring closure of heterocyclic systems [131,132]:

$$\qquad (47)$$

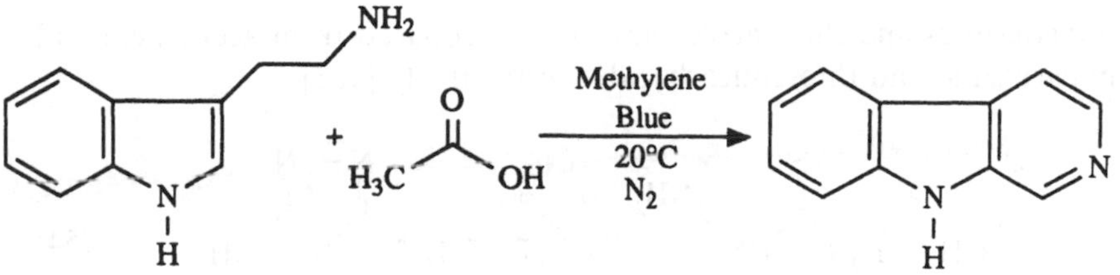

(48)

It can be used to prepare *N*-acylated compounds [133,134].

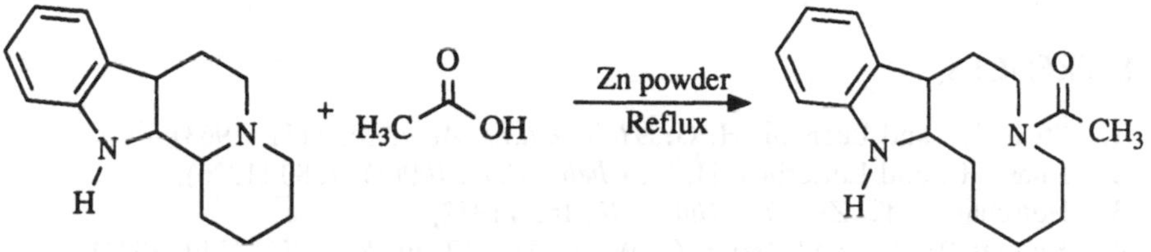

(49)

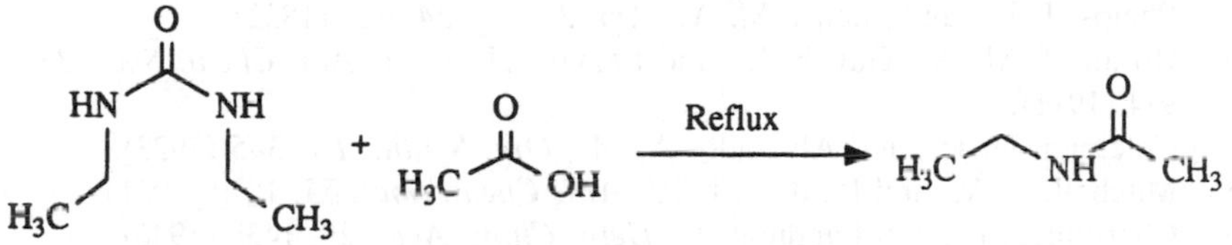

(50)

Substituted triazoles have been prepared using hydrazine as the nitrogen source [135].

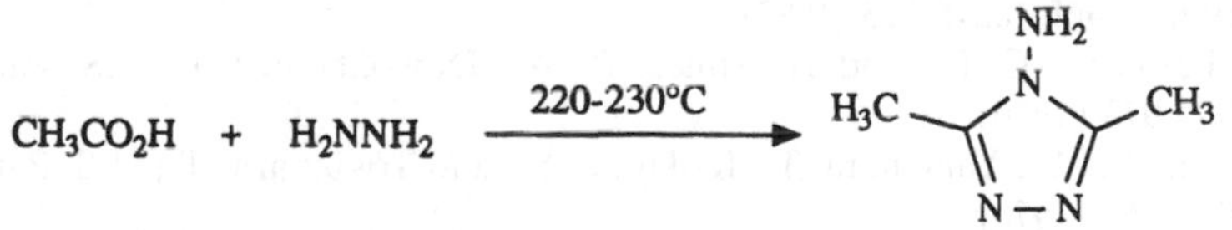

$$CH_3CO_2H + H_2NNH_2 \xrightarrow{220\text{-}230°C}$$

(51)

Tetrazoles have been converted to oxadiazoles [136].

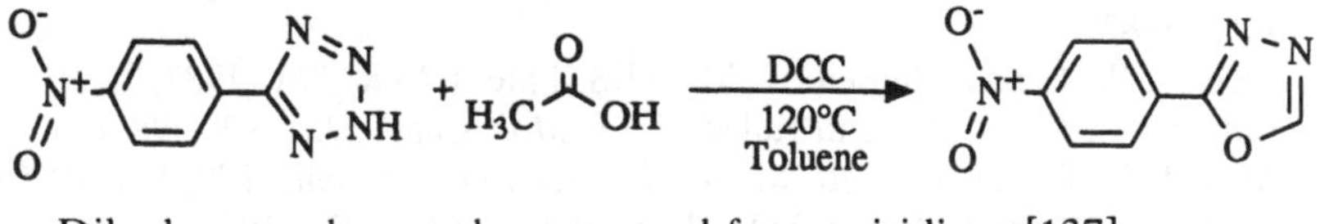

(52)

Dihydrooxazoles can be prepared from aziridines [137].

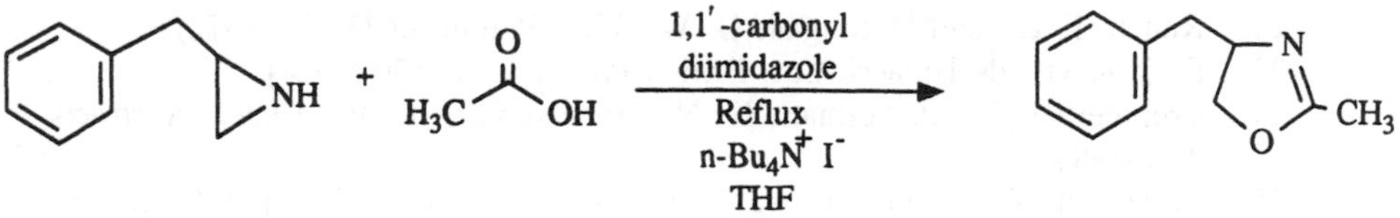

(53)

Oxadiazoles and thiadiazoles have been prepared from acetic acid and semicarbazide and thiosemicarbazide, respectively [138].

(54)

REFERENCES

1. Stich, K., and Leeman, H. G., *Helv. Chim. Acta, 46*, 1151 (1963).
2. Elias, H., and Lotterhos, H. F., *Chem. Ber., 109*(4), 1580 (1976).
3. Netschai, F. T., *Zh. Fiz. Khim., 31*, 165 (1957).
4. Kumler, W. D., and Porter, C. W., *J. Am. Chem. Soc., 56*, 2549 (1934).
5. Phelps, I. K., and Phelps, M. A., *Am. J. Sci., 24*, 429 (1832).
6. Rosanoff, M. A., Gulick, L., and Larkin, H. K., *J. Am. Chem. Soc., 33*, 974 (1911).
7. Coleman, G. H., and Alvarado, A. M., *Org. Synth., III*, 3–5 (1923).
8. Mitchell, J. A., and Reid, E. E., *J. Am. Chem. Soc., 53*, 1879 (1931).
9. Cherbuliez, E., and Landholt, F., *Helv. Chim. Acta, 29*, 1438 (1946).
10. Noyes, W. A., and Goebel, W. F., *J. Am. Chem. Soc., 44*, 2286 (1922).
11. Menschutkin, N., *J. Prakt. Chem., 29*, 422 (1884).
12. Benn, G., Farrar, D., and Karolia, S. A. M., Allied Colloids Ltd., European Patent Appl. 0,246,813 (1987).
13. Habermann, C. E., and Tefertiller, B. A., Dow Chem. Co., U.S. Patent 3,994,973 (1976).
14. Watanabe, Y., Yamahora, T., Inokuma, S., and Tokumaru, T., U.S. Patent 3,980,662 (1976).
15. Dockner, T., and Platz, R., BASF, U.S. Patent 3,928,439 (1975).
16. Ravindranathan, M., Kalyanam, N., and Sivaram, S., *J. Org. Chem., 47*, 4812 (1982).
17. Trogler, W. C., and Jensen, C. M., U.S. Patent 4,684,751 (1987).
18. Villain, G., Gaset, A., and Kalck, P., *J. Mol. Catal., 12*, 103 (1981).
19. Arnold, D. P., and Bennett, M. A., *J. Organomet. Chem., 199*, 119 (1980).
20. Goetz, R. W., and Medler, I. L., U.S. Patent 3,670,021 (1972).
21. Agence Nat. Valorisation, FR Patent 2,319,929 (1979).
22. Rauch, F. C., and Nachtigall, G. W., U.S. Patent 3,673,250 (1972).
23. R. J. Reynolds Tobacco Co., British Patent 1,351,530 (1974).
24. Eremeev, I. V., Zilberman, E. N., and Neman, L. B., *Chem. Abstracts* 97:1619864.
25. Zilberman, E. N., Vorontsova, N. B., Afonshin, G. N., Frenkel, R. S., and Romashenko, L. G., *Zhurnal Veso uznogo Khimichaskogo obshchestra D. I. Mendeleeva, 18*(6), 705 (1973).

26. Kawakami, K., Tanabe, T., and Nagano, O., European Patent Appl. 0,204,555 (1986).
27. Commeyras, A., Arnaud, A., L'Herault, C., Galzy, P., and Jallageas, J., U.S. Patent 4,001,081 (1977).
28. Agency of Ind. Sci. Tech., JP Patent 82,016,092 (1982).
29. Riecke, G., Sinow, D., and Jahn, K., East German Patent 90,139 (1972).
30. Masuko, F., and Katsura, T., U.S. Patent 4,536,599 (1985).
31. Kyowa Hakko Kogyo Co. Ltd., JP Patent Appl. 49,076,808 (1974).
32. Schoenbrunn, E. F., and Sinha, V. T., U.S. Patent 3,674,848 (1972).
33. Greene, J. L., and Clark, R. E., U.S. Patent 3,686,307 (1972).
34. Moore, L. D., U.S. Patent 3,670,020 (1972).
35. Rao, C. G., *Synth. Commun.*, *12*(3), 177 (1982).
36. Puga, J., Sanchez-Delgado, R. A., Ascanio, J., and Braga, D., *J. Chem. Soc. Chem. Commun.*, *22*, 1631 (1986).
37. Toray Inds. KK, JP Patent Appl. 52,128,302 (1977).
38. Asahi Chem. KK, JP Patent 75,020,043 (1975).
39. Chattopadhyaya, J. B., and Rama Rao, A. V., *Tetrahedron*, *30*, 2899 (1974).
40. Elmghari-Tabib, M., Laplanche, A., Venien, F., and Martin, G., *Water Res.*, *16*, 223–229 (1982).
41. Hitachi KK, JP Patent Appl. 57,194,027 (1982).
42. ICI, British Patent 1,173,773 (1968).
43. Oda, J., Horiike, M., and Inouye, Y., *Agr. Biol. Chem.*, *35*(10), 1648 (1971).
44. Gulliver, D. J., European Patent Appl. 0,190,937 (1986).
45. Migrdichian, V., *Organic Synthesis*, Vol. 1, 1st ed., Reinhold, New York, 1960, p. 379.
46. Husain, S. R., Ahmad, F., Ahmad, M., and Osman, S. M., *J. Am. Oil Chem. Soc.*, *61*, 954 (1984).
47. Schwartz, G., *Org. Synth.*, *25*, 35 (1945).
48. Galat, A., and Elion, G., *J. Am. Chem. Soc.*, *65*, 1566 (1943).
49. Lin, J. J., Knifton, J. F., and Yeakey, E. L., European Patent Appl. 0,170,830 (1986).
50. Bogoslovskii, V. E., Mikhalyuk, G. I., and Shamolin, A. I., *J. Appl. Chem. USSR* (Engl. Transl.), *45*, 1197 (1972).
51. Krommes, P., and Lorberth, J., *J. Organomet. Chem.*, *97*, 59 (1975).
52. Zaugg, H. E., *J. Am. Chem. Soc.*, *82*, 2903 (1960).
53. Bowden, F. L., Dronsfield, A. T., Haszeldine, R. N., and Taylor, D. R., *J. Chem. Soc. Perkin Trans.*, *1*, 516 (1973).
54. Kotov, V. I., Yakushkin, M. I., and Reshetov, V. A., *Zhur. Prakladhnoi Khimiiu*, *51*(10), 2287–2292 (1978).
55. Yakushkin, M. I., Kotov, V. I., and Kolesnikov, I. M., *Chem. Abstracts*, 92:93864x.
56. Neduv, M. B., Marxa, K., Bazakin, V. I., Komarenko, T. I., Zenchenko, S. M., Kirpichnikova, Z. F., and Gakh, I. G., U.S. Patent 4,139,557 (1979).
57. Kurnishi, M., Asano, S., and Isogai, N., U.S. Patent 3,580,968 (1971).
58. James, J. A., Jr., and Kramis, C. J., Fr. Patent 1,406,279 (1965).

59. Japan Gas Chem. Co. Inc., JP Patent 71,015,083 (1971).
60. Kirsanov, A. V., and Zolotov, Y. M., *Zhur. Obshchei. Khim.*, *21*, 1166 (1951).
61. Zadorskii, V. M., Solodovnik, V. V., and Shalakhman, Y. U., U.S.S.R. Patent 1,004,357 (1984).
62. Smith, W. E., Canadian Patent 1,073,468 (1975).
63. Smith, W. E., Canadian Patent 1,073,467 (1975).
64. Duffy, L. L., U.S. Patent 3,538,159 (1970).
65. Hibbs, F. M., and Pearson, W. B., British Patent 1,469,514 (1977).
66. Daughenbaugh, R. J., U.S. Patent 4,258,200 (1981).
67. Yakushkin, M. I., Kotov, V. I., and Smaeva, T. P., *Khim. Prom.*, *20*, 69 (1980).
68. Balasubramanian, S., and Murthy, P. S., *Man-made Textiles in India*, May, *23*(5), 1980.
69. Bellis, H. E., European Patent Appl. 0,185,823 (1986).
70. Nozaki, K., U.S. Patent 3,407,231 (1968).
71. Mitsubishi Gas Chem. Inc., JP Patent 79,011,286 (1979).
72. Takahashi, Y., Fukuoka, Y., Sasaki, K., and Senoo, S., U.S. Patent 3,751,465 (1973).
73. Hamamoto, K., and Yoshioka, M., Nippon Kagaku Zasshi, *80*, 326 (1959).
74. Takahashi, Y., and Fukuoka, Y., U.S. Patent 3,758,576 (1973).
75. Tamura, N., Fukuoka, Y., Nishikido, J., Yamamatsu, S., and Suzuki, Y., U.S. Patent 4,329,462 (1982).
76. Nishikido, J., Tamura, N., and Fukuoka, Y., U.S. Patent 4,304,937 (1981).
77. Kashiwagi, H., and Enomoto, S., *Nippon Kagaku Kaishi*, 279 (1980).
78. Smolanka, I. V., et al., Inventor's Certificate 183,731; *Buyl. izobret.*, no. 14 (1966).
79. Woodbury, R. P., and Rathke, M. W., *J. Org. Chem.*, *42*, 1688 (1977).
80. Brederick, H., Gompper, R., and Klemm, K., *Chem. Ber.*, *92*, 1456 (1959).
81. Arbuzov, B. A., Zobova, N. N., Agarov, A. V., and Sofronova, O. V., *Izv. Akad. Nauk SSSR, Ser. Khim.*, *3*, 712 (1975).
82. Bennetto, H. P., and Caldin, E. F., *J. Chem. Soc. A*, 2191 (1971).
83. Dennis, W. E., *J. Org. Chem.*, *35*, 3253 (1970).
84. Mauret, P., Fayet, J., and Jehl, C., *Bull. Soc. Chim. Fr.*, 429 (1976).
85. Vaughan, W. R., and Teegarden, D. M., *J. Am. Chem. Soc.*, *96*, 4902 (1974).
86. Kresze, G., and Uhlig, U., *Chem. Ber.*, *92*, 1048 (1959).
87. Hagemeyer, H. J., and Holmes, J. D., U.S. Patent 3,979,432 (1976).
88. Heinemann, H., Wert, R. W., and McCarter, W. S., *Ind. Eng. Chem.*, *41*, 2928 (1949).
89. Mitchell, J. A., and Reid, E. E., *J. Am. Chem. Soc.*, *53*, 321 (1931).
90. Van Epps, G. D., and Reid, E. E., *J. Am. Chem. Soc.*, *38*, 2128 (1916).
91. Abe, J., *Bull. Waseda Appl. Chem. Soc.*, *21*, 27 (1933).
92. Kobayashi, K., and Abe, J., *J. Soc. Chem. Ind., Jpn.*, *36*, 42 (1933).
93. Delmon, B. M., Portenart, M. C., and Viehe, H., U.S. Patent 4,203,917 (1980).

94. BASF AG., German Patent 2,310,184 (1974).
95. Bodrikov, I. V., Michurin, A. A., and Lyandaev, E. A., U.S.S.R. Patent 891,650 (1982).
96. Le Cloirec, C., Le Cloirec, P., Morvan, J., and Martin, G., Fr. Patent 2,543,137 (1983).
97. Nippon Kagaku KK, JP Patent 54,024,824 (1979).
98. Chang, C. D., Lang, W. H., and LaPierre, R. B., U.S. Patent 4,231,955 (1980).
99. Righetti, B., Hanni, H., and Morgenthal, K., Swiss Patent 642,942 (1984).
100. Hara, T., and Komatsu, S., Mem. Coll. Sci. Kyoto Imp. Univ., *8A*, 241 (1925).
101. Nicodemus, O., German Patent 463,123 (1928).
102. Olive, G., and Olive, S., U.S. Patent 4,179,462 (1979).
103. Auvil, S. R., and Penquite, C. R., U.S. Patent 4,272,452 (1981).
104. Gambell, J. W., and Auvil, S. R., U.S. Patent 4,272,451 (1981).
105. Shepelev, S. S., and Ione, K. G., U.S.S.R. Patent 1,321,722 (1987).
106. Aliev, S. M., Sokolovski, V. D., and Boreskov, G. K., U.S.S.R. Patent 740,753 (1980).
107. Aliev, S. M., Sokolovski, V. D., and Boreskov, G. K., U.S.S.R. Patent 721,421 (1980).
108. Aliev, S. M., Sokolovski, V. D., and Boreskov, G. K., U.S.S.R. Patent 738,657 (1980).
109. Aliev, S. M., Sokolovski, V. D., and Boreskov, G. K., U.S.S.R. Patent 612,929 (1978).
110. Pajonk, G., Teichner, S. J., and Zidan, F., Fr. Patent 2,407,021 (1979).
111. Leibscher, G., Eppert, G., Finn, C., Portsendo, M., and Stief, C., E. German Patent 243,492 (1987).
112. Brown, P. M., and Maselli, J. M., U.S. Patent 3,565,940 (1971).
113. McClain, D. M., and Mador, I. L., U.S. Patent 3,412,136 (1968).
114. Sanyo Chem. Ind., JP Patent 77,039,001 (1977).
115. Stuer, B. C., and Grob, W., U.S. Patent 1,421,743 (1922).
116. Stuer, B. C., and Grob, W., British Patent 147,067 (1920).
117. Farbenind, I. G., British Patent 332,258 (1929).
118. Farbenind, I. G., British Patent 295,276 (1927).
119. Stuer, B. C., German Patent 467,220 (1922).
120. Auger, V., *Compt. Rend.*, *145*, 1287–90 (1907).
121. Tashk Poly., U.S.S.R. Patent 503,856 (1976).
122. Olive, G., and Olive, S., U.S. Patent 4,058,548 (1977).
123. Ronzio, A. R., and Cook, W. B., *Org. Syntheses Collective*, Vol. 3, Wiley, New York, 1955, p. 71.
124. Bonnemann, H., Brinkmann, R., and Schenkluhn, H., *Synthesis*, *8*, 575 (1974).
125. Wakatsuki, Y., and Yamazaki, H., *Synthesis*, *1*, 26 (1976).
126. Tatone, D., Dich, T. C., Nacco, R., and Botteghi, C., *J. Org. Chem.*, *40*, 2987 (1975).

127. Pardy, R. B., British Patent 2,153,849 (1985).
128. Haggin, J., *Chem. Eng. News*, *62(42)*, *29 (1983)*.
129. *Eur. Chem. News*, *27*(678), 35 (1975).
130. Cannon, J. G., et al., *J. Med. Chem.*, *20*, 1111 (1977).
131. Cauzzo, G., and Jori, G., *J. Org. Chem.*, *37*, 1429 (1972).
132. Yamamoto, K., and Watanabe, H., *Chem. Lett.*, *9*, 1225 (1982).
133. Gaskell, A. J., and Joule, J. R., *Tetrahedron*, *24*, 5115 (1968).
134. Cherbuliez, E., and Landott, F., *Helv. Chim. Acta*, *29*, 1438 (1946).
135. Harbst, R. M., and Garrison, J. A., *J. Org. Chem.*, *18*, 872 (1953).
136. Povazanec, F., Kovac, J., and Svobada, J., *Collect. Czech. Chem. Commun.*, *45*(8), 1299 (1980).
137. Bates, G. S., and Varelas, M. A., *Can. J. Chem.*, *58*(23), 2562 (1980).
138. Grodno Medic Inst., U.S.S.R. Patent 1,092,156 (1984).

17
Specialty Chemicals

Philip C. Heidt and Ryan C. Schad
Eastman Chemical Company, Kingsport, Tennessee

I. INTRODUCTION

When using "Specialty Chemicals" as a chapter heading, one must bear in mind that this is an extremely broad category. Numerous chemicals, which could have been included if this were a book in itself, have been omitted by the author's choice. Instead, a representative or two was chosen from a broad list of product categories such as medicine, food and nutrition, and agriculture. Choice was based on a combination of quantities of acetic acid–based chemicals used in the production of these chemicals and on the importance of these chemicals in our everyday lives. However, various chemicals, such as calcium and sodium acetate and acetanilide used in the manufacture of dyestuffs, have been omitted owing to limited space. The acetates of mono-, di-, and triglycerides, which consume large volumes of acetic anhydride and are important chemicals used in the food industry, will not be discussed in this text because the technical issues that surround their manufacture do not warrant further discussion. A text entitled *Food Additives* provides in adequate detail the uses of the acetates of starch, mono-, di-, and triglycerides [1].

II. ANALGESICS

The analgesics market, created by the introduction of aspirin (1) in Germany in 1899 and later in the United States in 1900, has undergone significant changes over the past 35 years. Since its introduction in the 1950s, acetaminophen (2) has become a significant part of the analgesics market. The latest introduction to the analgesics market is ibuprofen (3). Although not entirely recognized around the world, ibuprofen's sales are growing at the fastest rate, while aspirin's sales have slowed. Depending on sources, the >2 billion (U.S.) dollar market for analgesics is currently divided as follows: aspirin (40–43%), acetaminophen (35–40%), and ibuprofen (20–22%) [2,3]. Another analysis indicates the three analgesics will each represent 33% of the market by 1993 [4].

Each of the three analgesics holds a niche in the market. Whereas some individuals cannot take aspirin, others (e.g., heart attack patients) may benefit more from using aspirin as opposed to either acetaminophen or ibuprofen.

Because this text focuses on the use of acetic acid and its derivatives, only aspirin and acetaminophen will be discussed in further detail.

A. Acetylsalicylic Acid

Acetylsalicylic acid, also known as aspirin, may be used as an analgesic, antipyretic, and/or anti-inflammatory agent. The early 1988 worldwide market for aspirin was 40,900 m.t./year, of which the United States represented some 13,600 m.t./year [5]. A more recent report revealed the U.S. production of aspirin in 1988 to be 10,200 m.t. [6]. The aforementioned is a good indicator that growth slowed in the United States owing in part to competition from other analgesics and in part to a mature market. Manufacturers of acetylsalicylic acid in the United States are Dow Chemical, Norwich, and Rhone-Poulenc. Most of the manufacturers outside the United States are located in Europe and Eastern Asia. One major change occurred in late 1989 when Monsanto sold its analgesic business to Rhone-Poulenc. The current market price for aspirin is $2.45/lb (approximately $5.40/kg) (U.S.) [7].

Table 1 Physical Properties of Acetylsalicylic Acid

Acetylsalicylic acid [50-78-2]
  Molecular weight: 180.16
  Density: 1.140
  Melting point (rapid heating): 135°C
  Decomposition point: 140°C
  Monoclinic tablets or needle-like crystals
  LD_{50} oral: mice, rats, 1.1, 1.5 g/kg
  Odorless; in moist air hydrolyzes to salicylic acid and acetic acid; stable in dry
    air
  Solubilities
    1 g/300 mL water at 25°C
    1 g/100 mL water at 37°C
    1 g/5 mL ethanol at 25°C
    1 g/17 mL chloroform at 25°C

The physical properties of acetylsalicylic acid are listed in Table 1 [8].

The first preparation of acetylsalicylic acid dates back to 1853 when Gerhardt reacted acetyl chloride with sodium salicylate [9]. The first reported medicinal use of acetylsalicylic acid was in Germany in 1899 and in the United States in 1900. The first U.S. patent for the preparation of acetylsalicylic acid was issued in 1900 [10]. Although it is still a registered trademark in some countries, aspirin is a generic name for acetylsalicylic acid in the United States. For purposes of discussion in this chapter, acetylsalicylic acid will now be referred to as aspirin.

A brief history of the synthesis of aspirin is given by Pelz [11]. Few significant changes in the preparation of aspirin have been reported in recent years. Most of these changes have focused on the elimination and/ or prevention of colored impurities, processes to increase isolated yields, and the use of different catalysts. Some of the different catalysts that have been used are phosphoric acid [12], amberlyst-15 [13], solid potassium hydroxide [14], perchloric acid [15], and pyridine [16]. Most of these changes have had little or no significant commercial applications.

Because unnecessary steps and impurities in the production of aspirin add costs, time, and equipment, one target has been to provide analytically pure aspirin in high yields that can be tableted directly from the initial reaction (an exemplary of this type of process was developed by Dow Chemical Company) [17].

Figure 1 is a flowsheet for the Dow process. First, a reactor is charged with acetic acid, salicylic acid, and excess acetic anhydride. The contents of the reactor are raised to 85°C ± 15°C and held there until the reaction

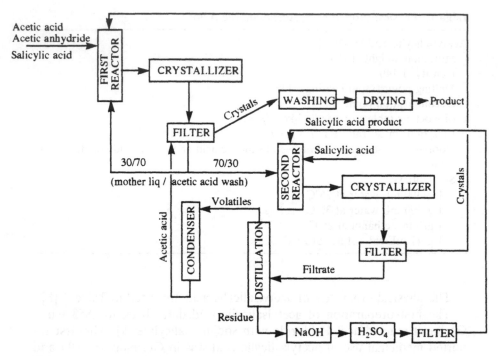

Figure 1 Flow process for production of acetylsalicylic acid.

is essentially complete (ca. 1.5 h). The reaction mixture is then cooled to allow crystallization to take place in the reaction vessel or in a separate crystallizer. The crystals are then filtered to remove the mother liquor, which is enriched with acetic anhydride. Finally, the crystals are washed with glacial acetic acid and then water before being dried. These crystals are now suitable for tableting. The filtrates (including the mother liquor) are combined and separated into two fractions. One fraction is returned to the initial reactor while the second fraction is fed into a second reactor. Salicylic acid is fed into the second reactor, and the reaction conditions mirror those of the first reactor. The crystals formed from the second reactor are not suitable for tableting and, therefore, must be recycled back into the first reactor. The filtrates are distilled to remove volatiles such as acetic acid, which can be recycled back into the washing process. The residue is hydrolyzed (sodium hydroxide) and reacidified to produce a salicylic acid product, which is returned to the second reactor.

In a vastly simplified preparation of aspirin, Robert T. Edmunds of Norwich developed a process whereby aspirin crystals in greater than 99.5% purity could be prepared in nearly theoretical yield without the need for filtration, washing and drying, or additional recovery from a mother liquor [18]. This process involves charging a glass-lined jacketed scale or weighing tank with salicylic acid, acetic acid, and a slight molar excess of acetic anhydride. Heated water is circulated throughout the jacket while the mixture is agitated. Once in solution, the salicylic acid–acetic anhydride mixture is filtered into a glass-lined rotary conical vacuum dryer. The pressure is gradually reduced to ca. 20–30 mmHg to remove volatiles. Trace odorants can be removed by a number of conventional methods.

B. Acetaminophen

N-Acetyl-*p*-aminophenol, also known as acetaminophen (APAP), is used as an analgesic and as an antipyretic agent. In early 1988, global capacities were reported at 31,800–34,100 m.t./year and consumption was 27,300 m.t./year [2]. By 1990, worldwide demand for APAP was 40,000 m.t./year, which represented a $250 million (U.S.) business [3]. APAP business in the United States is $90 million/year. Growth in the APAP business in the United States has slowed to 1–2% a year, while growth worldwide is 4–5% [3]. The U.S. capacity for APAP is 27,300 m.t./year [19]. A breakdown of U.S. capacity is shown in Table 2.

A current price of $7/kg (U.S.) represents little change over the past few years [7]. Unlike the aspirin producers discussed previously, the APAP industry has undergone some dramatic changes, such as Rhone-Poulenc's global acquisition of Monsanto's analgesic business operations, Mallinckrodt's purchase of RTZ Stavely Chemicals APAP business and of Penco's APAP business, and Hoechst-Celanese's entry into the U.S. APAP business. Hoechst's entry will increase U.S. capacity to almost double that of demand.

Physical properties for APAP are shown in Table 3 and U.S.P. specifications are shown in Table 4 [20,21]. Because of its antioxidant properties, APAP has been used in the manufacture of clear resins and plastics. APAP is also used as an intermediate in the preparation of dyes. And with its precursor, *p*-aminophenol, APAP is used in photographic areas as well.

As with most industrial processes, minor impurities are often formed in a reaction. Depending on end use, these minor impurities can create com-

Table 2 Manufacturers of Acetaminophen in the United States

Manufacturer (location)	Capacity (m.t./year)
Hoechst Celanese (Bishop, TX)	9,100
Mallinckrodt (Raleigh, NC)	11,400
Rhone-Poulenc (Luling, LA)	6,800

Table 3   Physical Properties of Acetaminophen

N-Acetyl-*p*-aminophenol [103-90-2]
  Molecular weight: 151.16
  Melting point: 169–170.5°C
  Density: 1.293
  Very slightly soluble in cold water; much more soluble in hot water. Soluble
    in methanol, ethanol, dimethylformamide, acetone, ethyl acetate, and
    ethylene dichloride. Slightly soluble in ether. Nearly insoluble in petroleum
    ether, pentane, benzene.
  LD_{50} oral: mice, 338 mg/kg

plications in future reactions and be expensive to remove. In the production of APAP, especially for use in pharmaceuticals, in which color and residues are closely regulated, the removal of impurities is critical.

There are three major starting routes for the commerical manufacturing of APAP. Two of these routes are related, one beginning with *p*-nitrophenol (PNP) and the other with *p*-aminophenol (PAP). The third, and most recent, route employs 4-hydroxyacetophenone (4-HAP) as a starting raw material. Each process has its own advantages and disadvantages as well as special equipment needs. Because PAP is obtained from PNP, those routes which utilize either one will be combined under one heading.

Color formation was one of the first problems addressed by corporate researchers. An early preparation of APAP employed boric acid as a catalyst and water-entraining solvents (e.g., benzene) to remove water as it was formed in the heating of PAP with acetic acid [22]. Despite care in starting with pure white PAP, color bodies still formed from the long heating time required and from the potential for contact with oxygen. A breakthrough in addressing color formation was the use of a blanket of sulfur dioxide in the acetylation process [23]. Although this method adequately addressed color formation, yields suffered, creating both expense and waste problems. The issue of yields then led to the use of catalysts such as boron trioxide and metaboric acid (of 176°C), instead of boric acid [24].

Table 4   U.S.P. Requirements for Acetaminophen

Contains not less than 98% and not more than 101.0%
  of $C_8H_9NO_2$, calculated on the anhydrous basis
Water, method I, not more than 0.5%
Residue on ignition, not more than 0.1%
Not more than 0.005% of free *p*-aminophenol
Melting range 168–172°C

A noncommercialized approach to the reduction of color formation and by-products was developed using electrolytically produced PAP from PNP and direct acetylation of the crude PAP produced without prior isolation [25]. In this method, PNP in sulfuric acid is electrolytically reduced to PAP. The sulfuric acid solution is partially neutralized to a pH of 1.5–4.9 with either calcium carbonate or calcium hydroxide. The precipitated sulfate is filtered off and the aqueous filtrate extracted with an organic solvent to remove impurities. The aqueous phase is then directly acetylated using acetic acid and acetic anhydride. This method of partial neutralization with calcium carbonate, or the like, eliminates the use of a more expensive base such as sodium hydroxide. Yields of 80–90% of APAP are obtained by this method.

Another dissatisfying problem in the production of APAP via PAP stems from by-product formation in the preparation and reduction of PNP. A process was developed whereby the hydrolysate-containing PNP, formed from the hydrolysis of *p*-nitrochlorobenzene, can be hydrogenated with hydrogen and 5% palladium on carbon in the presence of boric acid as catalyst and directly acetylated to form APAP [26,27]. A key factor in this process is maintaining the pH below 7.0. The overall production of APAP by this method is described as follows: A dispersion of the PNP previously prepared (from hydrolysis of *p*-nitrochlorobenzene) in water at 50°C is charged into a hydrogenator with a dispersion of boric acid and carbon in water. Five percent palladium on carbn is added and the hydrogenator charged with hydrogen to 4.82×10^6 dynes/cm^2 while cooling to 60°C. After approximately 60% of reduction has been obtained, half the required acetic anhydride is added. The hydrogenator, now at 70°C, is allowed to proceed to completion. The reaction material is filtered into a glass-lined autoclave, and sodium bisulfite is added to destroy any nitrites that may have formed. The remaining acetic anhydride is added and the acetylation is completed. Salts and other insoluble impurities are removed by centrifugation.

Because color can form in chemicals during storage (e.g., color formation on oxidation of PAP and related compounds), and because of the importance of having a moderately long shelf-life, a process has been developed to increase the color-free shelf-life of APAP [28]. This process involves dissolving the crude APAP in an appropriate solvent. This solution also contains formamidine-sulfinic acid and a reducing agent such as sodium metabisulfite. The solution is heated to near boiling for a period of time and then cooled, filtered, and washed. This procedure allows for the preparation of pharmaceutical-grade APAP.

Recently, a process has been developed to increase the recovery of APAP produced from the aqueous acetylation of PAP [29]. This process

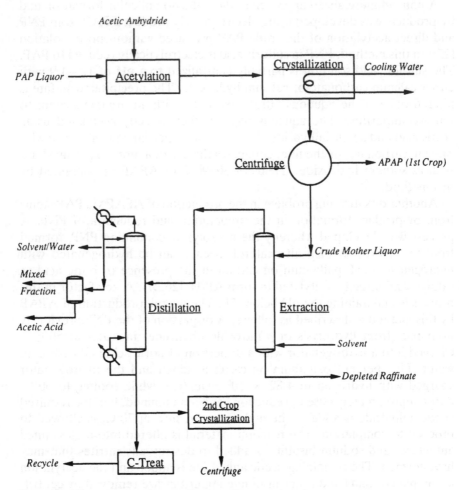

Figure 2 Flow process for production of acetaminophen.

is shown in Figure 2. The key factor in this process is the countercurrent extraction of the mother liquor and distillation of the solvent and acetic acid to recover a second crop of APAP. Yields may be greater than 90% by this method.

The third approach toward the preparation of APAP (and arguably the most effective route to a number of APAP-related products) involves 4-hydroxyacetophenone (4-HAP) as the starting material [30]. 4-HAP is conveniently prepared by the Fries rearrangement of phenyl acetate using acetic anhydride and anhydrous hydrogen fluoride as catalyst. The Fries rearrangement is shown in Eq. (1).

$$(1)$$

Once 4-HAP has been prepared, the corresponding oxime is formed and APAP is then prepared by the Beckman rearrangement [Eq. (2)] of the oxime using thionyl chloride in liquid sulfur dioxide. A process for the prevention of chlorinated by-products in the Beckman rearrangement has

$$(2)$$

also been developed by Hoechst-Celanese [31]. They found that the addition of small amounts (ca. 0.2 wt%) of potassium iodide to the mixture prior to rearrangement almost completely eliminated chlorinated by-products.

III. VITAMIN E ACETATE

α-Tocopheryl acetate (vitamin E acetate) is widely used to fortify numerous food items because of its ability to withstand oxidation compared to free vitamin E. Once ingested, α-tocopheryl acetate is hydrolyzed in the intestinal tract to liberate the free α-tocopherol. Vitamin E is an important antioxidant because of its lipid solubility.

α-Tocopheryl acetate is readily prepared by the acetylation of α-tocopherol with acetic anhydride. The major producers of *d*-α- and *d,l*-α-tocopheryl acetate are listed in Table 5. The market price for synthetic α-tocopheryl acetate has stabilized over the last decade, and it currently sells for $23/kg (U.S.) [7]. The acetate from *d*-α-tocopherol sells for approximately $57/kg (U.S.).

U.S. production of vitamin E in 1986 was reported at 8440 m.t., which is more than that produced in Western Europe and Japan [32]. Physical data on α-tocopheryl acetate are shown in Table 6 [33]. Biologically speaking, *d*-α-tocopheryl acetate is 1.4 times more active than *d,l*-α-tocopheryl acetate.

Of the four naturally occurring tocopherols that possess vitamin E activity (α, β, γ, δ), α-tocopherol has the highest activity. The natural isomer of α-tocopherol is (2R, 4'R, 8'R)-*d*-α-tocopherol (4). Tocopherols are widely distributed in vegetables, and their highest concentration is in the cereal grain oils. Other vegetables may be nearly devoid of tocopherols. Just as

Table 5 Manufacturers of d-α- and d,l-α-Tocopheryl Acetates in the United States, Western Europe, and Eastern Asia

---

d-α-Tocopheryl acetate
  Manufacturer (location)
    Eastman Kodak Company (Rochester, NY)
    Henkel Corporation (Kankakee, IL)
d,l-α-Tocopheryl acetate
  Manufacturer (location)
    United States
      BASF Corporation (Wyandotte, MI)
      Hoffmann-La Roche Inc. (Nutley, NJ)
    Western Europe
      BASF (Ludwigshafen, Germany)
      Bayer (Leverkusen, Germany)
      Hoffmann-La Roche (Grenzach, Germany)
      F. Hoffmann-La Roche (Basel, Switzerland)
      Roche (Sisseln, Switzerland)
    Eastern Asia
      Lucky Ltd. (Naju, S. Korea)

---

the concentration of tocopherols varies from vegetable oil to vegetable oil, so does the percentage of α-tocopherol contained. Safflower oil has the greatest percent composition of α-tocopherol (ca. 90%). Corn and soybean oils, which are more abundant, have roughly 20% α-tocopherol. β-Tocopherol (ca. 65%) predominates in wheat oil and γ-tocopherol predom-

Table 6 Physical Properties of d-α- and d,l-α-Tocopheryl Acetates

---

d,l-α-Tocopheryl acetate
  Molecular weight: 472.73
  Density: 0.9533
  Melting point: $-27.5°C$
  Boiling point$_{0.01}$: 184°C
  Soluble in: acetone, chloroform, ether
d-α-Tocopheryl acetate [58-95-7]
  Melting point: 28°C
  $[\alpha]_D^{25}$: $+0.25°$ (c = 10 in chloroform), $+3.2°$ (in ethanol)

---

inates in corn oil. γ- and δ-tocopherols are the most abundant in soybean oil [34].

(4)

α : R¹=R²=CH₃
β : R¹=CH₃, R²=H
ɣ : R¹=H, R²=CH₃
δ : R¹=R²=H

Tocopherols are obtained by vegetable oil distillate during the deodorizing step of refining and are concentrated by molecular distillation [35,36]. β-, γ-, and δ-tocopherols can be converted to α-tocopherol by methylation [37].

Apart from natural sources, α-tocopherol is also synthesized commercially. Some of these synthetic approaches will be mentioned later. Aqueous alkaline ferricyanide has been used in distinguish natural (*d*-α) from synthetic (*d,l*-α) tocopherol [38]. The *l*-epimer can be isolated from synthetic *d,l*-α-tocopherol via repeated fractionation of diastereomers by complexation with piperazine [39].

A process for the purification of tocopherols from vegetable oils has been developed [40]. This process employs caustic methanol to form two phases. The organic phase contains sterols, waxes, and the like. The methanolic phase contains the tocopherols that can be recovered by neutralization.

A process for the separation of the tocopherol homologs using selective deacylation (of the tocopheryl acetates) has been developed [41–43]. This process is sufficient for separating β-, γ-, and δ-tocopheryl acetates from α-tocopheryl acetate. Other processes that use either ion exchange chromatography or liquid chromatography require the use of large quantities of resin or adsorbents [44].

Because most attention in the synthesis of α-tocopheryl acetate is devoted to the synthesis of the α-tocopherol portion itself, the remaining portion of this section will deal with α-tocopherol..

Numerous variations using trimethylhydroquinone (TMHQ) to synthetically prepare chromane compounds have been developed. One major problem with this approach is the discoloration of TMHQ solutions caused by air oxidation. One process that minimizes discoloration uses the addition of small amounts of alkali metal borohydrides (e.g., $NaBH_4$) [45].

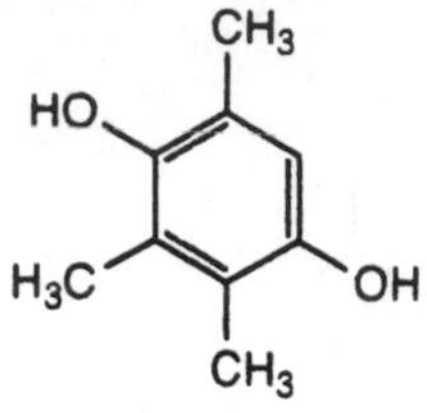

TMHQ

Chromanes have been prepared by the reaction of TMHQ with phytol in the presence of zinc chloride and acetic acid and in the presence of zinc and aqueous HCl [46–48]. They have also been prepared using TMHQ with phytyl halides (5) in the presence of zinc chloride; in a chlorinated lower aliphatic solvent in the presence of metallic tin and a Friedel-Crafts

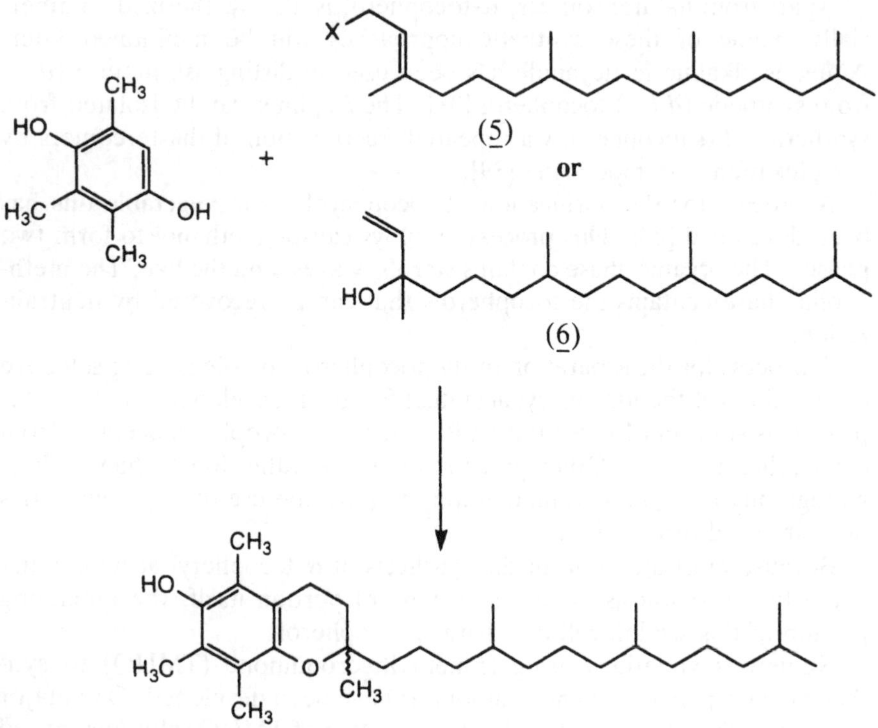

Figure 3   Preparation of α-tocopherol.

catalyst; and with BF_3 as a catalyst [49–52]. The currently preferred choice utilizes TMHQ and isophytol ($\underline{6}$). A number of developed conditions involving TMHQ and isophytol in combination include: (1) direct condensation using phosphorous pentoxide [53]; (2) employment of BF_3 as a condensation agent [54]; (3) zinc chloride in the presence of HCl [55]; (4) use of a strongly acidic sulfonic type cation exchange resin (e.g., amberlyst-15) [56]; (5) a combined acid condensation agent [57]; (6) the use of iron or ferrous chloride and HCl [58]; (7) a combined acid condensation agent and an amine salt of a nonoxidizing protic acid [59]; and (8) a combined acid condensating agent and an amine [60]. Still another preparatory approach for synthesizing chromanes via TMHQ involves dehydrophytol [61]. Each of the above processes produces d,l-α-tocopherol. The basic reaction sequence is shown in Figure 3.

In recent years, a number of synthetic methods have been developed for the stereoselective and/or stereospecific preparation of α-tocopherol. α-Tocopherol can be prepared with 90% retention of configuration from the cyclization of α-tocopherylquinone under acidic conditions [62]. Stereospecific preparation of α-tocopherol has been accomplished using 6-hydroxychroman-2-yl carboxylic acid derivatives ($\underline{7}$) or the corresponding aldehydes ($\underline{8}$) [63,64].

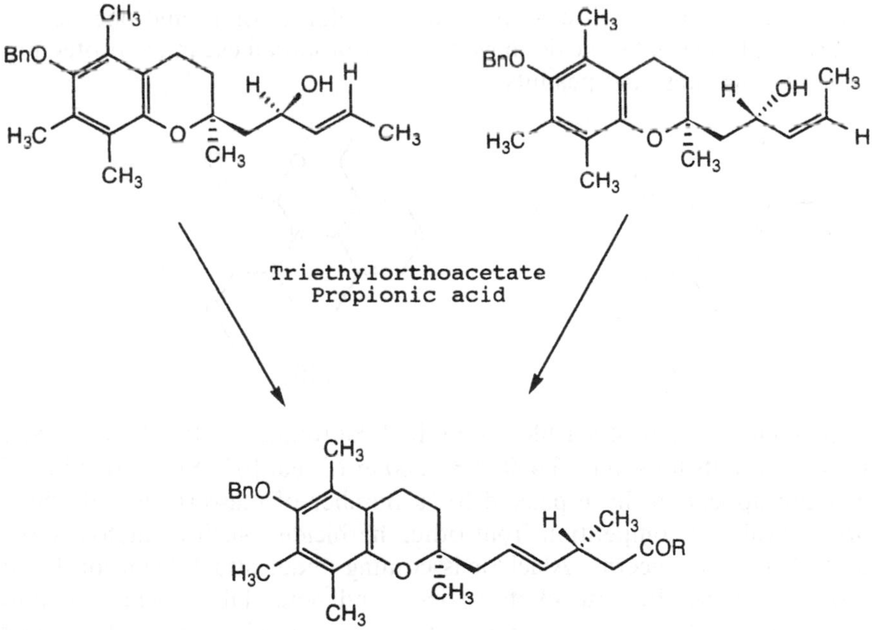

Figure 4 Claisen rearrangement.

An interesting synthesis of d-α-tocopheryl acetate uses a [3,3] sigmatropic (Claisen) rearrangement [65]. This route is shown in Figure 4.

The most complete synthetic work for the synthesis of all eight stereoisomers of α-tocopheryl acetate is described by Cohen and co-workers [66]. This work also includes physical data on each isomer.

IV. ALACHLOR AND METOLACHLOR

The agricultural industry holds a vast market potential for large end use of acetic acid derivatives in the manufacture of herbicides, pesticides, and insecticides. Alachlor (9) and metolachlor (10) are prime examples of major end use of acetic acid derivatives that serve the herbicide sector. Both these herbicides use α-chloroacetyl chloride or α-chloroacetic anhydride. These two herbicides have found widespread use in the protection of corn, soybeans, and peanuts.

Monsanto's sales of alachlor alone in 1988 totaled $500 million (U.S.), which accounts for some 36,400 to 38,200 m.t./year [67,68]. Production of alachlor appears to have peaked for a number of reasons, two of which are critical: (1) competition from other herbicides, such as metolachlor, and (2) health concerns. Alachlor is coming under attack from local and federal agencies because of its toxicity and potential to penetrate into groundwater. The Canadian government has already banned the use of alachlor-containing pesticides. However, because of the need for large

Table 7 Manufacturers of Alachlor in the United States and Eastern Asia

Monsanto Chemical Company (Muscatine, IA)
Comlets Chemical Industrial Co. (Taiwan)
Shen Hong Chemical (Taiwan)
Shinung Corporation (Taiwan)
Han-Nong Corporation (South Korea)
Korea Steel Chemical Co. (South Korea)
Oriental Chemical Industry Co. (South Korea)

volumes of pesticides in the United States and around the world, sales of alachlor should still be significant until a suitable cost-efficient and "nontoxic" replacement is developed.

Key manufacturers of alachlor in the United States and Eastern Asia are shown in Table 7; CIBA-GEIGY (St. Gabriel, LA) is the sole manufacturer of metolachlor.

Both alachlor and metolachlor are part of a series of acetanilides and acctamides of the general structures ($\underline{11}$ and $\underline{12}$) that possess phytotoxicity.

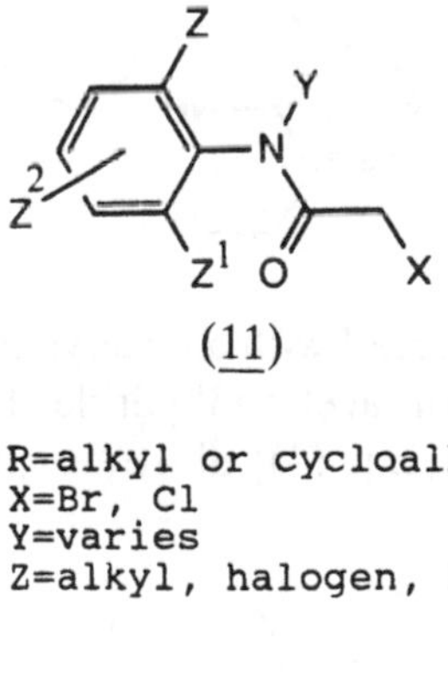

(11) 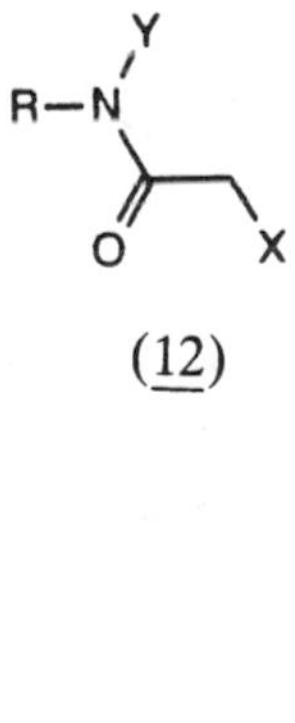   (12)

R=alkyl or cycloalkyl
X=Br, Cl
Y=varies
Z=alkyl, halogen, hydrogen

The physical properties of alachlor and metolachlor are shown in Table 8 [69–71].

The general structures for the acetanilides and acetamides are typically prepared in one of three ways, the first of which is most practical.

1. Acetylating a secondary amine [72–80]. In general, the synthesis of these α-haloacetanilides (or acetamides) begins with the desired substituted primary aniline and an appropriate alkylating agent under basic conditions. A base is usually necessary to react with acids of the type

Table 8 Physical Properties of Alachlor and Metolachlor

Alachlor ($C_{14}H_{20}ClNO_2$)
  Molecular weight: 269.77
  Density$_{15.6}$^{25}: 1.133
  Melting point: 40–41°C
  LD_{50} orally in rats: 1200 mg/kg
  Solubilities: soluble in ether, acetone, benzene, ethanol, ethyl acetate
  Hydrolyzed under strong acid or alkaline conditions
Metolachlor ($C_{15}H_{22}ClNO_2$)
  Molecular weight: 283.81
  Density^{20}: 1.12
  Index of refraction$_D$^{20}: 1.5301
  Boiling point$_{0.001}$: 100°C
  LD_{50} oral: rats, 2780 mg/kg
  Solubilities: soluble in most organic solvents

HX (X=Cl, Br), which are formed during the alkylation process. Such an example for the synthesis of metolachlor is shown in Eq. (3).

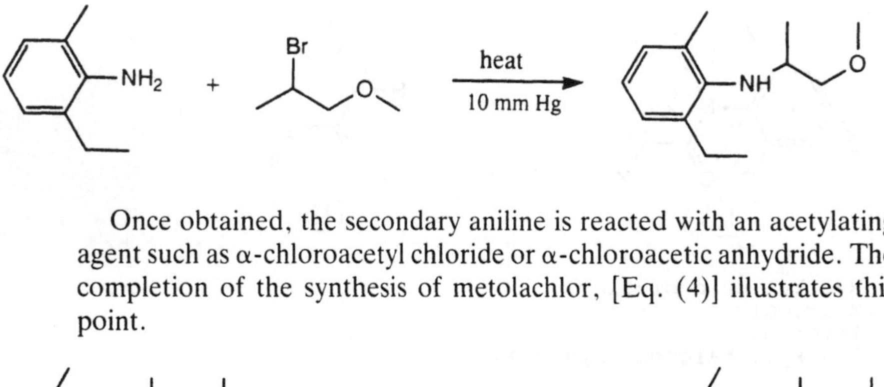

$$(3)$$

Once obtained, the secondary aniline is reacted with an acetylating agent such as α-chloroacetyl chloride or α-chloroacetic anhydride. The completion of the synthesis of metolachlor, [Eq. (4)] illustrates this point.

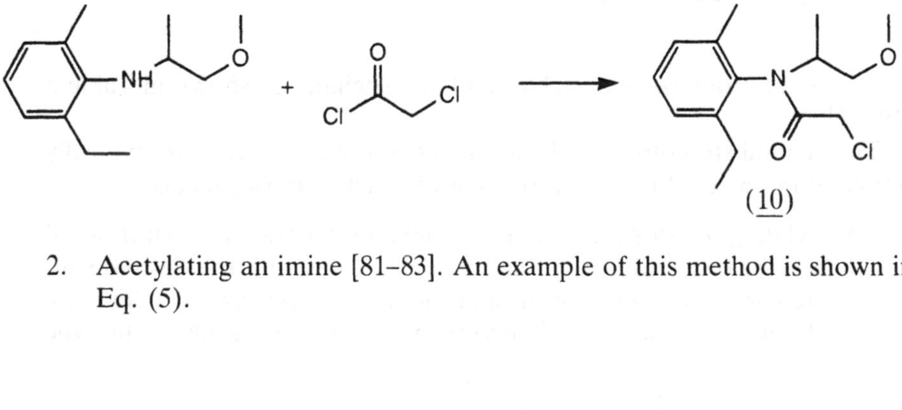

$$(4)$$

(**10**)

2. Acetylating an imine [81–83]. An example of this method is shown in Eq. (5).

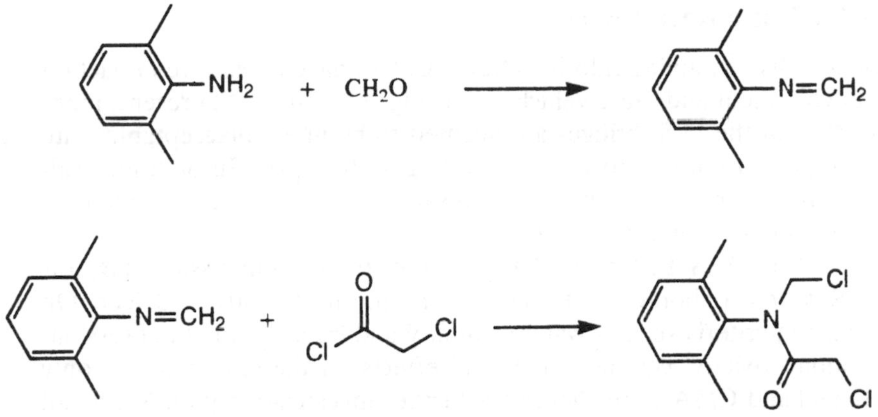

(5)

3. Alkylation of α-haloacetanilides (or acetamides) [84]. In this case, an anion is generated from the secondary amide under basic conditions and then alkylated. Extreme care and carefully selected conditions must be utilized to obtain usable quantities of the desired tertiary α-haloacetanilide. If conditions are not chosen carefully, significant amounts of by-products, such as aziridinones of type (13), will form [85,86].

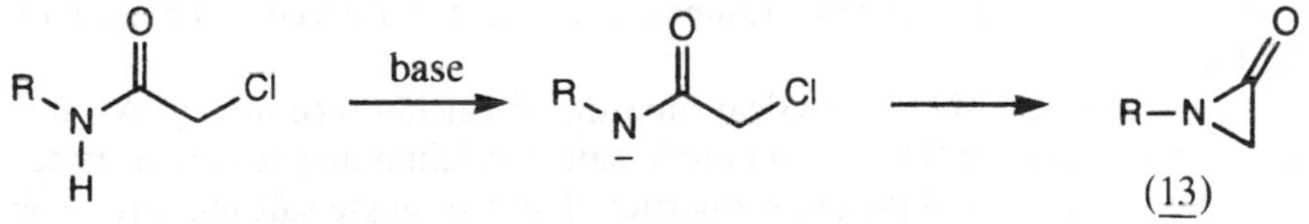

(13)

V. CALCIUM MAGNESIUM ACETATE

Calcium magnesium acetate (CMA) is an emerging bulk chemical in the world market. Two major uses of this chemical have been proposed that could increase demand for CMA into multimillion-ton quantities in the near future. One use is an alternative, noncorrosive deicing salt for roads and runways. The second major use for CMA is as an additive for coal-fired combustion units such as those used by electrical utilities, increasing combustion efficiency while independently reducing the amount of "acid rain" [87].

A. CMA as a Roadway Deicer

Deicing salts containing chlorine (NaCl, $CaCl_2$) have a negative impact on the environment and cause vehicle and bridge corrosion. At present, more than 40% of the U.S. bridges are deemed to be in an unacceptable state of disrepair, according to a federal rating system [88]. In addition, salt runoff from highways and unprotected salt storage areas can contaminate nearby drinking-water wells [89].

In 1976, the U.S. Federal Highway Administration commissioned Bjorksten Research Laboratories to develop an alternate roadway deicer. On the basis of criteria such as water solubility and freezing-point depression, corrosion, toxicity, cost potential, and effects on the environment, only methanol and CMA were chosen for further investigation [90]. Methanol was later eliminated from consideration because of its solvent nature and flammability, leaving CMA as the only acceptable alternative to chloride-containing deicer salts.

CMA is not only noncorrosive to metals; it is nondestructive to other highway materials, environmentally safe, and can be produced from acetic acid and dolomitic lime, raw materials readily available in the United States. Unfortunately, CMA's production costs are 10–20 times that of sodium chloride salt, making it prohibitively expensive in all but environmentally sensitive areas where regular salt cannot be used [89]. The effects of a winter maintenance program, however, extend far past the purchase price of a deicer. Several studies have shown the real cost of road salt, factoring in corrosion and environmental damage, far exceeds the price of CMA.

CMA can be either a physical mixture of calcium acetates (CA) and magnesium acetates (MA) or a double salt of calcium magnesium acetate. The double salt has different properties than the single-salt mixture. For example, the double salt is much less soluble in water than either CA or MA. Furthermore, deicing compositions of the double salt show improved ice-melting characteristics over what would be expected from a stoichiometric mixture of single salts [91]. Finally, the double salt is anhydrous, which permits high density and particle strength, no biodegradation in storage, exothermic dissolution, and low dustiness.

B. CMA as a Fossil Fuel Combustion Additive

As an additive to fossil fuel combustion units, CMA has dual functionality: (1) The calcium acts as a catalyst for combustion—more coal can be burned in the same size furnace, generating more electrical power without increased capital expenditure. (2) Calcium in CMA reacts with some of the sulfur in coal to form solid particle calcium sulfate, which is recovered from the stack gases by the precipitator [90]. Consequently, fewer sulfur oxides (SO_x) are formed during combustion, which could substantially reduce the amount of "acid rain."

C. CMA Production

As stated above, CMA is made by direct reaction of dolomitic lime (with MgO added to ensure the proper Mg:Ca ratio) and acetic acid. A generalized process flow diagram is shown in Figure 5 for continuous preparation of anhydrous CMA double salt. Note the addition of water in the mix tank to dilute the calcium and magnesium bases to about 40% water by weight [91]. Although glacial acid is generally used in the process (because it is readily available), it would be feasible to use a weak acetic acid stream as feedstock to the reactor. If too little acid is blended, or acid is added too slowly, side products such as calcium or magnesium acetates could form and precipitate out [91].

Following reaction and settling, the substrate is dried by direct contact and separated into suitable pellets, oversized material, and fines with the

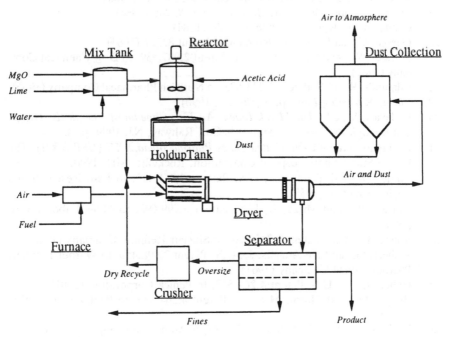

Figure 5 CMA process flow diagram. (From Ref. 91.)

latter two recycled to the dryer. Typically, the CMA substrate will make several cycles through the dryer before emerging as an acceptable product.

REFERENCES

1. Branen, A., Davidson, P., and Salminen, S., eds., *Food Additives*, Marcel Dekker, New York, 1990.
2. *Chem. Marketing Rep.*, p. 20 (Sept. 12, 1988).
3. *Chem. Week*, p. 32 (Nov. 21, 1990).
4. *Chem. Marketing Rep.*, p. 21 (March 7, 1988).
5. *Chem. Marketing Rep.*, p. 7, 28 (April 18, 1988).
6. *Synthetic Organic Chemicals, United States Production and Sales, 1989*, USTIC Publication 2338, U.S. Government Printing Office, Washington, DC, 1990, Section 6-2.
7. *Chem. Marketing Rep.* (Feb. 11, 1991), p. 28.
8. Budavari, S., ed., *The Merck Index: An Encyclopedia of Chemicals, Drugs, and Biologicals*, 11th ed., Merck & Co., Rahway, NJ, 1989, p. 134.
9. Gerhardt, C. *Annalen Der Chemie*, 87, 149 (1853).
10. Hoffman, F., U.S. Patent 644,077, to Farbenfabriken Bayer & Heyden (1900).
11. Pelz, J., *Pharmazie*, 41, 733 (1986).
12. Han, C., *Huaxue Tongbao*, 12, 30 (1989).
13. Akar, A., et al. *Bull. Tech. Univ. Istanbul*, 39, 149 (1986).
14. Zhang, G., and Zhang, M., *Huaxue Shiji*, 8, 245 (1986).
15. Candoros, F., Romanian Patent 85726 (1984).
16. Daescu, C., and Mate, I., Romanian Patent 83227 (1984).
17. Stoesser, W., and Surine, W., U.S. Patent 2,987,539, to Dow Chemical Company (1961).
18. Edmunds, R., U.S. Patent 3,235,583, to Norwich Pharmacal Company (1966).
19. *Chem. Marketing Rep.*, p. 3 (Sept. 3, 1990).
20. Budavari, S., ed., *The Merck Index: An Encyclopedia of Chemicals, Drugs, and Biologicals*, 11th ed., Merck & Co., Rahway, NJ, 1989, p. 8.
21. Permission granted *The United States Pharmacopeia XXII* (*USP XXII*), The United States Pharmacopeial Convention, Rockville, MD, 1990.
22. Cottle, D., and Young, D., U.S. Patent, 2,711,415, to Esso Research and Engineering Co. (1955).
23. Croxall, W., and Mirza, J., U.S. Patent 2,799,692, to Miles Laboratories, (1957).
24. Young, D., U.S. Patent 2,945,870, to Sinclair Refining Company (1960).
25. Wilbert, G., and De Angelis, J., U.S. Patent 2,998,450, to Warner-Lambert Pharmaceutical Company (1961).
26. Huber, J., Jr., U.S. Patent 4,264,525, to Penick Corporation (1981).
27. Ruopp, D., and Thorn, M., U.S. Patent 4,264,526, to Penick Corporation (1981).
28. Keel, B., et al., U.S. Patent 4,474,985, to Monsanto Company (1984).
29. Sathe, S., U.S. Patent 4,565,890, to Mallinckrodt, (1986).

30. Davenport, K., and Hilton, C., U.S. Patent 4,524,217, to Celanese Corporation (1985).
31. Fruchey, O., U.S. Patent 4,855,499, to Hoechst Celanese Corporation (1989).
32. *Synthetic Organic Chemicals, United States Production and Sales, 1986*, USTIC Publication 2009, U.S. Government Printing Office, Washington, DC, 1987, p. 79.
33. Budavari, S., ed., *The Merck Index: An Encyclopedia of Chemicals, Drugs, and Biologicals*, 11th ed., Merck & Co., Rahway, NJ, 1989, p. 1579.
34. Herting, D., ed., *Kirk-Othmer Encyclopedia of Chemical Technology*, Vol. 24, 3rd ed., Wiley, New York, 1984, p. 214.
35. Embree, N., *Chem. Rev.*, *29*, 317 (1941).
36. Stern, M., et al., *J. Am. Chem. Soc.*, *69*, 869 (1947).
37. Kasparek, S., and Machlin, L., eds., *Vitamin E: A Comprehensive Treatise*, Marcel Dekker, New York, 1980, Chapter 2.
38. Nelan, D., U.S. Patent 3,173,926, to Eastman Kodak Company (1965).
39. Nelan, D., and Robeson, C., *J. Am. Chem. Soc.*, *84*, 3196 (1962).
40. Willging, S., U.S. Patent 4,550,183, to Henkel Corporation (1985).
41. Foster, C., U.S. Patent 4,480,108, to Eastman Kodak Company (1984).
42. Foster, C., U.S. Patent 4,602,098, to Eastman Kodak Company (1986).
43. Foster, C., U.S. Patent 4,607,111, to Eastman Kodak Company (1986).
44. Kijima, S., and Nakamura, T., U.S. Patent 3,402,182, to Eisai Co. (1968).
45. Dodd, J., U.S. Patent 4,147,625, to Continental Oil Company (1979).
46. Bergel, F., et al., *Nature*, *142*, 36 (1938).
47. Smith, L., et al., *J. Am. Chem. Soc.*, *61*, 2615 (1939).
48. JP 31662, to Eisai Co. (1970).
49. Karrer, P., et al., *Helv. Chim. Acta*, *21*, 520 (1938).
50. Smith, L., et al., *Science 88*, 37 (1938).
51. Shiono, M., and Ninagawa, Y., U.S. Patent 4,252,726, to Kuraray Co. (1981).
52. Surmatis, J., and Weber, J., U.S. Patent 2,723,278, to Hoffman-La Roche (1955).
53. Werder, F., U.S. Patent 2,230,659, to Merck & Co. (1941).
54. Ehrmann, O., Ger. Patent 1,015,446 (1957).
55. Gloor, U., et al., *Helv. Chim. Acta*, *49*, 2303 (1966).
56. Moroe, T., et al., U.S. Patent 3,459,773, to Takasago Perfumery Co. (1969).
57. Greenbaum, S., et al., U.S. Patent 3,708,505, to Diamond Shamrock Corporation (1973).
58. Frick, H., et al., U.S. Patent 3,789,086, to Hoffmann-La Roche (1974).
59. Finnan, J., U.S. Patent 4,634,781, to BASF Corporation (1987).
60. Finnan, J., U.S. Patent 4,639,533, to BASF Corporation (1987).
61. Close, R., and Oroshnik, W., U.S. Patent 4,125,735, to SCM Corporation (1978).
62. Mayer, H., et al., U.S. Patent 3,455,959, to Hoffmann-La Roche (1969).
63. Scott, J., et al., U.S. Patent 4,026,907, to Hoffmann-La Roche (1977).
64. Chan, K., and Saucy, G., U.S. Patent 4,094,885, to Hoffmann-La Roche (1978).
65. Chan, K., et al., *J. Org. Chem.*, *43*, 3435 (1978).

66. Cohen, N., et al., *Helv. Chim. Acta*, *64*, 1158 (1981).
67. *Chem. Marketing Rep.*, pp. 4, 21 (May 9, 1988).
68. *St. Louis Post-Dispatch*, p. D-8 (May 6, 1988).
69. Budavari, S., ed., *The Merck Index: An Encyclopedia of Chemicals, Drugs, and Biologicals*, 11th Ed., Merck & Co., Rahway, NJ, 1989, p. 34.
70. Budavari, S., ed., *The Merck Index: An Encyclopedia of Chemicals, Drugs, and Biologicals*, 11th ed., Merck & Co., Rahway, NJ, 1989, p. 966.
71. Kidd, H., and James, D., eds., *The Agrochemicals Handbook*, 3rd ed., Royal Society of Chemistry, Cambridge, England, 1991.
72. Hamm, P., and Speziale, A., U.S. Patent 2,863,752, to Monsanto Chemical Company (1958).
73. Neth. Patent Appl. 6,602,564, to Monsanto Chemical Company (1967).
74. Olin, J., U.S. Patent 3,442,945, to Monsanto Chemical Company (1969).
75. Olin, J., U.S. Patent 3,547,620, to Monsanto Chemical Company (1970).
76. Vogel, C., and Aebi, R., U.S. Patent 3,937,730, to CIBA-GEIGY (1976).
77. Vogel, C., and Aebi, R., U.S. Patent 3,952,056, to CIBA-GEIGY (1976).
78. Hubele, A., U.S. Patent 4,025,648, to CIBA-GEIGY (1977).
79. Moser, H., U.S. Patent 4,032,657, to CIBA-GEIGY (1977).
80. Hubele, A., U.S. Patent 4,046,911, to CIBA-GEIGY (1977).
81. Chupp, J., U.S. Patent 3,574,746, to Monsanto Chemical Company (1971).
82. Olin, J., U.S. Patent 3,630,716, to Monsanto Chemical Company (1971).
83. Olin, J., U.S. Patent 3,637,847, to Monsanto Chemical Company (1972).
84. Chupp, J., and Goodin, R., U.S. Patent 4,258,196, to Monsanto Chemical Company (1981).
85. Sheehan, J., and Frankenfeld, J., *J. Am. Chem. Soc.*, *83*, 4792 (1961).
86. Sheehan, J., and Kurtz, R., *J. Am. Chem. Soc.*, *95*, 3415 (1973).
87. Wise, D., Levendis, Y., and Metghalchi, M., eds., *Calcium Magnesium Acetate, an Emerging Bulk Chemical for Environmental Applications*, Elsevier, Amsterdam, 1991, p. vii.
88. Sugarman, A., *Traveler*, p. 29 (September 1991).
89. Chollar, B., Smith, D., and Zenewitz, J., in *Calcium Magnesium Acetate, an Emerging Bulk Chemical for Environmental Applications* (D. Wise, Y. Levendis, and M. Metghalchi, eds.), Elsevier, Amsterdam, 1991, p. 1.
90. Schenk, R., in *Calcium Magnesium Acetate, an Emerging Bulk Chemical for Environmental Applications* (D. Wise, Y. Levendis, and M. Metghalchi, eds.), Elsevier, Amsterdam, 1991, p. 37.
91. Todd, H., Jr., and Walters, D., U.S. Patent 4,913,831, to Chevron Research Company (1990).

18

Chain Growth: Acrylics and Other Carboxylic Acids

J. Adrian Hawkins and Joseph R. Zoeller
Eastman Chemical Company, Kingsport, Tennessee

I. INTRODUCTION

The low cost of acetic acid, particularly when manufactured by carbonylation of methanol, makes it an attractive starting material for other, longer-chain-length carboxylic acids of commercial value. Unfortunately, there is little in the way of traditional methodology for the condensation or alkylation of the carboxylic acids. Therefore, at present, the commercially desirable carboxylic acids, such as propionic acid, butyric acid, isobutyric acid, acrylic acid, and crotonic acid, are all generated by processes that involve oxidation of the corresponding aldehyde, as either an intermediate or a starting material.

However, several changes in the research environment have led to new processes of potential future importance. All the aldehydes used at present ultimately trace their origin to petrochemical-based olefins. The petroleum shortages of the early 1970s prompted people to shift research efforts toward synthesis gas–based processes. Acetic acid, being a synthesis gas–based chemical, received considerable attention as an alternative starting material. Processes were discovered for the generation of acrylic acid, which is presently generated from propylene, and homologous acids, such as propionic acid and butyric acid, which are based on ethylene and propylene feedstocks, respectively. In addition, radical processes have now regained the interest of a number of organic chemists. As a result, processes were developed for the addition of olefins to acetic acid which represent interesting new approaches to γ-butyrolactones and longer-chain carboxylic acids and which may be practical in the future.

The future of these technologies is unknown because they are not yet practiced commercially. However, these processes are either presently in development or may be required if a shift from petroleum-based processes again becomes desirable. Therefore, they may well represent important future uses of acetic acid and, at this time, bear inclusion in this book.

II. ACETIC ACID–BASED ROUTES TO ACRYLIC ACID AND ACRYLATE ESTERS

A. Introduction

Glacial acrylic acid is a colorless liquid at room temperature with a pungent odor. To reduce its propensity for polymerization, it is usually stabilized with approximately 5 ppm hydroquinone monomethyl ether; however, care must also be taken to ensure that it is not allowed to freeze or be heated above approximately 40°C. During melting, stabilized glacial acrylic acid that has been allowed to freeze usually undergoes partial polymerization owing to zone refinement of the stabilizer. Polymerization will also occur at temperatures above approximately 40°C even in the presence of stabilizer.

The commercially important aliphatic esters of acrylic acid are all colorless liquids when pure and have odors generally described as unpleasant. The esters also tend to undergo polymerization and are usually stabilized with 50–200 ppm hydroquinone monomethyl ether or hydroquinone. The boiling points of the lower aliphatic esters increase and their specific gravities decrease with increasing molecular weights.

Acrylic acid is used primarily as the starting material for production of acrylate esters, including methyl, ethyl, butyl, 2-ethylhexyl, and methoxyethyl. These esters are in turn used in a variety of applications, including adhesives, coatings, resin modifiers, and fiber raw materials. Direct uses of acrylic acid in areas such as superabsorbent resin, detergent builders, and dispersing agents account for a relatively small amount of the acrylic acid consumed worldwide. These applications are, however, experiencing rapid growth in the United States and Western Europe.

Although the majority of acrylic acid and esters are currently produced via the air oxidation of propylene, the instability and upward trend in petrochemical feedstock prices, and the fact that most of the processes displaced by the propylene process were also based on petrochemical feedstock, served to spur research in the area of acrylic acid synthesis based on alternative raw materials. One of the more promising, and proven, acrylic acid feedstocks is acetic acid. Celanese and B. F. Goodrich both operated acetic acid–based acrylic acid plants until the mid-1970s. Production utilizing this process was discontinued primarily because of the toxicity of the propiolactone intermediate and the cost disadvantage compared to the propylene oxidation process. Although the prospect of higher prices and tighter supplies of propylene makes the use of acetic acid more attractive, the propiolactone toxicity issue must still be dealt with. One way of circumventing this problem, the direct condensation of formaldehyde with acetic acid, has been the subject of a flurry of recent research. Although much of this recent research has dealt with the improvement of catalyst selectivities and reaction rates, the data published thus far demonstrate the utility of this chemistry and should provide the impetus for further work in this area.

B. Synthesis via Ketene and Propiolactone

The preparation of ketene from acetic acid has been discussed in detail earlier and will not be covered here. It should be noted, however, that although ketene can be generated from acetone, higher selectivities and conversions to ketene are achieveable with the use of acetic acid, 92.5% selectivity at 38% conversion for acetic acid versus 78% selectivity at 11% conversion for acetone [1,2]. For this reason, the majority of commercial ketene/based acrylic acid production used acetic acid, rather than acetone, as the feedstock. The reaction of ketene with formaldehyde to produce propiolactone proceeds under extremely mild conditions and with a variety of catalysts and solvents.

$$H_2C{=}C{=}O + H_2C{=}O \xrightarrow[0-25°C]{Cat.} \quad \qquad (1)$$

Among the catalysts that have been used to effect this reaction are $ZnCl_2$, BF_3 etherate, $Zn(ClO_4)_2$, $Zn(SCN)_2$, $Zn(NO_3)_2$, silica-alumina, activated aluminas and clays, H_3BO_4, and $AlCl_3$ [3]. The catalysts of choice for commercial production were $ZnCl_2$ and $AlCl_3$ [4,5]. Although methanol, diethyl ether, and acetone are all effective solvents for the reaction, methanol was used in initial commercial production because of the higher catalyst solubility and resulting increase in throughput [5]. Methanol also provided for easier product isolation in the form of cleaner distillations. The most recent commercial production of acrylic acid via this route was conducted in propiolactone itself [5]. By-products from this reaction are largely acetic acid and anhydride produced via the reaction of ketene with water. Optimized propiolactone yields in excess of 90% are observed with these systems.

The conversion of propiolactone to acrylic acid can be effected either through the ring-opening polymerization of the propiolactone followed by thermal depolymerization of the propiolactone polymer or through rearrangement of the propiolactone directly to acrylic acid. Propiolactone can be quantitatively polymerized at 125°C in the presence of weak, solid base such as sodium carbonate [6]. Neat propiolactone polymer can be efficiently depolymerized at 200°C, in the absence of catalyst, allowing the liberated acrylic acid to be condensed rapidly and efficiently, thus minimizing losses to polymerization [7].

$$\underset{125^\circ C}{\xrightarrow{\text{base}}} \quad \xrightarrow{200^\circ C} \qquad (2)$$

The direct rearrangement of propiolactone to acrylic acid can be accomplished through the use of strong acid catalysts such as sulfuric or phosphoric acids. Commercially, 100% phosphoric acid was used because of its low cost, ruggedness, and efficiency. In 100% phosphoric acid at 170°C, quantitative conversion of propiolactone to acrylic acid is observed. Very little by-product formation is observed with either of these methods for the conversion of propiolactone to acrylic acid.

$$\underset{170^\circ C}{\xrightarrow{100\% \ H_3PO_4}} \qquad (3)$$

C. Synthesis from Methanol and Acetic Acid or Methyl Acetate

Although it has not been carried out on a commercial scale, the condensation of formaldehyde with acetic acid to produce acrylic acid has received interest from the industrial sector for over 30 years. Much of the earlier work in this area is recorded in the patent literature and, as such, does not provide a coherent, detailed overview. A number of detailed investigations of this reaction have been published recently, however, and have contributed greatly to our understanding of this and other reactions of C_1 chemicals of potential commercial value for the preparation of acrylic acids and esters. The overall reaction can be expressed by reaction (4).

$$\underset{\text{OH}}{\overset{\text{O}}{\parallel}}\ +\ H_2C{=}O\ \xrightarrow{\text{Cat.}}\ \underset{\parallel\quad\text{OH}}{\overset{\text{O}}{\parallel}}\ +\ H_2O \tag{4}$$

Studies of this and related reactions, e.g., the direct reaction of methanol with acetic acid or methyl acetate, have typically been conducted in the vapor phase at temperatures in the 200–400°C range when transition metal oxides were employed as catalysts and at temperatures of 350–550°C when nontransition metal oxides or nonmetal oxides were used. Depending on the reaction conditions, stoichiometry, catalysts, and reactants employed, the final products can be shifted toward acrylic acid or acrylate esters. The condensation of formaldehyde with acetic acid produces only acrylic acid and carbon dioxide when the reaction is run under the optimum conditions with an efficient catalyst. When formaldehyde is condensed with acetate esters or when the formaldehyde is generated from the in situ oxidation of methanol and subsequently condensed with either acetic acid or acetate ester, a mixture of acetic acid, acrylic acid, their esters, and formaldehyde results. A survey of the patent literature shows that acidic, basic, and amphoteric compounds have all been claimed as useful catalysts for these reactions [8–13]; however, recent investigations suggest that both acidic and basic sites are required [14–16]. In the publications of his studies of the reactions of formaldehyde or methanol with either acetic acid or methyl acetate, Ai refers to the reactions as aldol condensations and infers that both acidic and basic sites are required [14,15]. Support for this inference is provided by his catalyst composition optimization work. For example, when formaldehyde was reacted with acetic acid over a combination of vanadyl and titanium pyrophosphates, at 320°C in a continuous-flow, tube reactor, acrylic acid was produced in 96% yield. The catalyst activity could be maintained indefinitely by including 3.2% oxygen in the feed. In experiments conducted in which no oxygen was included in the feed, the catalytic activity decreased steadily with time on-stream, but could be totally regenerated by treatment of the catalyst with oxygen at 350°C.

The ruggedness of the catalyst, when employed in the presence of air, is attributed to its redox properties. Catalysts containing V/Ti ratios of 0.25–1.00 and P/metal ratios of 1.8–2.2 were found to be most effective. At V/Ti ratios lower than 0.25 or higher than 1.00, acrylic acid yields decreased. At lower P/metal ratios, acrylic acid yields decreased and carbon dioxide production increased. This observation led the author to conclude that the increase in basic sites induced by the decreased P content was responsible for the carbon dioxide formation. The effects of additives to this catalyst and the efficiencies of several other metal oxide and mixed metal oxides for this reaction were also studied. Acidic oxides, such as

WO_3, MoO_3, and V_2O_5, and amphoteric oxides, such as TiO_2, SnO_2, Fe_2O_3, and NiO, are not effective as catalysts for this reaction when used alone. The combination of WO_3 with an amphoteric oxide resulted in the formation of effective catalysts. Nontransition metal oxides such as SiO_2—B_2O_3 and silica-alumina were found to be effective in catalyzing the condensation reactions; however, they could not be easily regenerated. The throughput in optimum yield experiments reached 0.07 g of acrylic acid per gram of catalyst per hour. Increases in temperature increased the throughput slightly but also resulted in lower yields with a concomitant increase in carbon dioxide.

Ueda et al. found that the rate of reaction of methanol with the series acetophenone, acetone, acetonitrile, methyl propionate, and toluene showed decreasing reaction rates with increasing pKa over Cr— and Fe—MgO catalysts [16]. They also found that, of the other metals screened, the catalytic efficiency in converting methyl propionate to methyl methacrylate decreased as Mn > Cr > Cu > Ni > Al. In a separate study, these authors also found that the same metals, when incorporated in fluoro tetrasilicic mica, were also effective catalysts for these reactions as well as the condensation of methanol and methyl acetate. Under optimum conditions, the total selectivity for the acetate to acrylate moiety conversion was greater than 95% at conversions of greater than 13%. From the results in these studies they concluded that the probable mechanism in these reactions is simlar to the aldol condensation reactions and consists of the rate-determining proton abstraction from the acidic compound (ester, ketone, or acid) by a basic surface site, stabilization of the intermediate anion by metal-enhanced Lewis acid sites, formation of a formaldehyde intermediate from methanol via dehydrogenation, and, finally, condensation of the formaldehyde and anion intermediates.

III. HOMOLOGATION OF ACETIC ACID WITH SYNTHESIS GAS

In addition to acrylic acid, the homologous series of carboxylic acids, particularly propionic, n-butyric, and isobutyric acids, would be useful targets for a synthesis gas–based process if they could be derived from acetic acid. These materials are very useful in the form of their calcium salts as food and grain preservatives. The longer-chain carboxylic acids are very useful in modifying the properties of cellulose esters where they impart softness, flexibility, and increase hydrophobic properties. In general, owing to the higher cost, the longer-chain carboxylic acids are used only when improvements in performance warrant the extra cost, as in the case of the cellulose esters mentioned above.

In the early 1980s, John Knifton of Texaco reported a series of successful homologations of carboxylic acids using synthesis gas in the presence of ruthenium [17–20], rhodium [21], and palladium [22], which was subsequently extended by others [23–28]. When acetic acid was used as the carboxylic acid, the primary products were propionic acid, n-butyric acid, isobutyric acid, and traces of higher acids. Of these three catalysts, only the ruthenium catalyst system has been examined in detail. A number of mechanistic conjectures have appeared but, at present, no study has generated sufficient information to justify a choice among any of the various proposals.

However, work by Zoeller [26] and Luetgendorf et al. [28] has clearly demonstrated that, at least in the case of ruthenium, this is not a one-step process. Instead, the process is composed of a sequence of reactions initiated by a ruthenium-catalyzed hydrogenation of acetic acid to ethanol, followed by a conversion of the alcohol (or its corresponding ester) to the iodide, followed by carbonylation of the iodide. This sequence of events is depicted below. Higher carboxylic acids would result as the propionic acid product competes with acetic acid in the analogous reactions.

$$\textit{Step 1} \quad AcOH + 2H_2 \longrightarrow EtOH + H_2O \tag{5}$$

$$\textit{Step 2} \quad EtOH + HI \longrightarrow EtI + H_2O \tag{6}$$

$$\textit{Step 3} \quad EtI + CO + H_2O \longrightarrow EtCO_2H + HI \tag{7}$$

$$\text{Net reaction} \quad AcOH + CO + 2H_2 \longrightarrow EtCO_2H + 2H_2O \tag{8}$$

Assuming the difficult separation of the homologous series of acids could be accomplished, one unfortunate serious flaw will still limit the potential of this process. The water formed as coproduct in the reaction appears to inhibit the initial hydrogenation and, therefore, limits the ultimate conversion in this process to about 30–40%. Owing to this limitation, coupled with the difficulties inherent in separating such a diverse number of related products, this technology is not likely to be commercialized without some significant breakthroughs in both chemistry and engineering around this process.

IV. ADDITIONS TO OLEFINS: CHEMISTRY OF THE CARBOXYMETHYL RADICAL

A. Introduction

As mentioned earlier, there are few classic methods for the alkylation of acetic acid, and, as a result, acetic acid derivatives are more likely to be used for this purpose. However, in the 1960s and 1970s, several groups examined the generation of the carboxymethyl radical ($^{\cdot}CH_2CO_2H$) from acetic acid and its subsequent reaction with olefins [29–46]. When free carboxymethyl radicals are formed, these reactions lead to alkylation. However, when the radicals are generated as part of a metal-bound complex, γ-butyrolactones are formed instead. The processes involved in the two pathways and the end uses for the subsequent products are distinctly different and will be discussed separately.

B. Alkylation of the Carboxymethyl Radical with Olefins

The free radical–induced alkylation of acetic acid with olefins is one of the few methods available for the direct alkylation of acetic acid. More generally, one is forced to resort to more indirect processes using acetic acid derivatives or synthetic equivalents, such as malonic acid or its esters. The identification of methods for the direct alkylation of very inexpensive acetic acid has some economic incentive, particularly if the method can yield carboxylic acids of considerable chain length. These longer-chain carboxylic acids are very useful in generating a number of surfactants (soaps and detergents).

The most common method of generating the carboxymethyl radical has been the decomposition of di-*tert*-butyl peroxide [Eq. (9)], although a report in the Japanese patent literature indicates that peracetic acid is more efficient in this application [41]. Generation of the carboxymethyl radical with the one-electron oxidant manganese (III) acetate has also been reported as a means of initiating the process [43–46].

$$(CH_3)_3COOC(CH_3)_3 \longrightarrow 2(CH_3)_3CO^\cdot \longrightarrow 2CH_3^\cdot + 2CH_3COCH_3 \quad (9)$$

However, the use of manganese (III) acetate as a means of generating the carboxymethyl radical is somewhat complicated. Generally, γ-butyrolactones are formed as the preferred product, unless acetic anhydride is either added in significant quantities or, preferably, used to completely replace the acetic acid. Processes initiated by Mn(III) are distinctly different processes and will be discussed in more detail in the next section of this chapter.

Methyl acetate is a suitable replacement for acetic acid, if an ester product is preferred. However, using higher esters leads to olefination at the α-carbon of the alkyl portion of the ester, and therefore, higher esters are unsuitable for use in the reaction [29,39].

With the notable exception of ethylene, the reactions are fairly selective with 1-alkenes, giving selective addition to the terminal carbon for addition of a single olefin moiety to the carboxymethyl radical representing the prevailing process. This process is represented in reactions (10)–(12). Unfortunately, in general, the yields are insufficient for most large-scale commercial applications at this time. Although other acetic acid derivatives would need to be considered [29], this reaction could be useful in the generation of specialty acids. One such application has been the selective generation of terminal diacids from diolefins, where the product diacid was four carbons greater in length than the starting diene [42]. (These diacids are very useful polyester intermediates.)

Two secondary reactions are significant in this process and are particularly pronounced in the case of ethylene. In the first competing process, a second olefin moiety can react with the radical intermediate obtained from the addition of the carboxymethyl radical to the olefin before it is quenched by acetic acid. This reaction, which has been referred to as olefin telomerization, is represented in Eq. (13) and can happen several times prior to acetic acid quenching.

The second competing process is branching. This likely occurs by hydrogen abstraction from the α-carbon either via an intermolecular process in which the longer-chain carboxylic acid products compete with acetic acid for the radical initiator or by intramolecular transfer of the α-hydrogen atom. This process can be particularly pronounced when ethylene is used as the olefin source.

These side reactions can be useful. Since the telomerization is the predominant process when ethylene is used as the olefin, this process can be used as a method of generating the family of even-numbered carboxylic acids. Therefore, this process has been well studied [29,32–41] and an optimization of the ethylene process has been published [36].

$$R^{\cdot} + AcOH \longrightarrow {}^{\cdot}CH_2CO_2H \tag{10}$$

$$AcOH + H_2C{=}CHR \longrightarrow {}^{\cdot}C(R)HCH_2CH_2COOH \tag{11}$$

$$^{\cdot}C(R)HCH_2CH_2COOH + AcOH \longrightarrow$$
$$HC(R)HCH_2CH_2COOH + {}^{\cdot}CH_2CO_2H \tag{12}$$

$$^{\cdot}C(R)HCH_2CH_2COOH + n\ H_2C{=}CHR \longrightarrow$$
$$^{\cdot}C(R)HCH_2(RCHCH_2)_{n-1}CH_2COOH \tag{13}$$

C. Reaction of Olefins with Metal-Bound Carboxymethyl Radicals: Generation of γ-Butyrolactones

Olefination of the carboxymethyl radical generated from the interaction of one-electron oxidants represents a somewhat different process than the processes in which the radical is generated from organic free radical precursors. The predominant products in these processes are not the simple addition of an olefin, but are γ-butyrolactones instead. This reaction is represented by Eq. (14). By far, the predominant initiator has been

Mn(III), usually a manganese (III) acetate [47–56]. However, Ce(IV) and V(V) have also been shown to initiate the reaction [49,51].

$$RCH{=}CH_2 + CH_3COOH + 2M^{n+} \longrightarrow 2M^{(n-1)+} + \quad\quad\quad\quad (14)$$

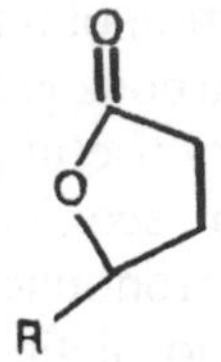

Very thorough studies of the generation of γ-butyrolactones using manganese (III) acetate have been undertaken, and the process has been reasonably well optimized. The yields can be quite good. For example, a yield of 88% was achieved with 1-octene [54]. The yield has been shown to be substantially improved by [54]:

1. The presence of an acetate salt (KOAc)
2. Generation of the manganese (III) acetate from hydrated manganese salts, as opposed to anhydrous forms of manganese (III) acetate precursors
3. Using excess oxidant, with an optimal yield achieved with 4.5 equiv. of Mn (III)

These observations, and several more specific experiments, led to a substantiated mechanism [55]. The manganese (III) acetates are generally believed to exist as oxo-centered clusters with three manganese atoms bound around the central oxygen atom as opposed to true stoichiometric complexes. The removal of a proton by acetate anion is apparently rate limiting and irreversible on the complex. This is based on the observation that the reaction is faster than proton exchange. This leads to a hypothetical manganese cluster–bound enolate, which rapidly leads to the manganese-bound carboxymethyl radical, which subsequently reacts with the olefin. Evidence indicates that neither the carboxymethyl radical nor the radical generated upon olefin addition dissociates from the cluster, but instead reacts in an intramolecular fashion using the manganese cluster as a pseudotemplate. This mechanism is best described pictorially and appears in Scheme (1). Other mechanistic conjectures were made previous to this study, but were unsubstantiated.

As mentioned earlier, addition of acetic anhydride shifts the selectivity toward simple alkylation. However, it does so at the expense of lactone formation [52–54]. The effect of acetic anhydride is well quantified in these studies, but the cause of the shift in selectivity is still not entirely clear. However, one might make the conjecture that the manganese acetate clus-

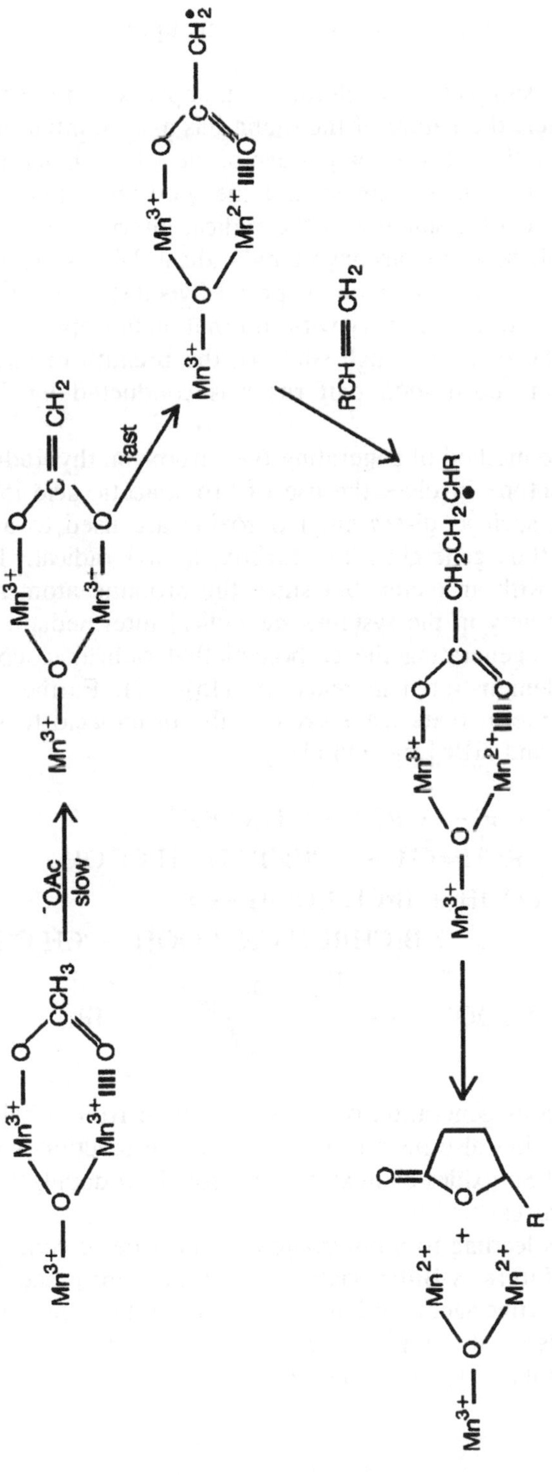

Scheme 1 Mechanism for the Manganese (III) assisted generation of γ-butyrolactones for acetic acid and olefins.

ters involved in the lactone formation are disrupted by the presence of acetic anhydride.

Based on the examples in the literature, it appears that most olefins are reactive. However, the nature of the olefin has a substantial influence on both the yield and the substitution pattern of the γ-butyrolactone product. The position of the substituents around the γ-butyrolactone products is reflective of the relative stability of the radical intermediate obtained by addition of the olefin to the carboxymethyl radical. With simple 1-alkenes, this means that the γ-butyrolactone is primarily substituted at the γ position. The relative stereochemistry with internal olefins appears to be sterically guided. The most thorough study of the breadth of usable olefins and the resultant substitution patterns was conducted by Fristad and Peterson [54].

An alternative method of generating the carboxymethyl radical to generate γ-butyrolactone involves the use of bromoacetic acid [57–59]. Ordinary initiators such as di-*tert*-butyl peroxide are used to abstract the bromine atom, thus generating the carboxymethyl radical. The radical reacts normally with an olefin, but since the bromine atom is the most easily reacted moiety in the system, the radical intermediate removes a bromine atom, regenerating the carboxymethyl radical to continue the chain. This is demonstrated in reactions (15)–(17). Further heating of the bromoacid intermediate leads cyclizes the bromoacid to the desired γ-butyrolactone and HBr [reaction (18)].

$$BrCH_2COOH + R^· \longrightarrow RBr + {}^·CH_2COOH \qquad (15)$$

$$^·CH_2COOH + RCH{=}CH_2 \longrightarrow {}^·CHRCH_2CH_2COOH \qquad (16)$$

$$^·CHRCH_2CH_2COOH + BrCH_2COOH \longrightarrow$$

$$BrCHRCH_2CH_2\ COOH + {}^·CH_2COOH \qquad (17)$$

$$BrCHRCH_2CH_2COOH \longrightarrow \quad \text{[structure]} + HBr \qquad (18)$$

A heterogeneous generation of γ-butyrolactone from ethylene, acetic acid, and oxygen has also been reported [60]. The reaction is carried out over U, As, and Sb oxides or mixtures thereof. Few details are available regarding the process.

The processes leading to γ-butyrolactones may be of considerable importance in the future. γ-Butyrolactones are important precursors to several biologically active agents and therefore represent potentially important starting materials for useful pharmaceuticals. In addition, these materials are useful in the flavorings and fragrance industries.

γ-Butyrolactone itself is used sparingly as a polymer intermediate, but its use would be more widespread if the cost could be reduced. The most important use for γ-butyrolactone is as a precursor for *N*-methyl pyrolli-dinone (NMP). NMP is a very useful polar aprotic solvent since it is generally regarded as nontoxic. As chlorinated solvents come under increasing scrutiny, this material has been the leading candidate for their replacement in a number of applications, particularly in the application of coatings in the industrial sector.

V. CONCLUSION

The future of the new technologies described in this chapter is uncertain since none are presently practiced and will depend heavily on changing economic factors. However, if petroleum-based feedstocks should again come under pressure, the syngas-based processes to acrylic acid and longer-chain carboxylic acids described in this chapter may become significant contributors to the chemical industry in the future. The carboxymethyl radical olefination processes, particularly the γ-butyrolactone, are still in the development stage and may begin to contribute to the overall supply of organic acids in the future.

REFERENCES AND FOOTNOTES

1. Froment, G., and Goethals, H. G., *Chem. Eng. Sci.*, *13*, 173–179 (1961).
2. Robeson, M. O., and Taylor, W. E., GB Patent 763,018 (1956).
3. Zaugg, H. E., *Organic Reaction*, Vol. VIII, Wiley, New York, 1954, Chapter 7, pp. 305–363, and references therein.
4. Kung, E. F., U.S. Patent 3,106,577 (1963).
5. Dunn, D. A., U.S. Patent 3,069,433 (1962).
6. Wearsch, N. C., and DePaola, A. J., U.S. Patent 3,002,017 (1962).
7. Schnizer, A. W., and Wheeler, E. N., U.S. Patent 3,176,042 (1965).
8. Holmes, J. D., U.S. Patent 4,085,143 (1978).
9. Ryu, J., U.S. Patent 4,560,790 (1985).
10. Hagenmeyer, H. J., Blood, A. E., and Snapp, T. C., U.S. Patent 3,928,458 (1975).
11. Leathers, J. M., and Woodward, G. E., U.S. Patent 3,051,747 (1962).
12. Sims, V. A., and Vitcha, J. F., U.S. Patent 3,247,248 (1966).
13. Schneider, R. A., U.S. Patent 4,165,438 (1979).
14. Ai, M., *Proceedings, 9th International Congress on Catalysis, Calgary, 1988*, Vol. 4, Chemical Institute of Canada, Ottawa, pp. 1562–1569.
15. Ai, M., *Appl. Catal.*, *59*, 227–235 (1990).
16. Ueda, W., Yokoyama, T., Kurokama, H., Moro-oka, Y., and Ikawa, T., *Sediyu Gakkaishi*, *29*(1), 72–79 (1986).

17. Knifton, J. F., *Chemtech*, *11*, 609 (1981).
18. Knifton, J. F., *Hydrocarbon Proc.*, 113 (1981).
19. Knifton, J. F., *ACS Symp. Ser.*, *152*, 225 (1981).
20. Knifton, J. F., *J. Mol. Catal.*, *11*, 91 (1981).
21. Knifton, J. F., U.S. Patent 4,334,092 (1982).
22. Knifton, J. F., U.S. Patent 4,334,094 (1982).
23. Dombek, B. D., U.S. Patent 4,897,473 (1982).
24. Drent, E., European Patent Appl. 64,287 (1983).
25. Drent, E., European Patent Appl. 75,335 (1983).
26. Zoeller, J. R., *J. Mol. Catal.*, *37*, 377 (1986).
27. Ono, H., Hashimoto, M., Fujiwara, K., Sugiyama, E., and Yoshida, K., *J. Organomet. Chem.*, *331*, 387 (1987).
28. Luetgendorf, M., Elvevoll, E., and Roper, M., *J. Organomet. Chem.*, *289*, 97 (1985).
29. The addition of α-carboxylic acid radicals has been reviewed several times. See (a) Vogel, H., *Synthesis*, 99 (1970). (b) Ramaiah, M., *Tetrahedron*, *43*, 3541 (1987). (c) Giese, B., *Angew. Chem. Int. Ed. Engl.*, *22*, 753 (1983). (d) Giese, B., *Angew. Chem. Int. Ed. Engl.*, *28*, 969 (1989).
30. Allen, J. C., Cadogan, J. I. G., and Hey, D. H., *Chem. Ind.*, 1621 (1962).
31. Allen, J. C., Cadogan, J. I. G., and Hey, D. H., *J. Chem. Soc.*, 1918 (1965).
32. Allen, J. C., Cadogan, J. I. G., and Hey, D. H., Belg. Patent 621,365 (1963).
33. Banes, F. W., Fitzgerald, W. P., and Nelson, F. F., U.S. Patent 2,585,723 (1966).
34. Seitz, A., Jr., and Moote, T. P., Jr., *Ind. Eng. Chem. Process Res. Dev.*, *1*, 132 (1962).
35. Roe, E., Konen D.A., and Swern, D., *J. Am. Oil Chem. Soc.*, *42*, 457 (1965).
36. Suhara, Y., *Bull. Chem. Soc. Jpn.*, *46*, 990 (1973).
37. Suhara, Y., *Yukagaku*, *25*, 522 (1976).
38. Freidlina, R. K., Terent'ev, A. B., Ikonnikov, N. S., and Churilova, A. M., *Dokl. Akad. Nauk. SSSR*, *208*, 1366 (1973).
39. Ikonnikov, N. S., Terent'ev, A. B., Churilova, M. A., and Freidlina, R. K., *Izv. Akad. Nauk. SSSR, Ser. Khim.*, 2479 (1972).
40. Grinevich, I. A., Zagorets, P. A., Shostenko, A. G., Dodonov, A. M., and Myshkin, V. E., *Chem. Abstr.*, *87*, 53670w (1975).
41. Miyazaki, T., Munemiya, S., and Tasaka, A., Jpn. Kok. 48078115 (1973).
42. Bradney, M. A., Forbes, A. D., and Wood, J., *Am. Chem. Soc.*, *Div. Petrol. Chem.*, *Prepr.*, *16*, B20 (1971).
43. Klein, W. J., *J. Roy. Neth. Chem. Soc.*, *94*, 48 (1975).
44. Okano, M., *Chem. Ind.*, 423 (1972).
45. Klein, W. J., U.S. Patent 4,014,910 (1977).
46. Simonnot, R. S., Fr. Demande 2,010,295 (1970).
47. Bush, J. B., and Finkbeiner, H. L., *J. Am. Chem. Soc.*, *90*, 5903 (1968).
48. Heiba, E. I., Dessau, R. M., and Koehl, W. J. *J. Am. Chem. Soc.*, *90*, 5905 (1968).
49. Heiba, E. I., Dessau, R. M., and Rodewald, P. G., *J. Am. Chem. Soc.*, *96*, 7977 (1974).

50. Dessau, R., and Heiba, E., U.S. Patent 3,992,417 (1976).
51. Heiba, E. I., and Dessau, R. M., U.S. Patent 4,175,089 (1979).
52. Okano, M., *Bull. Chem. Soc. Jpn.*, *49*, 1041 (1976).
53. Midgley, G., and Thomas, C. B., *J. Chem. Soc. Perkin Trans.*, *II*, 1537 (1984).
54. Fristad, W. E., and Peterson, J. R., *J. Org. Chem.*, *50*, 10 (1985).
55. Fristad, W. E., Peterson, J. R., Ernst, A. B., and Urbi, G. B., *Tetrahedron*, *42*, 3429 (1986).
56. Heiba, E. I., and Dessau, R. M., U.S. Patent 4,285,868 (1981).
57. Kharasch, M. S., Skell, P. S., and Fisher, P., *J. Am. Chem. Soc.*, *70*, 1055 (1948).
58. Nakano, T., Kayama, M. and Nagai, Y., *Bull. Chem. Soc. Jpn.*, *60*, 1049 (1987).
59. Nakano, T., Kayama, M., Matsumoto, H., and Nagai, Y., *Chem. Lett.*, 415 (1981).
60. Murib, J. H., U.S. Patent 4,247,467 (1981).

19

Key Properties and Hazards

Peter N. Lodal
Eastman Chemical Company, Kingsport, Tennessee

I. INTRODUCTION

This chapter deals with the key safety properties and special hazards associated with acetic acid and derivatives discussed in other chapters of this book. In particular, reactive hazards associated with acetic anhydride and diketene are discussed in detail. Material Safety Data Sheets (MSDSs) for some of the largest-volume derivatives are provided at the end of this chapter.

II. ACETIC ACID

Acetic acid is a monofunctional organic acid with a pKa of about 4.5. Although not as strong as most mineral acids or formic acid, it is still sufficiently corrosive to warrant significant safety precautions.

A. Flammability

Acetic acid has a TCC flash point of 39°C (103°F), making it borderline between flammable and combustible. Its autoignition point is 516°C (960°F), making it readily ignitable by open flame and heat. However, flammability decreases significantly when water is added to the mixture. Table 1 summarizes experimental flash point data for various mixtures of acetic acid and water in air. Beyond 56% acid (weight), no flash point was discernible.

Table 1 Flash Points of
Acetic Acid/Water Mixtures

Acetic acid in water, wt%	Flash point, °C (°F)
100	39 (103)
95	56 (133)
90	61 (142)
80	65 (149)
70	83 (183)
56	91 (196)
40	No flash
30	No flash

Source: Ref. 13.

Fenlon [1] has used similar data to generate the flammability chart shown in Figure 1. Shraer [2] has published similar data for mixtures at elevated pressures.

It should be noted that use of solvents for separation of acetic acid and water can change the flash point significantly. For example, 30% acetic acid in water showed no flash point per the data in Table 1. However, additions of small amounts of isopropyl acetate (flash point 2°C, 35°F) changed the measured mixture flash point dramatically, as shown in Table 2.

Other physical safety hazards and handling precautions are given in the MSDS Appendix. In addition, numerous publications deal with the handling precautions and prevention and mitigation of accidental spills [3–8]. The information contained in the MSDS Appendix is up to date as of the publication date; readers are cautioned to contact their chemical supplier for any more recent information available.

A related hazard in the manufacture of acetic acid and acetic anhydride is the nature of the catalyst system. Use of methyl iodide and metal carbonyls may require the use of a secondary containment system to minimize the probability of, and provide for mitigation of, a potential toxic vapor cloud release.

III. ACETIC ANHYDRIDE

Acetic anhydride is much like acetic acid in its physical appearance and hazards. It is significantly different in one major respect: it is only partly miscible with water at low temperatures (<45°C) in the absence of acetic acid. Figure 2 shows a ternary phase diagram [9] illustrating this point.

Table 2 Flash Points of a 30%
Acetic Acid/Water Mixture with
Isopropyl Acetate

Isopropyl acetate in mixture, %	Flash point, °C (°F)
0.3	67 (152)
1.0	52 (126)
3.0	22 (72)

Source: Ref. 14.

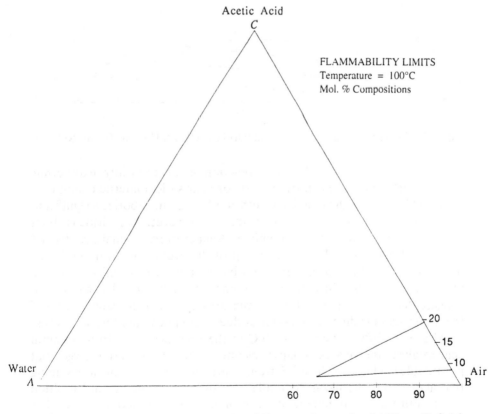

Figure 1 Flammability limits for acetic acid/water mixtures in air. (From Ref. 1.)

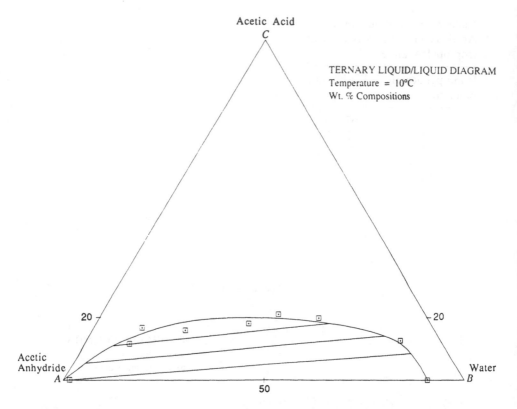

Figure 2   Ternary diagram for acetic anhydride/acetic acid/water. (From Ref. 9.).

Since the acetic anhydride/water reaction is energetically exothermic [~25,000 Btu/lbmol), at stoichiometric or near stoichiometric conditions, the heat liberated in the reaction is sufficiently large to vaporize a significant fraction of the resulting acetic acid product. Since acetic anhydride is about 5% more dense than water near ambient temperatures, a small, controlled addition of water may layer out on top of the acetic anhydride until sufficient acetic acid has been generated by the partial miscibility to tie the two phases together in a nearly uncontrolled manner. This has led to catastrophic eruptions of hot acidic mixtures upon violent vaporization of the acetic acid product, with fatalities due to chemical and thermal burns.

    Above 45–50°C, or above −20°C in the presence of a strong mineral acid catalyst, the phases collapse instantly, with the resulting large (but predictable) release of heat. Extreme care should be taken in assuming that above 50°C the hazard potential is insignificant; the actual transition temperature is nearly impossible to reproduce in the laboratory and has been observed to go as high as 65°C with no appreciable reaction [10].

The comments on water reactivity for acetic anhydride also apply in a general way to diketene; however, diketene is far more reactive at a given temperature than acetic anhydride, so the layering potential is far less. In many cases, adequate vessel protection devices will be difficult or impossible to design with an acceptable degree of reliability, even with DIERS test data availability [11,12]. Therefore, prevention, rather than protection, systems may need to be used to avoid hazardous situations.

To adequately assess this potential, the cases of liquid water versus steam leakage will be considered separately.

A. Steam Leak

The most common cause of steam leakage into a process would be a ruptured heater tube. Calculations show that, if full steam input is assumed, and all the water reacts instantaneously with the acetic anhydride, extremely large multiple relief devices would be needed to adequately protect the vessel [7,11,12,15]. However, calculated steam leakages are usually 2–5 times greater than the normal maximum steam flow allowed by the trim on the steam flow control valve.

One prevention system design employs redundant column base pressure measurement as the diagnostic for uncontrolled water addition into the column. As the pressure rises, the steam flow can be halted by triggering a column emergency shutdown procedure (e.g., stop steam and feed flow, maximize cooling water to the condenser). This solution requires that all affected steam loops have a steam flow controller. Additionally, the condensate from the reboiler should be diverted to prevent contamination of the steam system from backleakage of process through the severed tube.

B. Liquid Water Leak

This section covers suggested protection/prevention methods for systems where there is the potential for liquid water and anhydride to mix. These systems are primarily large storage vessels with upstream cooling or vent condensers, smaller pressure vessels (reflux accumulators, sidedraw tanks), and their associated condensers and coolers.

The reason for handling these situations separately is the partial miscibility potential of the water/anhydride system; the potential for water added to anhydride to "layer out" until sufficient acetic acid has been formed (through slow reaction via the small mutual solubility) to collapse the phases. When the phases collapse, water is added to the anhydride at an extremely high molar flow rate. Since the water/anhydride reaction is very exothermic, this high, uncontrolled addition rate may result in a large temperature transient, with potential eruption due to rapid vapor generation.

Three specific systems will be considered: large, atmospheric storage vessels, smaller pressure vessels, and the heat exchangers associated with these vessels.

1. Large Atmospheric Storage Vessels

Large atmospheric storage vessels present the greatest potential risk, because they contain the largest volume of acetic anhydride and are rated to contain little or no pressure. To prevent a potentially hazardous situation from developing, an early warning system using a small resin bed in a sidestream from the normal tank feed stream, downstream of the product coolers and/or condensers, should be installed. This bed, containing a cationic ion exchange resin, mimics the addition of a strong mineral acid catalyst (such as sulfuric acid), which promotes the anhydride/water reaction at ambient temperatures. Coupled with temperature detectors up and downstream of the bed, this device can give early warning of a tube leak and initiate action to find and isolate the leak source.

It is recommended that no automatic action be initiated as a result of a temperature alarm. This is due to the fact that, depending on the magnitude of the leak, temperature could go either up (i.e., a small leak with reaction) or down (i.e., a massive leak where the temperature detectors are swamped with cold water). Also, if automatic action shutting off cooling water to a suspected leaking exchanger were taken, a bad situation could be made worse by increasing tank temperature and promoting phase collapse of water already in the tank.

For these same reasons, the temperature measurements should have both positive and negative deviation alarms (in addition to a delta T alarm), so that any drastic change from the norm is alarmed.

2. Small Pressure Vessels

These vessels, primarily reflux accumulators and other similar small tanks, present a smaller risk for two reasons: (1) the inventory of anhydride is much smaller, and (2) they can withstand much more of a pressure transient (which would likely occur in the event of a phase collapse) without rupturing. Because of these differences, the resin bed monitoring system is not recommended for these vessels. Rather, a temperature indicator with alarms in the tank itself is recommended to call attention to any sudden temperature change, whether positive or negative.

3. Heat Exchangers

The heat exchangers associated with storage tanks and pressure vessels are the most likely source for the water. Potential failures could be due to construction errors (poor tube to tubesheet joints and split tubes, in the case of shell-and-tube heat exchangers), corrosion, or mechanical damage. To minimize the potential for water leakage into the process, seal-welded tubes for all exchangers in anhydride service are recommended. More frequent inspection of anhydride service exchangers may also be warranted, and leak checking of these exchangers after every major pressure vessel inspection shutdown, by running cooling water through the unit before chemicals are reintroduced, visually checking for leaks to the process, is also recommended.

Another potential solution for exchangers is to always guarantee that the acetic anhydride or diketene is at a pressure above that of the cooling water, so that any leakage occurs from organic to aqueous side, thus limiting the molar rate of addition.

C. Emergency Vent Discharges

Emergency vent (rupture disks and relief valves) discharge should be kept entirely separate from the normal process vent header system. Off-normal discharges do not require an air permit; therefore, safety and personnel protection issues dictate the design.

D. Vapor Relief of Acetic Acid and Anhydride

For acid and anhydride (neither of which represents a major vapor cloud potential), vapor discharge directed up and away from the relief device should be sufficient. Rainout (entrained liquid) and limited-access elevated locations (such as building roofs and distillation column platforms) in the downstream plume path should be considered in determining the location and elevation of the relief discharge.

Particularly for anhydride, a vent into a scrubbed standpipe, with water seal at the base, may be necessary to prevent noxious fumes from escaping into an occupied area (control room, etc.).

IV. DIKETENE

Diketene, unlike acetic anhydride, has a significant potential to autopolymerize. This is potentially dangerous for two reasons: First, a large amount of carbon monoxide gas is released during the polymerization reaction, which can overpressure the containment vessel in a short period of time. Second, the polymerization reaction proceeds rapidly at ambient temperatures, thus requiring a highly reliable refrigeration source to prevent its initiation. Again, as is the case with acetic anhydride, reliable design of vents and relief devices is tenuous at best [11,12,15]. Protection mechanisms, including a liquid dump into an unconfined area, are the best method for prevention of vessel rupture.

MATERIAL SAFETY DATA SHEETS

MATERIAL SAFETY DATA SHEET

EASTMAN CHEMICAL PRODUCTS, INC.

Kingsport, Tennessee 37662

For Emergency Health and Safety Information, Call: (615) 229-6094

For Other Information, Call Your Eastman Representative

Eastman Operator: (615) 229-2000   Date of Preparation: April 19, 1990

Replaces Previous Editions

IDENTIFICATION AND USE

—Name: Glacial Acetic Acid

—Chemical Name: acetic acid
—Synonyms: PM 21, PM 3598
—Molecular Formula: $C_2H_4O_2$
—Molecular Weight: 60.05
—Product Use: solvent

COMPOSITION

Component(s)	Approx Weight %	CAS Reg No	Eastman Kodak No.
Acetic acid	100	64-19-7	900763

HAZARD SUMMARY

DANGER! CAUSES SEVERE SKIN AND EYE BURNS
VAPOR IRRITATING TO THE EYES AND RES-
PIRATORY TRACT
HARMFUL IF SWALLOWED
COMBUSTIBLE LIQUID AND VAPOR

PHYSICAL DATA

—Physical State: liquid
—Color: colorless

Reprinted with permission from Eastman Chemical Company, Kingsport, Tennessee.

—Odor: pungent
—Odor Threshold: 0.48 ppm
—Specific Gravity (water = 1): 1.02
—Vapor Pressure at 20°C (68°F): 11.4 mmHg
—Volatile Fraction by Weight: 100%
—Vapor Density (air = 1): 2.1
—Evaporation Rate (*n*-butyl acetate = 1): 0.97
—Boiling Point: 115°C (239°F)
—Melting Point: 17°C (63°F)
—Solubility in Water: complete
—pH: not available
—Octanol/Water Partition Coefficient: log P = −0.30, P = 0.5

FIRE AND EXPLOSION HAZARD

—Flashpoint by Tag Closed Cup: 39°C (103°F)
—Lower Explosive Limit: 6.6 vol%
—Upper Explosive Limit: 19.3 vol%
—Autoignition Temperature by ASTM E 659-78: 516°C (960°F)
—Hazardous Combustion Products: carbon dioxide, carbon monoxide
—Means of Extinction: water spray, dry chemical, carbon dioxide, or "alcohol" foam
—Special Fire-Fighting Procedures: Wear self-contained breathing apparatus and protective clothing to prevent contact with skin and eyes. Use water spray to keep fire-exposed containers cool.
—Sensitivity to Mechanical Impact: insensitive
—Explosive Power: not available
—Sensitivity to Static Discharge: not available
—Unusual Fire and Explosion Hazards: Material is combustible. Keep away from heat and flame.

REACTIVITY DATA

—Stability: stable
—Stability calculated by ASTM CHETAH 4.3: insensitive
   —Heat of decomposition: −0.19 kcal/g
   —Heat of combustion: −3.48 kcal/g
—Incompatibility—Material can react with: bases, oxidizers
—Hazardous Polymerization: will not occur

TOXICOLOGICAL PROPERTIES
—Exposure Limits

　—ACGIH Threshold Limit Value (TLV): 10 ppm-TWA, 15 ppm-STEL
　—OSHA (USA) Permissible Exposure Limit (PEL): 10 ppm-TWA

—Effects of Acute Exposure

　—Inhalation: Vapor irritating.

　—Eyes: Contact with liquid causes severe burns. Vapor irritating.

　—Skin: Contact with liquid causes severe burns.

　—Ingestion: May be corrosive to the gastrointensinal tract if swallowed.

—Carcinogenicity Classification (components present at 0.1% or more)

　—IARC: not listed
　—NTP (USA): not listed
　—OSHA (USA): not listed
　—ACGIH: not listed

—Toxicity Data

—Acute Toxicity Data	Species	Result
Oral LD$_{50}$	rat	3310 to 3530 mg/kg
Oral LD$_{50}$	mouse	4960 mg/kg
Inhalation LC$_{50}$	mouse	5620 ppm/1 h
Dermal LD$_{50}$	rabbit	1060 mg/kg
Skin irritation	rabbit	corrosive
Eye irritation	rabbit	corrosive

PREVENTIVE MEASURES, VENTILATION, AND PERSONAL PROTECTION

—Preventive Measures: Do not get in eyes, on skin, on clothing. Avoid breathing vapor at concentrations greater than the exposure limits. Keep container closed. Use with adequate ventilation. Wash thoroughly after handling.

—Ventilation: Good general ventilation (typically 10 air changes per hour) should be used. Ventilation rates should be matched to conditions. local exhaust ventilation or an enclosed handling system may be needed to control airborne levels below recommended exposure limits.

—Respiratory Protection: An approved respirator must be worn if exposure is likely to exceed recommended exposure limits. Respirator type: acid gas

—Eye Protection: Wear safety glasses with side shields (or goggles) and a face shield.

—Hand Protection: Wear impermeable gloves.

—Footwear: Wear impermeable boots.

—Clothing: Wear an impermeable apron or smock.

—Recommended Decontamination Facilities: eye bath, safety shower and washing facilities

—Note: Recommendations for personal protection are for industrial handling of material; requirements for laboratories should be in accordance with good laboratory practices.

FIRST AID

—Inhalation: Remove to fresh air. Treat symptomatically. Get medical attention.

—Eyes: In case of contact, immediately flush with plenty of water for at least 15 min. Get medical attention immediately.

—Skin: Immediately flush with plenty of water for at least 15 min while removing contaminated clothing and shoes. Get medical attention immediately.

—Wash clothing before reuse. Destroy contaminated shoes.

—Ingestion: Call a physician or poison control center immediately. Do NOT induce vomiting. Give victim a glass of water. Never give anything by mouth to an unconscious person.

SPILL, LEAK, AND DISPOSAL PROCEDURES

—Steps to be taken in case material is spilled or released: Eliminate all ignition sources. Wear a self-contained breathing apparatus and appropriate personal protective equipment (see above). For small spills, absorb spill with inert material, then place in a chemical waste container. For large spills, flush spill area with water spray. Prevent runoff from entering drains, sewers, or streams. Neutralize spill and/or washings with soda ash or lime. Clean Water Act (USA) and Superfund (USA) reportable quantity (RQ): 2270 kg (5000 lb)

—Waste Disposal (Observe all laws concerning health and environment.): incineration

SPECIAL STORAGE AND HANDLING PRECAUTIONS

Keep from contact with oxidizing materials. Keep container closed. Since emptied containers retain product residue, follow label warnings even after container is emptied.

ENVIRONMENTAL EFFECTS DATA

—Summary: This environmental effects summary is written to assist in addressing emergencies created by an accidental spill which might occur during the shipment of this material, and, in general, it is not meant to address discharges to sanitary sewers or publically owned treatment works. This material is a moderately acidic aqueous solution, and this property may cause adverse environmental effects. It has the following properties: a high biochemical oxygen demand and a potential to cause oxygen depletion in aqueous systems, a low potential to affect aquatic organisms, a low potential to affect secondary waste treatment microbial metabolism, a low potential to affect the growth of some plants, a moderate to high potential to affect the germination of some plants, ready biodegradability by unacclimated microorganisms, and a low potential to bioconcentrate. When diluted with water, this material released directly or indirectly into the environment is not expected to have a significant impact.

—Oxygen Demand Data
   —COD: 1.0 g oxygen/g
   —BOD-5: 0.34–0.88 g oxygen/g

—Acute Aquatic Effects:
  —96-h LC_{50}—fathead minnow: >100 mg/L
  —96-h LC_{50}—daphnid: approx. 100 mg/L
  —48-h LC_{50}—mosquito fish: 251 mg/L
  —48-h LC_{50}—golden orfe: 410 mg/L

—Secondary Waste Water Treatment Effects: 5-h IC_{50}: >5000 mg/L

—7-Day Plant Germination Effects—No-adverse-effect concentration:
  —Ryegrass: 1 mg/L
  —Radish: 10 mg/L
  —Lettuce: 10 mg/L

—7-Day Plant Seedling Effects—No-adverse-effect concentration:
  —Marigold: 100 mg/L
  —Radish: 100 mg/L
  —Corn: 100 mg/L
  —Lettuce: 100 mg/L

TRANSPORTATION INFORMATION

—DOT (U.S.A.) Classification: corrosive material (liquid)
—DOT Shipping Name and Number: acetic acid, glacial, UN 2789
—DOT Reportable Quantity: 2270 kg (5000 lb)

—TDG (Canada) Classification: class 8, packing group II
—TDG Shipping Name and Number: acetic acid, glacial, UN 2789
—TDG Regulated Limit: 50 kg (110 lb)
—TDG Subsidiary Risk: 9.2

—International Civil Aviation Organization (ICAO): class 8, packing group II
—ICAO Shipping Name and Number: acetic acid, glacial, UN 2789

—International Maritime Dangerous Goods (IMDG): class 8
—IMDG Shipping Name and Number: acetic acid, glacial, UN 2789

OTHER INFORMATION

—OSHA (U.S.A.) hazardous chemical(s) according to 29 CFR 1910.1200: acetic acid

—WHMIS (Canada) controlled material(s): acetic acid
—WHMIS (Canada) controlled product: yes; classification(s): B/3, E

—Material(s) known to the State of California (USA) to cause cancer: none
—Material(s) known to the State of California (USA) to cause adverse reproductive effects: none

—Chemical(s) subject to the reporting requirements of Section 313 of Title III of the Superfund Amendments and Reauthorization Act (SARA) of 1986 and 40 CFR Part 372 (USA): none
—SARA (USA) Sections 311 and 312 hazard classification(s): immediate (acute) health hazard, fire hazard

—HMIS Hazard Ratings: Health—3, Flammability—2, Chemical Reactivity—0
—NFPA Hazard Ratings: Health—2, Flammability—2, Chemical Reactivity—1
NOTICE: HMIS and NFPA ratings involve data and interpretations that may vary from company to company and are intended only for rapid, general identification of the magnitude of the specific hazard. To deal adequately with the safe handling of this material, all the information contained in this MSDS must be considered.

—TSCA: This material is listed in the TSCA (U.S.A.) inventory.

—EINECS: This material is listed in the EINECS (EEC inventory).

The information contained herein is furnished without warranty of any kind. Users should consider these data only as a supplement to other information gathered by them and must make independent determinations of suitability and completeness of information from all sources to assure proper use and disposal of these materials and the safety and health of employees and customers and the protection of the environment.

MATERIAL SAFETY DATA SHEET

EASTMAN CHEMICAL PRODUCTS, INC.
EASTMAN KODAK COMPANY
Kingsport, Tennessee 37662

For Health Hazard Information, Call: (615) 229—6094

For Other Information, Call Your Eastman Representative

Eastman Operator: (615) 229-2000   Date of Preparation: 05-24-89

SECTION I. IDENTIFICATION

—Name:

Acetic Anhydride

—Synonyms: PM 9, PM 6355, Acetyl Oxide
—Formula: $C_4H_6O_3$
—Molecular Weight: 102.09

SECTION II. COMPONENT AND PRECAUTIONARY DATA

A. Component:	Approx Weight %	CAS Reg No	Eastman Kodak No
Acetic anhydride*	100	108-24-7	900004

See Section VI-A for information on exposure limits.
*Hazardous material as defined by OSHA, 29 CFR 1910.1200.

B. Precautionary Label Statements:

DANGER! CAUSES SKIN AND EYE BURNS
         VAPOR IRRITATING
         WATER REACTIVE
         COMBUSTIBLE

Do not get in eyes, on skin, on clothing.
Avoid breathing vapor.
Keep container closed.
Use with adequate ventilation.

Reprinted with permission from Eastman Chemical Company, Kingsport, Tennessee.

Wash thoroughly after handling.
Keep away from heat and flame.

FIRST AID: In case of contact, immediately flush eyes or skin with plenty of water for at least 15 min while removing contaminated clothing and shoes. Thoroughly clean shoes before reuse. Call a physician. In case of irritation of eyes, nose, and throat, remove from exposure, treat symptomatically, and call a physician if symptoms persist.

IN CASE OF FIRE: Use water spray, dry chemical, "alcohol" foam, or CO_2. Use water with caution. Material reacts with water. Do not add water to a closed container. Use water spray to keep fire-exposed containers cool.

IN CASE OF SPILL: Eliminate all ignition sources. Use water spray to disperse vapors. Flush spill area with large volume of water spray applied quickly to entire spill. Prevent runoff from entering drains, sewers, and streams. Neutralize washing with soda ash or lime.

Since emptied containers retain product residue, follow label warnings even after container is emptied.

FOR MANUFACTURING USE ONLY

SECTION III.   PHYSICAL DATA

—Appearance and Odor: Colorless liquid; pungent, acidic odor.
—Boiling Point: 139°C (282°F).
—Melting Point: −73.1°C (−100°F).
—Specific Gravity: (H_2O = 1): 1.08 at 15.6°/15.6°C.
—Vapor Pressure: 10^2 mmHg at 36°C.
—Solubility in Water: Decomposes.

SECTION IV.   FIRE AND EXPLOSION HAZARD DATA (1)

—Flash Point: 53°C (127°F), Method Used: Tag Closed Cup.
—Autoignition Temperature: 332°C (630°F), Method Used: ASTM D 2155.
—Flammable Limits: LEL 2.8% at 81°C
                                    UEL 12.4% at 129°C

—Extinguishing Agent: Water Spray, Dry Chemical, CO_2, or "Alcohol" Foam.
—Special Fire-Fighting Procedures: Wear self-contained breathing apparatus and protective clothing to prevent contact with skin and eyes. Use water with caution. Material reacts with water. Do not add water to a closed container since the reaction may result in violent rupture of the container. Use water spray to disperse vapors and keep fire-exposed containers cool.
—Unusual Fire and Explosion Hazards: Reacts with water (see Special Fire-Fighting Procedures).

SECTION V. REACTIVITY DATA

—Stability: Stable.
—Incompatibility: Oxidizing materials can cause a vigorous reaction.
—Incompatibility: Acetic anhydride is slowly soluble in water, but may react violently after a period of time to form acetic acid.
—Hazardous Decomposition Products: As with any other organic material, combustion will produce carbon dioxide and probably carbon monoxide.
—Hazardous Polymerization: Will not occur.

SECTION VI. TOXICITY AND HEALTH

A. Exposure Limits

—Threshold Limit Value (TLV): 5 ppm-C, ACGIH, 1988–1989.
—OSHA Permissible Exposure Limit (PEL): 5 ppm-Ceiling.
—A NIOSH industrial hygiene analytical method is available.

B. Exposure Effects

Inhalation: Vapor irritating to nose, throat, and lungs.

Eyes: Liquid causes burns. Vapor irritating and may cause eye burns.

Skin: Liquid causes burns.

C. First Aid

Inhalation: Remove from exposure, treat symptomatically, and call a physician if symptoms persist.

Eyes: Immediately flush with plenty of water for at least 15 min. Call a physician immediately.

Skin: Immediately flush with plenty of water for at least 15 min while removing contaminated clothing and shoes. Call a physician immediately. Wash contaminated clothing before reuse. Thoroughly clean shoes before reuse.

D. Toxicity Data

Test	Species	Result	Acute Toxicity Classification (3)
Acute oral LD_{50}	Rat	1780 mg/kg (4)	Slightly toxic
Dermal LD_{50}	Rabbit	4000 mg/kg (5)	Practically nontoxic
Inhalation LC_{50}	Rat	1000 ppm/4 h (4)	
Inhalation LC_{100}	Rat	2000 ppm/4 h (4)	
Skin irritation	Rabbit	Slight (4)	
Skin irritation	Guinea pig	Strong (6)	
Skin sensitization	Guinea pig	Slight (7)	
Eye irritation	Rabbit	Corrosive (4)	

Exposures to vapor concentrations above 5 ppm have resulted in acute eye and upper respiratory tract irritation, while exposure to high vapor concentrations may cause ulcerations of the nasal mucosa and even bronchospasm. Both the liquid and vapor can produce severe eye damage. Workers exposed to the vapor may develop conjunctivitis and associated photophobia. (7) There are no known cumulative effects.

SECTION VII.   VENTILATION AND PERSONAL PROTECTION

A.   Ventilation

Good general ventilation (typically 10 air changes per hour) should be used. Ventilation rates should be matched to conditions. Local exhaust ventilation or an enclosed handling system may be needed to control airborne levels below recommended exposure limits (see Section VI-A).

B.   Respiratory Protection

An appropriate full-face NIOSH-approved respirator for acid gas must be worn if exposure is likely to exceed recommended exposure limits (see Section VI-A). If respirators are used, a program should be established to assure compliance with OSHA Standard 29 CFR 1910.134.

C. Skin and Eye Protection

Wear safety glasses with side shields (or goggles) and a face shield. Impermeable gloves should be worn. An impermeable apron or smock and boots should be worn to minimize skin contact. A safety shower, an eye bath, and washing facilities should be available. Wash thoroughly after handling.

SECTION VIII. SPECIAL STORAGE AND HANDLING PRECAUTIONS

Material is classified as a Combustible Liquid. Keep away from heat and flame. Since emptied containers retain product residue, follow label warnings even after container is emptied.

SECTION IX. SPILL, LEAK, AND DISPOSAL PRACTICES

Steps to be Taken in Case Material is Released or Spilled: Eliminate all ignition sources. Use water spray to disperse vapors. Flush spill area with large volume of water spray applied quickly. Prevent runoff from entering drains, sewers, or streams. Neutralize spill and/or washings with soda ash or lime.

Waste Disposal Method: Incineration. Observe all federal, state, and local laws concerning health and environment.

SECTION X. ENVIRONMENTAL EFFECTS DATA

A. Summary

This product has been tested for environmental effects. Data (6, 9, 10, 11) for both acetic acid and acetic anhydride are available, and these data have been used to provide the following estimate of environmental impact:

Acetic acid has a high biological oxygen demand, and it is expected to cause significant oxygen depletion in aquatic systems. Acetic anhydride has a moderate potential to affect aquatic organisms. Acetic anhydride is expected to be readily biodegradable and is not likely to bioconcentrate. If diluted with a large amount of water, a small quantity of this product released directly or indirectly into the environment is not expected to have a significant impact.

B. Oxygen Demand Data (for acetic acid):

—COD: 1.03 g/g (6)
—BOD_5:   0.74 g/g (6); 0.34–0.88 g/g (9)
—BOD_{20}:   0.9 g/g (9)

C. Acute Aquatic Effects (for acetic anhydride):

—96-h EC_{50}*; Water flea: 3200 mg/L (with neutralization); 55 mg/L (without neutralization) (10)
—48-h LC_{50}; Golden orfe (minnow): 265 mg/L; 279 mg/L (11)**

*Fifty percent immobilization concentration.
**Results of identical tests carried out at two different laboratories.

SECTION XI. TRANSPORTATION

DOT Hazard Classification: Corrosive material.
Flashpoint: See Section IV.
Proper DOT Shipping Name: Acetic Anhydride.
UN Number: 1715.

SECTION XII. REFERENCES

1. File data, Material Safety Program, Eastman Chemicals Division, Eastman Kodak Company, Kingsport, Tennessee.

2. NIOSH MANUAL OF ANALYTICAL METHODS, 2nd Edition, Volume 3. Issued by the National Institute for Occupational Safety and Health. U.S. Government Printing Office, Washington, 1977, Method S170.

3. AM IND HYG ASSOC Q 10, 93–96 (1949).

4. AMA ARCH IND HYG OCCUP MED 4, 119–122 (1951).

5. REGISTRY OF TOXIC EFFECTS OF CHEMICAL SUB-STANCES. Issued by the National Institute for Occupational Safety and Health. U.S. Government Printing Office, Washington, 1982.

6. Unpublished data, Health and Environment Laboratories, Eastman Kodak Company, Rochester, New York.

7. N. H. Proctor and J. P. Hughes. CHEMICAL HAZARDS OF THE WORKPLACE. Philadelphia, J. B. Lippincott, 1978, p. 81.

8. F. A. Patty, Editor. INDUSTRIAL HYGIENE AND TOXICOLOGY, 2nd Revised Edition, Volume II. Wiley-Interscience, New York, 1963, pp. 1817–1818.

9. K. Verschueren. HANDBOOK OF ENVIRONMENTAL DATA ON ORGANIC CHEMICALS. Van Nostrand Reinhold Company, New York, 1977, pp. 60–63.

10. Z WASSER ABWASSER FORSCH 15, 1–6 (1982).

11. Z WASSER ABWASSER FORSCH 11, 161–164 (1978).

SECTION XIII. HAZARD RATINGS

	Health	Flammability	Reactivity
HMIS* Rating:	3	2	2
NFPA** Rating:	3	2	0W

NOTICE: These ratings involve data and interpretations that may vary from company to company and are intended only for rapid, general identification of the magnitude of the specific hazard. TO DEAL ADEQUATELY WITH THE SAFE HANDLING OF THIS MATERIAL, ALL THE INFORMATION CONTAINED IN THIS MSDS MUST BE CONSIDERED. The customer is responsible for determining the proper personal protective equipment needed for its particular use of this material.
*Hazardous Materials Identification System's [HMIS] Revised RAW MATERIALS RATING MANUAL. National Paint & Coatings Association, Fall 1984.
**NFPA 704 Standard System for the Identification of the Fire Hazards of Materials, National Fire Protection Association, 1985.

The information contained herein is furnished without warranty of any kind. Users should consider these data only as a supplement to other information gathered by them and must make independent determinations of suitability and completeness of information from all sources to assure proper use and disposal of these materials and the safety and health of employees and customers.

MATERIAL SAFETY DATA SHEET

EASTMAN CHEMICAL PRODUCTS, INC.
EASTMAN KODAK COMPANY
Kingsport, Tennessee 37662

For Health Hazard Information, Call: (615) 229-6094

For Other Information, Call Your Eastman Representative

Eastman Operator: (615) 229-2000   Date of Preparation: 05-23-89

SECTION I. IDENTIFICATION

—Name:

   Acetaldehyde

—Synonyms: PM 160; Ethanal
—Formula: CH_3CHO
—Molecular Weight: 44.05

SECTION II. COMPONENT AND PRECAUTIONARY DATA

A. Component:	Approx weight %	CAS reg no	Eastman Kodak no
Acetaldehyde* **	100	75-07-0	900468

See Section VI-A for information on exposure limits.
*Hazardous chemical as defined by OSHA, 29 CFR 1910.1200.
**Chemical subject to the reporting requirements of section 313 of Title III of the Superfund Amendments and Reauthorization Act of 1986 and 40 CFR Part 372.

B. Precautionary Label Statements:

DANGER! EXTREMELY FLAMMABLE LIQUID AND
         VAPOR
         FORMS EXPLOSIVE PEROXIDES—STORE
         AWAY FROM HEAT AND LIGHT
         CAUSES SKIN AND EYE IRRITATION
         MAY POLYMERIZE

Keep away from heat, sparks, and flame.
Do not allow to evaporate to near dryness.

Reprinted with permission from Eastman Chemical Company, Kingsport, Tennessee.

Avoid contact with eyes, skin, and clothing.
Avoid breathing vapor.
Do not contaminate.
Keep container closed.
Use with adequate ventilation.
Wash thoroughly after handling.

FIRST AID: In case of contact, immediately flush eyes with plenty of water for at least 15 min. Get medical attention. Flush skin with water. Wash clothing before reuse. In case of vapor irritation of eyes, nose, and throat, remove from exposure, treat symptomatically, and get medical attention if symptoms persist.

IN CASE OF FIRE: Use "alcohol" foam, dry chemical, CO_2, or water spray. Water may be ineffective in fighting the fire. Use water spray to keep fire-exposed containers cool.

IN CASE OF SPILL: Eliminate all ignition sources. Emergency personnel should wear self-contained breathing apparatus. Use water spray to protect personnel attempting to stop the leak. Use water spray to disperse vapors and to dilute spill to a nonflammable mixture. Prevent runoff from entering drains, sewers, and streams.

Since emptied containers retain product residue, follow label warnings even after container is emptied. Do not cut, drill, grind, or weld on or near this container.

FOR MANUFACTURING USE ONLY

SECTION III. PHYSICAL DATA (1)

—Appearance and Odor: Colorless liquid; pungent odor.
—Boiling Point: 21°C (70°F).
—Specific Gravity (H_2O = 1): 0.788 (20°/20°C).
—Vapor Pressure: 400 mmHg at 5.0°C (41°F).
—Volatile Fraction by Weight: 1.0.
—Vapor Density (Air = 1): 1.52.
—Evaporation Rate (ethyl ether = 1): 3.0.
—Solubility in Water: Complete.

SECTION IV.　FIRE AND EXPLOSION HAZARD DATA (1)

—Flash Point: $-39°C$ ($-38°F$); Method Used: Tag Closed Cup.
—Autoignition Temperature: 149°C (300°F); Method Used: ASTM D 2155.
—Cool Flame: 141°C (286°F), Method Used: ASTM E 659-78.
—Flammable Limits: LEL 4%.
　　　　　　　　　　UEL 57%.
—Extinguishing Agent: Water Spray, Dry Chemical, CO_2, "Alcohol" Foam
—Special Fire-Fighting Procedures: Wear self-contained breathing apparatus and protective clothing to prevent contact with skin and eyes. In advanced fires, firefighting should be done from a safe distance or protected locations. Water may be ineffective for fire fighting. Use water spray to keep fire-exposed containers cool.
—Unusual Fire and Explosion Hazards: DANGER, EXTREMELY FLAMMABLE, VAPORS MAY CAUSE FLASH FIRE. Extremely reactive. Can be oxidized and reduced readily. Can be easily polymerized. These reactions can be violent. Contact with air may result in the formation of explosive peroxides.

SECTION V.　REACTIVITY DATA (2)

—Stability: Stable, however forms unstable peroxides with exposure to air.
—Incompatibility: Oxidizing and reducing materials, organic acids, acid anhydrides, alcohols, halogens, ketones, phenols, amines, ammonia, hydrogen cyanide, or mercury oxosalts can cause a vigorous reaction.
—Hazardous Decomposition Products: As with any other organic material, combustion will produce carbon dioxide and probably carbon monoxide.
—Hazardous Polymerization: May occur. Conditions to Avoid: Contact with air and heat, acids or bases.

SECTION VI.　TOXICITY AND HEALTH

A.　Exposure Limits

—Threshold Limit Value (TLV): 100 ppm—TWA, 150 ppm—STEL, ACGIH, 1988–89.
—OSHA Permissible Exposure Limit (PEL): 100 ppm—TWA, 150 ppm—STEL.

—A NIOSH industrial hygiene analytical method is available (3).
—An odor threshold of 0.05 ppm has been reported (4).

B. Exposure Effects

Carcinogenicity Status: This chemical has been listed as a carcinogen or potential carcinogen for hazard communication purposes by: International Agency for Research on Cancer (IARC) Monographs (5). Acetaldehyde has also been listed by the State of California as a chemical known to the state to cause cancer (6).

Acetaldehyde was tested for carcinogenicity in rats and hamsters by inhalation exposure at high concentrations and in hamsters by intratracheal instillations. By the inhalation route, an increased incidence of carcinomas was induced in the nasal mucosa of rats, and laryngeal carcinomas were induced in hamsters. In another inhalation study in hamsters, using a lower exposure level, and in an intratracheal installation study, no increased incidence of tumors was observed. In its final evaluation, IARC stated: "There is sufficient evidence for the carcinogenicity of acetaldehyde to experimental animals. There is inadequate evidence for the carcinogenicity of acetaldehyde to humans."

In the inhalation studies, levels of acetaldehyde that caused clear evidence of carcinogenicity also caused severe necrosis of the respiratory tract, resulting in respiratory distress, lack of weight gain and increased mortality. Due to these effects, the high dose levels had to be reduced in both the rat and hamster studies. In the rat study, where somewhat lower concentrations were used, degenerative changes of the olfactory epithelium were still observed: the lowest concentration (a) was about twice the minimally toxic level determined in a subchronic study (7), (b) exceeds the TLV-TWA (Section VI-A) by a factor of 7.5, and (c) exceeds the level used by the ACGIH to define an "industrial substances of low carcinogenic potency" (8) by a factor of 135.

Acetaldehyde is a natural constituent of alfalfa, apples, blueberries, broccoli, coffee, cotton, grapefruit, grapes, lemons, mushrooms, onions, oranges, peaches, pears, pineapples, raspberries, and strawberries. It is found in the essential oils of rosemary, balm, clary sage, daffodil, bitter orange, camphor, angelica, fennel, mustard, and peppermint. It is a metabolic intermediate in higher plants and a product of alcohol fermentation. (5)

The Health and Environment Laboratories of Eastman Kodak Company have reviewed all the available data relating to the carcinogenicity of acetaldehyde, including mutagenicity data and carcinogenicity data on compounds of similar structure. Based on this information, Eastman Kodak Company does not believe that acetaldehyde, used with good industrial hygiene practices, presents a cancer hazard to humans.

Inhalation: Nose and throat irritation were not produced at 200 ppm in one study (9) while 134 ppm was described as "mildly irritating" in another study (10). High vapor concentrations cause upper respiratory tract irritation and narcosis ("drunkenness," sleepiness, dizziness, etc.). Eyes: A vapor level of 50 ppm can cause some degree of irritation (8). A splash of liquid can be expected to cause painful but superficial injury to the cornea, with rapid healing (11).

Skin: Contact may cause the skin to become reddened, then white and wrinkled. This may be followed by peeling. (1)

C. First Aid

Inhalation: Remove from exposure, treat symptomatically, and get medical attention if symptoms persist.

Eyes: Immediately flush with plenty of water for at least 15 min and get medical attention.

Skin: Immediately flush with plenty of water for at least 15 min. Get medical attention if symptoms persist. Wash contaminated clothing before reuse.

D. Toxicity Data

Test	Species	Result	Acute toxicity classification (12)
Acute oral LD_{50}	Rat	1930 mg/kg (13)	Slightly toxic
Inhalation LC_{50}	Rat	13,300 ppm/4 h (6)	

Inhalation Studies: In a 4-week study in rats, 400 ppm was near a no-observed-effect level (NOEL), producing slight degeneration of the nasal olfactory epithelium (6); whereas, 390 ppm was a NOEL in a 90-day study in Syrian golden hamsters (14). Hyperplastic and metaplastic changes in the rspiratory epithelium were produced at the next test level in each study: 1000 ppm for rats and 1340 ppm for hamsters.

Inhalation Study: Four groups of male and female rats were exposed to 0, 750, 1500, or 3000 ppm, 6 h/day, 5 days/wk, for a maximum of 27 months. (The highest dose was reduced progressively over a period of 11 months to 1000 ppm due to toxicity.) To date, only results obtained in the first 15 months of the study have been reported. Major lesions occurred in the nose and larynx. The nasal lesions comprised: (a) degenerative changes of the olfactory epithelium at all dose levels, (b) stratified squamous metaplasia of the respiratory epithelium often accompanied by severe keratinization and occasionally papillomatous hyperplasia, almost exclusively observed in the top-dose level, and (c) malignant tumors (squamous cell carcinomas and adenocarcinomas) at all dose levels. Hyperplasia and keratinized stratified squamous metaplasia of the laryngeal epithelium were seen at the 2 highest dose levels. (15)

SECTION VII. VENTILATION AND PERSONAL PROTECTION

A. Ventilation

Good general ventilation (typically 10 air changes per hour) should be used. Ventilation rates should be matched to conditions. Local exhaust ventilation or an enclosed handling system may be needed to control airborne levels below recommended exposure limits (see Section VI-A).

B. Respiratory Protection

An appropriate full-face NIOSH-approved respirator for organic vapor must be worn if exposure is likely to exceed recommended exposure limits (see Section VI-A). If respirators are used, a program should be established to assure compliance with OSHA Standard 29 CFR 1910.134.

C. Skin and Eye Protection

Wear safety glasses with side shields (or goggles). A face shield is recommended. Impermeable gloves should be worn. A safety shower, an eye bath, and washing facilities should be available. Wash thoroughly after handling.

SECTION VIII. SPECIAL STORAGE AND HANDLING PRECAUTIONS

Material is extremely flammable. Vapors may ignite explosively. Prevent buildup of vapors to explosive concentrations; use with adequate ventilation. Keep away from heat, sparks, and flame. Do not smoke; extinguish all flames and pilot lights; and turn off stoves, heaters, electric motors, and other sources of ignition when explosive concentrations of vapor are present or likely. Keep container closed. Peroxide former. Do not allow to evaporate to near dryness. Store away from heat and light. After opening, purge container with nitrogen before reclosing. Periodically test for peroxide formation on long-term storage. Storage under an inert gas blanket (nitrogen) is recommended. May polymerize. Do not contaminate. Comply with all federal, state, and local codes pertaining to the storage, handling, dispensing, and disposal of flammable liquids.

Since emptied containers retain product residue, follow label warnings even after container is emptied. Do not cut, drill, grind, or weld on or near this container.

SECTION IX. SPILL, LEAK, AND DISPOSAL PRACTICES

Steps to Be Taken in Case Material Is Released or Spilled: Eliminate all ignition sources. Emergency personnel should wear self-contained breathing apparatus. Use water spray to protect workers attempting to stop leak. Use water spray to disperse vapors and to dilute spill to a nonflammable mixture. Prevent runoff from entering drains, sewers or streams. Clean Water Act and Superfund reportable quantity (RQ): 1000 lb.

Waste Disposal Method: Mix with compatible chemical which is less flammable and incinerate. Observe all federal, state, and local laws concerning health and environment.

SECTION X. ENVIRONMENTAL EFFECTS DATA

A. Summary

Data for this material (16–19) have been used to evaluate the following properties and provide the following estimate of environmental impact: this material has a high biochemical oxygen demand and significant potential to cause oxygen depletion in aqueous systems, a high potential to affect secondary waste treatment microorganisms, a moderate potential to affect some aquatic organisms, ready biodegradability, a low potential to persist in the environment, and a low potential to bioconcentrate. The direct, instantaneous discharge to a receiving body of water of an amount of this material which will rapidly produce by dilution a final concentration of 5 mg/L or less is not expected to cause adverse environmental effects. After dilution with a large amount of water, followed by secondary waste treatment, this material is not expected to cause adverse environmental effects.

B. Oxygen Demand Data

—ThOD: 1.82 g/g (16)
—BOD$_5$: 1.3 g/g; 1.27 g/g (16)

C. Acute Aquatic Effects

—96-h LC$_{50}$; Bluegill sunfish: 53 mg/L (16,17,18)
—48-h LC$_{50}$; Golden orfe (minnow): 124 mg/L; 140 mg/L (19)*
—24-h LC$_{50}$; Pinperch: 70 mg/L (16,17,18)

*Results of the same test carried out at two different laboratories.

D. Secondary Waste Treatment Effects

—230 mg/L of acetaldehyde produced 50% inhibition of oxygen utilization of sewage organisms as compared to controls. (17,18)

E. Bioconcentration Potential

—Octanol/water partition coefficient: Log P = 0.43 (calculated); P = 2.7 (16)

SECTION XI. TRANSPORTATION

DOT Hazard Classification: Flammable Liquid.
Flash Point: See Section IV.
Proper DOT Shipping Name: Acetaldehyde.
UN Number: 1089.

SECTION XII. REFERENCES

1. File data, Material Safety Program, Eastman Chemicals Division, Eastman Kodak Company, Kingsport, Tennessee.

2. L. Bretherick. HANDBOOK OF REACTIVE CHEMICAL HAZ-ARDS, 2nd Edition. Butterworths, Boston, 1979, p. 375.

3. NIOSH MANUAL OF ANALYTICAL METHODS, 2nd Edition, Volume 5. Issued by the National Institute for Occupational Safety and Health. U.S. Government Printing Office, Washington, 1979, Method S 345.

4. J APPL TOXICOL 3, 272-290 (1983).

5. IARC MONOGRAPHS ON THE EVALUATION OF THE CAR-CINOGENIC RISK OF CHEMICALS TO HUMANS, Vol. 36. International Agency for Research on Cancer, Lyon, France, February 1985, pp. 101–131.

6. California's Safe Drinking Water and Toxic Enforcement Act of 1986. Listed April 1, 1988.

7. TOXICOLOGY 23, 293-307 (1982).

8. THRESHOLD LIMIT VALUES and BIOLOGICAL EXPOSURE INDICES for 1985–86. American Conference of Governmental Industrial Hygienists, Cincinnati, 1985, pp. 43–46.

9. J IND HYG TOXICOL 28, 262–266 (1946).

10. J AM MED ASSOC 165, 1908–1913 (1957).

11. W. M. Grant. TOXICOLOGY OF THE EYE, 2nd Edition. Charles C Thomas, Springfield, Illinois, 1974, p. 76.

12. AM IND HYG ASSOC Q 10, 93–96 (1949).

13. AMA ARCH IND HYG OCCUP MED 4, 119–122 (1951).

14. ARCH ENVIRON HEALTH 30, 449–452 (1975).

15. TOXICOLOGY 31, 123–133 (1984).

16. K. Verschueren. HANDBOOK OF ENVIRONMENTAL DATA ON ORGANIC CHEMICALS, 2nd Edition. Van Nostrand Reinhold Company, New York, 1983, pp. 139–141.

17. Batelle's Columbus Laboratories. WATER QUALITY CRITERIA DATA BOOK—VOLUME 3—EFFECTS OF CHEMICALS ON AQUATIC LIFE, Selected Data from the Literature Through 1968. U.S. Environmental Protection Agency, Washington, Project No. 18050 GWV, Contract No. 68-01-007, May 1971.

18. J. E. McKee and H. W. Wolf, Editors. WATER QUALITY CRITERIA, Publication No. 3-A. State of California, 1963.

19. Z WASSER ABWASSER FORSCH 11, 161–164 (1978).

SECTION XIII.   HAZARD RATINGS

	Health	Flammability	Reactivity
HMIS* Rating:	2	4	2
NFPA** Rating:	2	4	2

NOTICE: These ratings involve data and interpretations that may vary from company to company and are intended only for rapid, general identification of the magnitude of the specific hazard. TO DEAL ADEQUATELY WITH THE SAFE HANDLING OF THIS MATERIAL, ALL THE INFORMATION CONTAINED IN THIS MSDS MUST BE CONSIDERED. The customer is responsible for determining the proper personal protective equipment needed for its particular use of this material.

*Hazardous Materials Identification System's [HMIS] Revised RAW MATERIALS RATING MANUAL, National Paint & Coatings Association, Fall 1984.

**NFPA 704 Standard System for the Identification of the Fire Hazards of Materials, National Fire Protection Association, 1985.

The information contained herein is furnished without warranty of any kind. Users should consider these data only as a supplement to other information gather by them and must make independent determinations of suitability and completeness of information from all sources to assure proper use and disposal of these materials and the safety and health of employees and customers.

MATERIAL SAFETY DATA SHEET

EASTMAN CHEMICALS DIVISION
EASTMAN KODAK COMPANY
Kingsport, Tennessee 37662

For Health Hazard Information, Call: 615/229-6094
Date of Preparation: 06-08-89

SECTION I. IDENTIFICATION

—Name: Ketene

—Formula: C_2H_2O
—Molecular Weight: 42.04

SECTION II. PRODUCT AND COMPONENT HAZARD DATA

A. **Component:**	Approx weight %	CAS reg no	Eastman Kodak no
Ketene*	>90	463-51-4	038756

See Section VI-A for information on exposure limits.
*Hazardous chemical as defined by OSHA, 29 CFR 1910.1200.

B. Precautionary Label Statements:

DANGER! HIGHLY REACTIVE
FLAMMABLE GAS
MAY BE FATAL IF INHALED
MAY CAUSE DELAYED LUNG INJURY
GAS EXTREMELY IRRITATING
MAY POLYMERIZE

Keep away from heat, sparks, and flame.
Keep container closed.
Do not breathe gas.
Do not get in eyes, on skin, on clothing.
Use with adequate ventilation.

POISON call a physician immediately

FIRST AID: If inhaled, remove to fresh air. If breathing is difficult, give oxygen. GET MEDICAL ATTENTION EVEN IF SYMPTOMS

OF IRRITATION ARE MILD OR QUICKLY SUBSIDE AS LUNG INJURY MAY HAVE OCCURRED. In case of contact, immediately flush eyes and skin with plenty of water for at least 15 min. Wash clothing before reuse.

IN CASE OF FIRE: Use water spray, dry chemical, or CO_2. Stop flow of gas. Use water spray to keep fire-exposed containers cool.

IN CASE OF SPILL: Eliminate all ignition sources. Use water spray to disperse vapors. Neutralize spill area and washings with soda ash or lime. Prevent runoff from entering drains, sewers, and streams.

Since emptied containers retain product residue, follow label warnings even after container is emptied. Do not cut, drill, grind, or weld on or near this container.

FOR MANUFACTURING USE ONLY

SECTION III.   PHYSICAL DATA (1)

—Appearance and Odor: Colorless, irritating gas.
—Boiling Point: $-56°C$ ($-69°F$).
—Vapor Density (Air = 1): 0.81 at 25°C.
—Vapor Pressure: Gas.
—Solubility in Water: Reacts slowly.

SECTION IV.   FIRE AND EXPLOSION HAZARD DATA (1)

—Flash Point: Gas.
—Extinguishing Agent: Water spray, dry chemical, or CO_2.
—Special Fire-Fighting Procedures: Wear self-contained breathing apparatus and protective clothing to prevent contact with skin and eyes. Stop flow of gas. Use water spray to keep fire-exposed containers cool.
—Unusual Fire and Explosion Hazards: DANGER, EXTREMELY FLAMMABLE, VAPORS MAY CAUSE FLASH FIRE. (SEE SECTION VIII.)

SECTION V.   REACTIVITY DATA (1)

—Stability: Unstable, readily polymerizes.
—Stability calculated by ASTM CHETAH 4.3: Sensitive.

—Incompatibility: Will react with molecules containing oxygen, nitrogen, or sulfur.
—Hazardous Decomposition Products: As with any other organic material, combustion will produce carbon dioxide and probably carbon monoxide.
—Hazardous Polymerization: Will occurr. Conditions to Avoid: Can polymerize on heating or in the presence of acids or bases.

SECTION VI.   TOXICITY AND HEALTH

A.   Exposure Limits

—ACGIH Threshold Limit Value and OSHA Permissible Exposure Limit: 0.5 ppm—TWA, 1.5 ppm—STEL.
—A NIOSH industrial hygiene analytical method is available. (2)

B.   Exposure Effects

Inhalation: Gas causes severe upper respiratory tract irritation. May be fatal if inhaled. May cause lung injury. Symptoms of lung injury may be delayed for many hours.

Eyes: Gas causes irritation. Lacrimator.

Skin: Gas irritating.

C.   First Aid

POISON CALL A PHYSICIAN IMMEDIATELY

Inhalation: Remove to fresh air. If breathing is difficult, give oxygen. Call a physician immediately. GET MEDICAL ATTENTION EVEN IF SYMPTOMS OF IRRITATION ARE MILD OR QUICKLY SUBSIDE AS LUNG INJURY MAY HAVE OCCURRED.

Eyes: Immediately flush with plenty of water for at least 15 min.

Skin: Immediately flush with plenty of water. Wash clothing before reuse.

D.   Toxicity Data

Ketene is one of the most highly irritant gases to the respiratory tract. Its toxicity appears to be of the same order of magnitude as that of phosgene and also resembles phosgene in its delayed action on the respiratory system. A concentration of 17 ppm ketene for 10 min resulted in an LC_{50} for mice, but 1 ppm was tolerated without apparent chronic injury for 6 months by a number of animal species on an exposure schedule of 6 h daily, 5 days/week. (3)

SECTION VII. VENTILATION AND PERSONAL PROTECTION

A. Ventilation

Good general ventilation (typically 10 air changes per hour) should be used. Ventilation rates should be matched to conditions. Normally, local exhaust ventilation or an enclosed handling system will be needed to control air contamination below recommended exposure limits (see Section VI-A).

B. Respiratory Protection

An appropriate full-face NIOSH-approved respirator for organic vapor must be worn if exposure is likely to exceed recommended exposure limits (see Section VI-A). If respirators are used, a program should be established to assure compliance with OSHA Standard 29 CFR 1910.134.

C. Skin and Eye Protection

Wear safety glasses with side shields (or goggles) and a face shield. An impermeable apron or smock and boots should be worn to minimize skin contact. A safety shower, an eye bath, and washing facilities should be available. Wash thoroughly after handling.

SECTION VIII. SPECIAL STORAGE AND HANDLING PRECAUTIONS

Ketene cannot be shipped or stored in a gaseous state.

SECTION IX. SPILL, LEAK, AND DISPOSAL PRACTICES

Steps to Be Taken in Case Material Is Released or Spilled: Wear appropriate protective clothing. See Section VII. Eliminate all ignition sources. Use water spray to disperse vapors. Neutralize spill and/or washings with soda ash or lime. Prevent runoff from entering drains, sewers, or streams. Waste Disposal Method: Mix with compatible chemical which is less flammable and incinerate. Observe all federal, state, and local laws concerning health and environment.

SECTION X.   ENVIRONMENTAL EFFECTS DATA

This material has not been tested for environmental effects.

SECTION XI.   TRANSPORTATION

Should not be transported outside of plant boundaries.

SECTION XII.   REFERENCES

1.   File data, Material Safety Program, Eastman Chemicals Division, Eastman Kodak Company, Kingsport, Tennessee.

2.   NIOSH MANUAL OF ANALYTICAL METHODS, 2nd Edition, Volume 2. Issued by the National Institute for Occupational Safety and Health. U.S. Government Printing Office, Washington, 1977, Method S92.

3.   DOCUMENTATION OF THE THRESHOLD LIMIT VALUES, 4th Edition. American Conference of Governmental Industrial Hygienists, Cincinnati, 1980, pp. 241.

The information contained herein is furnished without warranty of any kind. Users should consider these data only as a supplement to other information gathered by them and must make independent determinations of suitability and completeness of information from all sources to assure proper use and disposal of these materials and the safety and health of employees and customers.

MATERIAL SAFETY DATA SHEET

EASTMAN CHEMICAL COMPANY
Kingsport, Tennessee 37662

For Emergency Health and Safety Information, Call: 800-EASTMAN

For Other Information, Call Your Eastman Representative

Eastman Operator: (615) 229-2000   Date of Preparation: April 3, 1991
MSDS No. 10,388A   Replaces Previous Editions

IDENTIFICATION AND USE

—Name: Diketene

—Chemical Name: 4-Methylene-2-oxetanone
—Synonyms: PM 1423; Ketene dimer; 3-Butene-beta-lactone
—Molecular Formula: C4H402
—Molecular Weight: 84.08
—Product Use: chemical intermediate

COMPOSITION

Component(s)	Approx weight %	CAS reg no	Eastman Kodak no
Diketene	100	674-82-8	909807

HAZARD SUMMARY

DANGER! FLAMMABLE LIQUID AND VAPOR
MAY BE FATAL IF INHALED
HIGHLY REACTIVE
MAY POLYMERIZE RESULTING IN HAZARDOUS
CONDITION
CAUSES SKIN AND EYE BURNS
VAPOR EXTREMELY IRRITATING TO THE EYES
AND RESPIRATORY TRACT
LACRIMATOR—CAUSES EYE IRRITATION

Reprinted with permission from Eastman Chemical Company, Kingsport, Tennessee.

PHYSICAL DATA

—Physical State: liquid
—Color: colorless
—Odor: sharp
—Odor Threshold: not available
—Specific Gravity (water = 1) at 20°C (68°F): 1.096
—Vapor Pressure at 20°C (68°F): 7.9 mmHg
—Volatile Fraction by Weight: 100%
—Vapor Density (air = 1): 2.9
—Evaporation Rate: not available
—Boiling Point: 127°C (260°F) (decomposition may occur)
—Melting Point: −7.5°C (18.5°F)
—Decomposes: 98°C (208°F)
—Solubility in Water: slight
—pH: not available
—Octanol/Water Partition Coefficient: not available

FIRE AND EXPLOSION HAZARD

—Flashpoint by Tag Closed Cup: 34°C (93°F)
—Lower Explosive Limit: not available
—Upper Explosive Limit: not available
—Autoignition Temperature by ASTM D 2155: 310°C (590°F)
—Hazardous Combustion Products: carbon dioxide, carbon monoxide
—Means of Extinction: water spray or carbon dioxide. Avoid using foam and dry chemical since they can react with diketene.
—Special Fire-Fighting Procedures: Wear self-contained breathing apparatus and protective clothing to prevent contact with skin and eyes. Fight fire from a protected location. USE WATER WITH CAUTION. Material reacts with water. Do not add water to a closed container since the reaction may result in violent rupture of the container. Use water spray to keep fire-exposed containers cool.
—Sensitivity to Mechanical Impact: Impact or elevated temperature can cause violent decomposition. Material does not contribute to the explosion in a heavy confinement test.
—Explosive Power: not available
—Sensitivity to Static Discharge: not available
—Unusual Fire and Explosion Hazards: Thermally unstable; see Reactivity Data. Reacts violently with water. Material is flammable. Vapors may

travel considerable distance to a source of ignition and flash back. Keep away from heat, sparks, and flame. Keep container closed. Use with adequate ventilation.

REACTIVITY DATA

—Stability: Unstable. Material can decompose violently above 98°C (208°F). Unstable. Conditions to avoid: contamination and excessive temperature. Temperatures in excess of 0°C (32°F) lead to deterioration of product quality.

—Incompatibility—Material can react violently with: acids, bases, amines, oxidizers, and Friedel-Craft catalysts. Pressure may develop in container if contents are exposed to water.

—Hazardous Polymerization: May occur. Conditions to avoid: contamination and temperature. No known inhibitor for diketene.

TOXICOLOGICAL PROPERTIES

—Exposure Limits

—ACGIH Threshold Limit Value (TLV) and OSHA (USA) Permissible Exposure Limit (PEL): not established

—Effects of Acute Exposure

—General: Animal experiments and human experience indicate that the major effect of the liquid and vapor is that of irritation. Diketene is a strong lacrimator and is an irritant of the eyes, nose, and throat. In two episodes of corneal burn in humans, recovery was complete within 48 h.

—Inhalation: May be fatal if inhaled. Vapor extremely irritating.

—Eyes: Contact with liquid causes severe burns. Vapor extremely irritating. Lacrimator. Causes eye irritation.

—Skin: Contact with liquid causes severe burns.

—Ingestion: May cause irritation of the gastrointestinal tract if swallowed.

—Carcinogenicity Classification (components present at 0.1% or more)

   —IARC: not listed
   —NTP (USA): not listed
   —OSHA (USA): not listed
   —ACGIH: not listed

—Toxicity Data

Oral LD$_{50}$	rat	400 to 800 mg/kg
Oral LD$_{50}$	rat	540 mg/kg
Oral LD$_{50}$	mouse	800–1600 mg/kg
Inhalation LC$_{50}$	rat	551 ppm/1 h
Dermal LD$_{50}$	guinea pig	10–29 mL/kg
Dermal LD$_{50}$	rabbit	6.73 g/kg
Skin irritation	guinea pig	moderate
Skin irritation	rabbit	moderate
Eye irritation	rabbit	severe

PREVENTIVE MEASURES, VENTILATION, AND PERSONAL PROTECTION

—Preventive Measures: Do not breathe vapor. Do not get in eyes, on skin, on clothing. Keep container closed. Use only with adequate ventilation. Wash thoroughly after handling.

—Ventilation: Good general ventilation (typically 10 air changes per hour) should be used. Ventilation rates should be matched to conditions. Normally, local exhaust ventilation or an enclosed handling system will be needed to control airborne levels. Maintain air concentrations below irritating levels.

—Respiratory Protection: An approved respirator should be worn if needed. Respirator type: organic vapor

—Eye Protection: Wear safety glasses with side shields (or goggles) and a face shield.

—Hand Protection: Wear impermeable gloves.

—Footwear: Wear impermeable boots.

—Clothing: Wear an impermeable apron or smock.

—Recommended Decontamination Facilities: eye bath, safety shower and washing facilities.

—Note: Recommendations for personal protection are for industrial handling of material; requirements for laboratories should be in accordance with good laboratory practices.

FIRST AID

—Inhalation: Remove to fresh air. If not breathing, give articifical respiration. If breathing is difficult, give oxygen. Call a physician or poison control center immediately.

—Eyes: In case of contact, immediately flush with plenty of water for at least 15 min. Call a physician or poison control center immediately. In case of irritation by vapor, remove from exposure, treat symptomatically, and get medical attention if symptoms persist.

—Skin: Immediately flush with plenty of water for at least 15 min while removing contaminated clothing and shoes. Call a physician or poison control center immediately. Wash clothing before reuse. Destroy contaminated shoes.

—Ingestion: Call a physician or poison control center immediately. Do NOT induce vomiting. Give victim a glass of water. Never give anything by mouth to an unconscious person.

SPILL, LEAK, AND DISPOSAL PROCEDURES

—Steps to be taken in case material is spilled or released: Eliminate all ignition sources. Wear a self-contained breathing apparatus and appropriate personal protective equipment (see above). For large spills, use water spray to disperse vapors and dilute spill to a nonflammable mixture. Prevent runoff from entering drains, sewers, or streams.

—Waste Disposal (Observe all laws concerning health and environment): incineration

SPECIAL STORAGE AND HANDLING PRECAUTIONS

Keep material below 0°C (32°F). Do not allow water to get into container because of reaction. Contents may develop pressure upon prolonged exposure to heat. Avoid heat or contamination. Keep container closed. Since emptied containers retain product residue, follow label warnings even after container is emptied. Residual vapors may explode on ignition; do not cut, drill, grind, or weld on or near this container.

TRANSPORTATION INFORMATION

—DOT (USA) Classification: flammable liquid
—DOT Shipping Name and Number: flammable liquid, corrosive, n.o.s. (diketene) (Note: This material is a poison inhalation hazard)
—DOT Shipping Number: UN 2924

(Note: For domestic surface transportation only, it is management's decision to not ship this material internationally or by air.)

OTHER INFORMATION

—OSHA (USA) hazardous chemical(s) according to 29 CFR 1910.1200: diketene
—WHMIS (Canada) Ingredient Disclosure List: diketene
—WHMIS (Canada) controlled material(s): diketene
—WHMIS (Canada) controlled product: yes; classification(s): B/2, D/1/ A, F

—Material(s) known to the State of California to cause cancer: none
—Material(s) known to the State of California to cause adverse reproductive effects: none
—Massachusetts Substance List: diketene
—New Jersey Workplace Hazardous Substance List: diketene
—Pennsylvania Hazardous Substance List: diketene
—Chemical(s) subject to the reporting requirements of Section 313 of Title III of the Superfund Amendments and Reauthorization Act (SARA) of 1986 and 40 CFR Part 372 (USA): none
—SARA (USA) Sections 311 and 312 hazard classification(s): immediate (acute) health hazard, fire hazard, reactive hazard.

—HMIS Hazard Ratings: Health-3, Flammability-3, Chemical Reactivity-2

—NFPA Hazard Ratings: Health-3, Flammability-3, Chemical Reactivity-2

NOTICE: HMIS and NFPA ratings involve data and interpretations that may vary from company to company and are intended only for rapid, general identification of the magnitude of the specific hazard. To deal adequately with the safe handling of this material, all the information contained in this MSDS must be considered.

—TSCA: This material is listed in the TSCA (USA) inventory.

—EINECS: This material is listed in the EINECS (EEC inventory).

The information contained herein is furnished without warranty of any kind. Users should consider these data only as a supplement to other information gathered by them and must make independent determinations of suitability and completeness of information from all sources to assure proper use and disposal of these materials and the safety and health of employees and customers and the protection of the environment.

MATERIAL SAFETY DATA SHEET

EASTMAN CHEMICAL PRODUCTS, INC.
Eastman Kodak Company
Kingsport, Tennessee 37662

For Emergency Health and Safety Information, Call: (615) 229-6094

For Other Information, Call Your Eastman Representative

Eastman Operator: (615) 229-2000   Date of Preparation: 12/18/90
MSDS No. 10,833A   Replaces Previous Editions

IDENTIFICATION AND USE

—Name: Methyl Acetate

—Chemical Name: Methyl acetate
—Synonyms: PM 9051; acetic acid, methyl ester
—Molecular Formula and Weight: $C_3H_6O_2$; 74.09
—Product Use: solvent

COMPOSITION

Component(s)	Approx weight %	CAS reg no	Eastman Kodak no
Methyl acetate	100	79-20-9	900520

HAZARD SUMMARY

DANGER!   EXTREMELY FLAMMABLE LIQUID AND VA-
POR—VAPOR MAY CAUSE FLASH FIRE
HARMFUL IF INHALED, ABSORBED THROUGH
SKIN, OR SWALLOWED
VAPOR IRRITATING TO THE EYES AND RESPI-
RATORY TRACT
CAUSES EYE IRRITATION
HIGH VAPOR CONCENTRATIONS MAY CAUSE
DROWSINESS

PHYSICAL DATA

—Physical State: liquid
—Color: colorless
—Odor: pleasant
—Odor Threshold: 0.1–200 ppm
—Specific Gravity (water = 1) at 20°C (68°F): 0.934
—Vapor Pressure at 20°C (68°F): 170 mmHg
—Vapor Density (air = 1): 2.6
—Evaporation Rate: not available
—Boiling Point: 57°C (135°F)
—Melting Point: −99°C (−146°F)
—Solubility in Water: appreciable
—pH: not available
—Octanol/Water Partition Coefficient: not available

FIRE AND EXPLOSION HAZARD

—Flashpoint by Tag Closed Cup: −10°C (14°F)
—Lower Explosive Limit: 3.1 vol%
—Upper Explosive Limit: 16.0 vol%
—Autoignition Temperature: 502°C (936°F)
—Hazardous Combustion Products: carbon dioxide, carbon monoxide
—Means of Extinction: water spray, dry chemical, carbon dioxide, or "alcohol" foam
—Special Fire-Fighting Procedures: Wear self-contained breathing apparatus and protective clothing to prevent contact with skin and eyes. Use water spray to keep fire-exposed containers cool. Water may be ineffective in fighting the fire.
—Sensitivity to Mechanical Impact: not applicable
—Explosive Power: not available
—Sensitivity to Static Discharge: not available
—Unusual Fire and Explosion Hazards: Material is extremely flammable. Vapors may cause a flash fire or ignite explosively. Vapors may travel considerable distance to a source of ignition and flash back. Prevent buildup of vapors to explosive concentrations; use with adequate ventilation. Keep away from heat, sparks, and flame. Do not smoke; extinguish all flames and pilot lights; and turn off stoves, heaters, electric motors, and other sources of ignition when explosive concentrations of vapor are present or likely. Keep container closed.

REACTIVITY DATA

—Stability: Stable
    Heat of decomposition: -0.36 kcal/g
    Heat of combustion: -4.83 kcal/g
—Incompatibility—Material can react violently with: oxidizers
—Hazardous Polymerization: will not occur

TOXICOLOGICAL PROPERTIES

—Exposure Limits
  —ACGIH Threshold Limit Value (TLV) and OSHA (USA) Permissible
    Exposure Limit (PEL): 200 ppm-TWA, 250 ppm-STEL

—Effects of Exposure

    General: Methyl acetate is less narcotic than some of the higher an-
    alogs such as amyl acetate. It has been suggested that methyl acetate
    may resemble methyl alcohol in producing atrophy of the optic nerve.
    This claim has not been adequately substantiated, and no cases of
    irritation or systemic injury have been reported from industrial ex-
    posures at or below 200 ppm. High concentrations have been reported
    to cause intoxication due to hydrolysis of methyl acetate to methanol
    and acetic acid.

—Inhalation: Harmful if inhaled. High vapor concentrations may cause
  drowsiness and irritation.

  —Eyes: Contact with liquid causes irritation. High vapor concentrations
    may cause irritation.

  —Skin: Harmful if absorbed through skin.

  —Ingestion: Harmful if swallowed.

—Carcinogenicity Classification (components present at 0.1% or more)

  —IARC: not listed
  —NTP (USA): not listed
  —OSHA (USA): not listed
  —ACGIH: not listed

—Toxicity Data

– Acute Toxicity Data	Species	Result
Oral LD_{50}	rat	>5000 mg/kg
Inhalation LC_{50}	—	not available
Dermal LD_{50}	guinea pig	>20 mL/kg
Skin irritation	guinea pig	slight to moderate
Eye irritation	rabbit	slight to moderate

PREVENTIVE MEASURES, VENTILATION, AND PERSONAL PROTECTION

—Preventive Measures: Avoid breathing vapor at concentrations greater than the exposure limits. Avoid contact with eyes, skin, and clothing. Keep container closed. Use with adequate ventilation. Wash thoroughly after handling.

—Ventilation: Good general ventilation (typically 10 air changes per hour) should be used. Ventilation rates should be matched to conditions. Local exhaust ventilation or an enclosed handling system may be needed to control airborne levels below recommended exposure limits.

—Respiratory Protection: An approved respirator must be worn if exposure is likely to exceed recommended exposure limits. Respirator type: organic vapor

—Eye Protection: Wear safety glasses with side shields (or goggles).

—Hand Protection: Wear impermeable gloves.

—Footwear: Wear impermeable boots.

—Clothing: Wear an impermeable apron or smock.

—Recommended Decontamination Facilities: eye bath, safety shower and washing facilities.

—Note: Recommendations for personal protection are for industrial handling of material; requirements for laboratories should be in accordance with good laboratory practices.

FIRST AID

—Inhalation: Remove to fresh air. Treat symptomatically. Get medical attention if symptoms persist.

—Eyes: In case of contact, immediately flush with plenty of water for at least 15 min. Get medical attention.

—Skin: Immediately flush with plenty of water for at least 15 min while removing contaminated clothing and shoes. Get medical attention if symptoms persist. Wash clothing before reuse. Destroy contaminated shoes.

—Ingestion: Call a physician or poison control center immediately. Induce vomiting as directed by medical personnel. Never give anything by mouth to an unconscious person.

SPILL, LEAK, AND DISPOSAL PROCEDURES

—Steps to be taken in case material is spilled or released: Eliminate all ignition sources. Wear self-contained breathing apparatus and protective clothing to prevent contact with skin and eyes. For small spills, absorb spill with inert material, then place in a chemical waste container. For large spills, use water spray to disperse vapors and dilute spill to a nonflammable mixture. Prevent runoff from entering drains, sewers, or streams.

—Waste Disposal (Observe all laws concerning health and environment): incineration

SPECIAL STORAGE AND HANDLING PRECAUTIONS

Keep from contact with oxidizing materials. Comply with all national, state or provincial, and local codes pertaining to the storage, handling, dispensing, and disposal of flammable liquids. Since emptied containers retain product residue, follow label warnings even after container is emptied. Residual vapors may explode on ignition; do not cut, drill, grind, or weld on or near this container.

TRANSPORTATION INFORMATION

—DOT (USA) Classification: flammable liquid
—DOT Shipping Name and Number: methyl acetate, UN 1231

—TDG (Canada) Classification: class 3.2, packing group II
—TDG Shipping Name and Number: methyl acetate, UN 1231

—International Civil Aviation Organization (ICAO) Classification: class 3, packing group II
—ICAO Shipping Name and Number: methyl acetate, UN 1231

—International Maritime Dangerous Goods (IMDG) Classification: class 3.2
—IMDG Shipping Name and Number: methyl acetate, UN 1231

OTHER INFORMATION

—OSHA (USA) hazardous chemical(s) according to 29 CFR 1910.1200: methyl acetate

—WHMIS (Canada) Ingredient Disclosure List: methyl acetate
—WHMIS (Canada) controlled material(s): none
—WHMIS (Canada) controlled product: yes; classification(s): B/2, D/2/B

—Material(s) known to the State of California (USA) to cause cancer: none
—Material(s) known to the State of California (USA) to cause adverse reproductive effects: none
—Massachusetts Substance List: none
—New Jersey Workplace Hazardous Substance List: none
—Pennsylvania Hazardous Substance List: methyl acetate

—Chemical(s) subject to the reporting requirements of Section 313 of Title III of the Superfund Amendments and Reauthorization Act (SARA) of 1986 and 40 CFR Part 372 (USA): none
—SARA (USA) Sections 311 and 312 hazard classification(s): immediate (acute) health hazard, fire hazard

—HMIS Hazard Ratings: Health—2, Flammability—3, Chemical Reactivity—0
—NFPA Hazard Ratings: Health—1, Flammability—3, Chemical Reactivity—0
NOTICE: HMIS and NFPA ratings involve data and interpretations that may vary from company to company and are intended only for rapid,

general identification of the magnitude of the specific hazard. To deal adequately with the safe handling of this material, all the information contained in this MSDS must be considered.

—TSCA: This material is listed in the TSCA (USA) inventory.
—EINECS: This material is listed in the EINECS (EEC inventory).

MATERIAL SAFETY DATA SHEET

EASTMAN CHEMICAL PRODUCTS, INC.
Kingsport, Tennessee 37662

For Emergency Health and Safety Information, Call: (615) 229-6094

For Other Information, Call Your Eastman Representative

Eastman Operator: (615) 229-2000   Date of Preparation: August 3, 1990
Replaces Previous Editions

IDENTIFICATION AND USE

—Name: "EASTMAN" Ethyl Acetate

—Chemical Name: Acetic Acid, ethyl ester
—Synonyms: PM 54; PM 54-UG; PM 10577, EASTMAN Ethyl Acetate,
    Urethane Grade; EASTMAN Ethyl Acetate, Food Grade
—Molecular Formula and Weight: CH3C00C2H5 88.11
—Product Use: solvent, chemical intermediate, food additive

COMPOSITION

Component(s):	Approx weight %	CAS reg no	Eastman Kodak no
Ethyl acetate	100	141-78-6	900300

HAZARD SUMMARY

WARNING! FLAMMABLE LIQUID AND VAPOR
HIGH VAPOR CONCENTRATIONS MAY CAUSE
DROWSINESS AND IRRITATION

PHYSICAL DATA

—Physical State: liquid
—Color: colorless
—Odor: sweet, ester

Reprinted with permission from Eastman Chemical Company, Kingsport, Tennessee.

—Odor Threshold: 0.016 ppm (lowest detectable) to 50 ppm (100% identifiable)
—Specific Gravity (water = 1) at 20°C (68°F): 0.902
—Vapor Pressure at 20°C (68°F): 86 mmHg
—Vapor Density (air = 1): 3.04
—Evaporation Rate (*n*-butyl acetate = 1): 4.1
—Boiling Point: 78°C (172°F)
—Melting Point: −83°C (−117°F)
—Viscosity at 20°C (68°F): 0.4519 cP
—Solubility in Water: moderate
—pH: not available
—Octanol/Water Partition Coefficient: log P = 0.73, P = 5.47

FIRE AND EXPLOSION HAZARD

—Flashpoint by Tag Closed Cup: −4°C (24°F)
—Lower Explosive Limit at 38°C (100°F): 2.02 vol%
—Upper Explosive Limit at 38°C (100°F): 10.7 vol%
—Autoignition Temperature by ASTM D 2155: 485°C (905°F)
—Hazardous Combustion Products: carbon dioxide, carbon monoxide
—Means of Extinction: water spray, dry chemical, carbon dioxide, or "alcohol" foam
—Special Fire-Fighting Procedures: Wear self-contained breathing apparatus and protective clothing to prevent contact with skin and eyes. Use water spray to keep fire-exposed containers cool. Water may be ineffective in fighting the fire.
—Sensitivity to Mechanical Impact: insensitive
—Explosive Power: not available
—Sensitivity to Static Discharge: not available
—Electrical Resistance: 650 megohn
—Unusual Fire and Explosion Hazards: Material is flammable. Vapors may travel considerable distance to a source of ignition and flash back. Keep away from heat, sparks, and flame. Keep container closed. Use with adequate ventilation.

REACTIVITY DATA

—Stability: stable
—Incompatibility—Material can react violently with: oxidizers
—Hazardous Polymerization: will not occur

TOXICOLOGICAL PROPERTIES

—Exposure Limits

  —ACGIH Threshold Limit Value (TLV) and OSHA (USA) Permissible Exposure Limit (PEL): 400 ppm

—Effects of Acute Exposure

  —Inhalation: High vapor concentrations may cause drowsiness and ir-ritation.

  —Eyes: Low hazard for usual industrial handling. However, any material that contacts the eye may be irritating or may cause mechanical injury.

  —Skin: Prolonged or repeated contact may cause drying, cracking, or irritation.

  —Ingestion: Expected to be a low ingestion hazard.

—Carcinogenicity Classification (components present at 0.1% or more)

  —IARC: not listed
  —NTP (USA): not listed
  —OSHA (USA): not listed
  —ACGIH: not listed

—Toxicity Data

—Acute Toxicity Data	Species	Result
Oral LD$_{50}$	rat	5.60 g/kg
Inhalation LC$_{50}$	rat	>8,000 ppm
Dermal LD$_{50}$	rabbit	>20 mL/kg
Skin irritation	rabbit	slight
Eye irritation	rabbit	slight
Skin sensitization	human	none

—Subchronic Toxicity Data: Guinea pigs exposed to 2000 ppm ethyl acetate for 4 h/day, 6 days/week showed no evidence of harm after 65 exposures. Mice were narcotized but recovered from 3–4 h exposure to 5000 ppm. Cats exposed to 9200 ppm demonstrated salivation and coughing, and exposure to 17,000 ppm was lethal. Deaths were attributed to pulmonary edema, hemorrhage, and hyperemia of the respiratory tract. Repeated exposure of rabbits to 4450 ppm for 1 h/day for 40 days resulted in secondary anemia with leukocytosis, hyperemia, cloudy swelling, and fatty degeneration of various organs.

Human subjects have found ethyl acetate irritating at 400 ppm, but industrial experience indicates that higher concentrations can be tolerated. No anesthetic symptoms occur at 400–600 ppm even with 2–3 h exposure. A concentration of 8600–20,000 ppm has been considered dangerous to humans for short exposures.

PREVENTIVE MEASURES, VENTILATION, AND PERSONAL PROTECTION

—Preventive Measures: Avoid breathing vapor at concentrations greater than the exposure limits. Avoid prolonged or repeated contact with skin. Keep container closed. Use with adequate ventilation.

—Ventilation: Good general ventilation (typically 10 air changes per hour) should be used. Ventilation rates should be matched to conditions. Local exhaust ventilation or an enclosed handling system may be needed to control airborne levels below recommended exposure limits.

—Respiratory Protection: An approved respirator must be worn if exposure is likely to exceed recommended exposure limits. Respirator type: organic vapor

—Eye Protection: Safety glasses with side shields (or goggles) are recommended for any type of industrial chemical handling.

—Hand Protection: For operations where prolonged or repeated skin contact may occur, impermeable gloves are recommended.

—Recommended Decontamination Facilities: eye bath and washing facilities

—Note: Recommendations for personal protection are for industrial handling of material; requirements for laboratories should be in accordance with good laboratory practices.

FIRST AID

—Inhalation: In case of irritation by vapor, remove from exposure, treat symptomatically, and get medical attention if symptoms persist.

—Eyes: Any material that contacts the eye should be washed out immediately and medical attention obtained if symptoms persist.

—Skin: Wash with soap and plenty of water.

—Ingestion: Call a physician or poison control center immediately. Induce vomiting as directed by medical personnel. Never give anything by mouth to an unconscious person.

SPILL, LEAK, AND DISPOSAL PROCEDURES

—Steps to be taken in case material is spilled or released: Eliminate all ignition sources. Wear a self-contained breathing apparatus and appropriate personal protective equipment (see above). For large spills, use water spray to disperse vapors and dilute spill to a nonflammable mixture. Prevent runoff from entering drains, sewers, or streams. Superfund (USA) reportable quantity (RQ): 5000 lb

—Waste Disposal (Observe all laws concerning health and environment): incineration

SPECIAL STORAGE AND HANDLING PRECAUTIONS

Keep from contact with oxidizing materials. Comply with all national, state or provincial, and local codes pertaining to the storage, handling, dispensing, and disposal of flammable liquids. Since emptied containers retain product residue, follow label warnings even after container is emptied. Residual vapors may explode on ignition; do not cut, drill, grind or weld on or near this container.

ENVIRONMENTAL EFFECTS DATA

—Summary: This environmental effects summary is written to assist in addressing emergencies created by an accidental spill which might occur during the shipment of this material, and, in general, it is not meant to address discharges to sanitary sewers or publically owned treatment works. Data for this material have been used to estimate its environmental impact. It has the following properties: a high biochemical oxygen demand and a potential to cause oxygen depletion in aqueous systems, a low potential to affect aquatic organisms, ready biodegradability by unacclimated microorganisms, and a low potential to bioconcentrate. When diluted with water, this material released directly or indirectly into the environment is not expected to have a significant impact.

—Oxygen Demand Data
  —COD: 1.54 g oxygen/g
  —BOD-5: 1.24 g oxygen/g

—Acute Aquatic Effects:
  —96-h LC_{50}—daphnid: 2500 mg/L or microL/L
  —48-h LC_{50}—golden orfe: 270 mg/L or microL/L

TRANSPORTATION INFORMATION

—DOT (USA) Classification: Flammable Liquid
—DOT Shipping Name and Number: ethyl acetate, UN 1173
—DOT Reportable Quantity: 5000 lb

—TDG (Canada) Classification: Class 3.2, Packing Group II
—TDG Shipping Name and Number: ethyl acetate, UN 1173

—International Civil Aviation Organization (ICAO) Classification: Class 3, Packing Group II
—ICAO Shipping Name and Number: ethyl acetate, UN 1173

—International Maritime Dangerous Goods (IMDG) Classification: Class 3.2
—IMDG Shipping Name and Number: ethyl acetate, UN 1173

OTHER INFORMATION

—OSHA (USA) hazardous chemical(s) according to 29 CFR 1910.1200: Ethyl acetate

—WHMIS (Canada) controlled material(s): none
—WHMIS (Canada) controlled product: yes; classification(s): B/2

—Material(s) known to the State of California (USA) to cause cancer: none

—Material(s) known to the State of California (USA) to cause adverse reproductive effects: none

—Chemical(s) subject to the reporting requirements of Section 313 of Title III of the Superfund Amendments and Reauthorization Act (SARA) of 1986 and 40 CFR Part 372 (USA): none

—SARA (USA) Sections 311 and 312 hazard classification(s): fire hazard

—HMIS Hazard Ratings: Health—1, Flammability—3, Chemical Reactivity—0

—NFPA Hazard Ratings: Health—1, Flammability—3, Chemical Reactivity—0

NOTICE: HMIS and NFPA ratings involve data and interpretations that may vary from company to company and are intended only for rapid, general identification of the magnitude of the specific hazard. To deal adequately with the safe handling of this material, all the information contained in this MSDS must be considered.

—TSCA: This material is listed in the TSCA (USA) inventory.

—EINECS: This material is listed in the EINECS (EEC inventory).

The information contained herein is furnished without warranty of any kind. Users should consider these data only as a supplement to other information gathered by them and must make independent determinations of suitability and completeness of information from all sources to assure proper use and disposal of these materials and the safety and health of employees and customers and the protection of the environment.

MATERIAL SAFETY DATA SHEET

EASTMAN CHEMICAL PRODUCTS, INC.
Eastman Kodak Company
Kingsport, Tennessee 37662

For Emergency Health and Safety Information, Call: (615) 229-6094

For Other Information, Call Your Eastman Representative

Eastman Operator: (615) 229-2000 Date of Preparation: October 2, 1990
Replaces Previous Editions

IDENTIFICATION AND USE

—Name: EASTMAN Propyl Acetate

—Chemical Name: Propyl acetate
—Synonyms: PM 117; acetic acid, propyl ester; n-propyl acetate
—Molecular Formula and Weight: CH300C3H7; 102.13
—Product Use: solvent, chemical intermediate, food additive

COMPOSITION

Component(s)	Approx weight %	CAS reg no	Eastman Kodak no
Propyl acetate	100	109-60-4	900747

HAZARD SUMMARY

WARNING! FLAMMABLE LIQUID AND VAPOR
HIGH VAPOR CONCENTRATIONS MAY CAUSE
DROWSINESS AND IRRITATION

PHYSICAL DATA

—Physical State: liquid
—Color: colorless
—Odor: sweet, ester
—Odor Threshold: 20 ppm
—Specific Gravity (water = 1) at 20° C (68°F): 0.855

—Vapor Pressure at 20°C (68°F): 23 mmHg
—Vapor Density (air = 1): 3.52
—Evaporation Rate (n-butyl acetate = 1): 2.3
—Boiling Point: 102°C (216°F)
—Melting Point: −92°C (−134°F)
—Solubility in Water: moderate
—pH; not available
—Octanol/Water Partition Coefficient: not available

FIRE AND EXPLOSION HAZARD

—Flashpoint by Tag Closed Cup: 13°C (55°F)
—Lower Explosive Limit at 38°C (100°F): 1.71 vol%
—Upper Explosive Limit at 93°C (199°F): 7.95 vol%
—Autoignition Temperature by ASTM D 2155: 457°C (855°F)
—Hazardous Combustion Products: carbon dioxide, carbon monoxide
—Means of Extinction: water spray, dry chemical, carbon dioxide, or "alcohol" foam
—Special Fire-Fighting Procedures: Wear self-contained breathing apparatus and protective clothing to prevent contact with skin and eyes. Use water spray to keep fire-exposed containers cool. Water may be ineffective in fighting the fire.
—Sensitivity to Mechanical Impact: not available
—Explosive Power: not available
—Sensitivity to Static Discharge: not available
—Electrical Resistivity: 6400 megohm-cm
—Unusual Fire and Explosion Hazards: Material is flammable. Vapors may travel considerable distance to a source of ignition and flash back. Keep away from heat, sparks, and flame. Keep container closed. Use with adequate ventilation.

REACTIVITY DATA

—Stability: stable
   Heat of decomposition: −0.30 kcal/g
—Incompatibility—Material can react violently with: oxidizers
—Hazardous Polymerization: will not occur

TOXICOLOGICAL PROPERTIES
—Exposure Limits

—ACGIH Threshold Limit Value (TLV) and OSHA (USA) Permissible
Exposure Limit (PEL): 150 ppm-TWA, 200 ppm-STEL

—Effects of Acute Exposure

—Inhalation: High vapor concentrations may cause drowsiness and irritation.

—Eyes: Low hazard for usual industrial handling. However, any material
that contacts the eye may be irritating or may cause mechanical injury.

—Skin: Low hazard for usual industrial handling.

—Ingestion: Expected to be a low ingestion hazard.

—Carcinogenicity Classification (components present at 0.1% or more)

—IARC: not listed
—NTP (USA): not listed
—OSHA (USA): not listed
—ACGIH: not listed

—Toxicity Data

—Acute Toxicity Data	Species	Result
Oral LD$_{50}$	rat	9370 mg/kg
Oral LD$_{50}$	mouse	8300 mg/kg
Inhalation LC$_{50}$	—	not available
Inhalation LC	rat	8000 ppm for 4 h killed 4/6
Dermal LD$_{50}$	rabbit	>20 mL/kg
Dermal LD$_{50}$	guinea pig	>10 mL/kg
Skin irritation	guinea pig	slight
Skin irritation	rabbit	slight
Eye irritation	rabbit	slight

Cats exposed to 5300 ppm, 6 h/day for 5 days, showed moderate irritation
and salivation. Exposure at 7400 ppm produced staggering within 30–
45 min and deep narcosis in 4.5–5.5 h; exposure at 24,000 ppm produced

narcosis in 13–18 min. However, concentrations of 7400 ppm for 5.5 h and 24,000 ppm for 0.5 h were fatal to one-fourth of the animals tested (total number of animals unspecified).

While experimental inhalation of rather high concentrations of the propyl acetates have been shown to produce irritation, narcosis, and death in certain cases, no permanent industrial injury has been caused in workmen exposed to these acetates. However, it has been found that human exposures up to 14,000 ppm within 1 week cause conjunctival irritation, cough, and a feeling of chest oppression; after discontinuing exposure to the ester, recovery is prompt.

PREVENTIVE MEASURES, VENTILATION, AND PERSONAL PROTECTION

—Preventive Measures: Avoid breathing gas at concentrations greater than the exposure limits. Keep container closed. Use with adequate ventilation.

—Ventilation: Good general ventilation (typically 10 air changes per hour) should be used. Ventilation rates should be matched to conditions. Local exhaust ventilation or an enclosed handling system may be needed to control airborne levels below recommended exposure limits.

—Respiratory Protection: An approved respirator must be worn if exposure is likely to exceed recommended exposure limits. Respirator type: organic vapor

—Eye Protection: Safety glasses with side shields (or goggles) are recommended for any type of industrial chemical handling.

—Recommended Decontamination Facilities: eye bath and washing facilities.

—Note: Recommendations for personal protection are for industrial handling of material; requirements for laboratories should be in accordance with good laboratory practices.

FIRST AID

—Inhalation: Remove to fresh air. Treat symptomatically. Get medical attention if symptoms persist.

—Eyes: Any material that contacts the eye should be washed out immediately and medical attention obtained if symptoms persist.

—Skin: Wash with soap and water and get medical attention if symptoms occur.

—Ingestion: Call a physician or poison control center immediately. Induce vomiting as directed by medical personnel. Never give anything by mouth to an unconscious person.

SPILL, LEAK, AND DISPOSAL PROCEDURES

—Steps to be taken in case material is spilled or released: Eliminate all ignition sources. Wear a self-contained breathing apparatus and appropriate personal protective equipment (see above). For large spills, use water spray to disperse vapors and dilute spill to a nonflammable mixture. Prevent runoff from entering drains, sewers, or streams.

—Waste Disposal (Observe all laws concerning health and environment): incineration

SPECIAL STORAGE AND HANDLING PRECAUTIONS

Keep from contact with oxidizing materials. Comply with all national, state or provincial, and local codes pertaining to the storage, handling, dispensing, and disposal of flammable liquids. Since emptied containers retain product residue, follow label warnings even after container is emptied. Residual vapors may explode on ignition; do not cut, drill, grind, or weld on or near this container.

ENVIRONMENTAL EFFECTS DATA

—Summary: This environmental effects summary is written to assist in addressing emergencies created by an accidental spill which might occur during the shipment of this material, and, in general, it is not meant to address discharges to sanitary sewers or publicly owned treatment works.

Data for this material have been used to estimate its environmental impact. It has the following properties: a high biochemical oxygen demand and a potential to cause oxygen depletion in aqueous systems, a low potential to affect aquatic organisms, and a low potential to bioconcentrate. When diluted with a large amount of water, this material released directly or indirectly into the environment is not expected to have a significant impact.

—Oxygen Demand Data
  —BOD-5: 1.34 g oxygen/g

—Acute Aquatic Effects:
  —48-h LC_{50}—golden orfe: 97–194 mg/L
  —96-h LC_{50}—daphnid: 511 mg/L

TRANSPORTATION INFORMATION

—DOT (USA) Classification: flammable liquid
—DOT Shipping Name and Number: propyl acetate, UN 1276

—TDG (Canada) Classification: Class 3.2, Packing Group II
—TDG Shipping Name and Number: *n*-propyl acetate, UN 1276
—TDG Subsidiary Risk: not applicable

—International Civil Aviation Organization (ICAO) Classification: Class 3, Packing Group II
—ICAO Shipping Name and Number: *n*-propyl acetate, UN 1276
—ICAO Subsidiary Risk: not applicable

—International Maritime Dangerous Goods (IMDG) Classification: Class 3.2
—IMDG Shipping Name and Number: *n*-propyl acetate, UN 1276
—IMDG Subsidiary Risk: not applicable

OTHER INFORMATION

—OSHA (USA) hazardous chemical(s) according to 29 CFR 1910.1200: propyl acetate

—WHMIS (Canada) controlled material(s): none
—WHMIS (Canada) controlled product: yes; classification(s): B/2

—Material(s) known to the State of California to cause cancer: none
—Material(s) known to the State of California to cause adverse reproductive effects: none

—Massachusetts Substance List: none
—New Jersey Workplace Hazardous Substance List: none
—Pennsylvania Hazardous Substance List: propyl acetate

—Chemical(s) subject to the reporting requirements of Section 313 of Title
  III of the Superfund Amendments and Reauthorization Act (SARA) of
  1986 and 40 CFR Part 372 (USA): none
—SARA (USA) Sections 311 and 312 hazard classification(s): fire hazard

—HMIS Hazard Ratings: Health—1, Flammability—3, Chemical Reactivity—0
—NFPA Hazard Ratings: Health—1, Flammability—3, Chemical Reactivity—0
NOTICE: HMIS and NFPA ratings involve data and interpretations that
may vary from company to company and are intended only for rapid,
general identification of the magnitude of the specific hazard. To deal
adequately with the safe handling of this material, all the information
contained in this MSDS must be considered.

—TSCA: This material is listed in the TSCA (USA) inventory.

—EINECS: This material is listed in the EINECS (EEC inventory).

The information contained herein is furnished without warranty of any
kind. Users should consider these data only as a supplement to other
information gathered by them and must make independent determinations
of suitability and completeness of information from all sources to assure
proper use and disposal of these materials and the safety and health of
employees and customers and the protection of the environment.

MATERIAL SAFETY DATA SHEET

EASTMAN KODAK COMPANY
343 State Street
Rochester, New York 14650

For Emergency Health, Safety, and Environmental Information, call 716-722-5151
For all other purposes, call 800-225-5352, in New York State call 716-458-4014

Date of Preparation: 06/12/86   Kodak Accession Number: 904650

SECTION I. IDENTIFICATION

—Product Name: Cellulose Acetate
—Formula: Polymer of various molecular weights
—CAT No(s): 117 3244; 117 3251; 117 3269
—Chem. No(s): 04650
—Kodak's Internal Hazard Rating Codes: R: 1 S: 1 F: 1 C: 0

SECTION II. PRODUCT AND COMPONENT HAZARD DATA

COMPONENT(S):	Percent	ACGIH TLV(R)	CAS reg. no.
Cellulose Acetate	ca. 100	—	9004-35-7

SECTION III. PHYSICAL DATA

—Appearance: Fine white powder
—Melting Pont: 260°C (500°F)
—Vapor Pressure: Negligible
—Evaporation Rate (n-butyl acetate = 1): Negligible
—Volatile Fraction by Weight: Negligible
—Specific Gravity (Water = 1): 1.27
—Solubility in Water (by Weight): Negligible

SECTION IV. FIRE AND EXPLOSION HAZARD DATA

—Flash Point: Not Applicable
—Extinguishing Media: Water spray; Dry chemical; Carbon dioxide

---

Reprinted with permission from Eastman Chemical Company, Kingsport, Tennessee.

—Special Fire Fighting Procedures: None
—Unusual Fire and Explosion Hazards: This material in sufficient quantity and reduced particle size is capable of creating a dust explosion.

SECTION V. REACTIVITY DATA

—Stability: Stable
—Incompatibility: Strong oxidizers
—Hazardous Decomposition Products: Combustion will produce carbon dioxide and probably carbon monoxide.
—Hazardous Polymerization: Will not occur.

SECTION VI. TOXICITY AND HEALTH HAZARD DATA

A. Exposure Limits:

Not established.

B. Exposure Effects:

Inhalation: Low hazard for usual industrial handling.
Skin: Low hazard for usual industrial handling.
Eye: No specific hazard known. Contact may cause transient irritation.
Ingestion: Expected to be a low ingestion hazard.

C. First Aid:

Inhalation: None should be needed.
Skin: None should be needed.
Eye: Flush eyes with plenty of water.
Ingestion: None should be needed.

SECTION VII. VENTILATION AND PERSONAL PROTECTION

A. Ventilation and Respiratory Protection:

Good ventilation should be sufficient.

B. Skin and Eye Protection:

Safety glasses recommended in industrial operations involving chemicals.

SECTION VIII.   SPECIAL STORAGE AND HANDLING PRECAUTIONS

Keep from contact with oxidizing materials.

SECTION IX.   SPILL, LEAK, AND DISPOSAL PROCEDURES

Sweep up material and package for safe feed to an incinerator. Dispose by incineration or contract with licensed chemical waste disposal agency. Discharge, treatment, or disposal may be subject to federal, state, or local laws.

For transportation information regarding this material, please phone the Eastman Kodak Distribution Center nearest you: Rochester, NY (716) 254-1300; Oak Brook, IL (312) 654-5300; Chamblee, GA (404) 455-0123; Dallas, TX (214) 241-1611; Whittier, CA (213) 945-1255; Honolulu, HI (808) 833-1661.

The information contained herein is furnished without warranty of any kind. Users should consider these data only as a supplement to other information gathered by them and must make independent determinations of the suitability and completeness of information from all sources to assure proper use and disposal of these materials and the safety and health of employees and customers.

MATERIAL SAFETY DATA SHEET

EASTMAN KODAK COMPANY
343 State Street
Rochester, New York 14650

For Emergency Health, Safety, and Environmental Information, call 716-722-5151
For all other purposes, call 800-225-5352, in New York State call 716-458-4014

Date of Revision: 12/31/90   Kodak Accession Number: 900002

SECTION I. IDENTIFICATION

—Product Name: Acetamide
—Synonym(s): Acetic Acid Amide
—Formula: C_2H_5NO
—CAT No(s): 100 0256; 100 2047; 100 2880; 100 3391; 100 3458; 100 5081; 156 0952
—Chem. No(s): 00002
—Kodak's Internal Hazard Rating Codes: R: 2 S: 2 F: 1 C: 0

SECTION II. PRODUCT AND COMPONENT HAZARD DATA

COMPONENT(S):	Percent	ACGIH TLV(R)	CAS reg. no.
Acetamide	ca. 100	—	60-35-5

SECTION III. PHYSICAL DATA

—Appearance and Odor: White crystalline solid; mousy odor
—Melting Point: 80°C (176°F)
—Vapor Pressure: Negligible
—Evaporation Rate (n-butyl acetate = 1): Negligible
—Volatile Fraction by Weight: Negligible
—Specific Gravity (Water = 1): 1.16
—Solubility in Water (by Weight): Appreciable

Reprinted with permission from Eastman Chemical Company, Kingsport, Tennessee.

SECTION IV. FIRE AND EXPLOSION HAZARD DATA

—Flash Point: Not Applicable
—Extinguishing Media: Water spray; Dry chemical; Carbon dioxide
—Special Fire Fighting Procedures: Wear self-contained breathing apparatus and protective clothing.
—Unusual Fire and Explosion Hazards: Fire or excessive heat may produce hazardous decomposition products. This material, like most organic materials in powder form, is capable of creating a dust explosion.

SECTION V. REACTIVITY DATA

—Stability: Stable
—Incompatibility: Strong oxidizers; metals; halogenated materials
—Hazardous Decomposition Products: Combustion will produce carbon dioxide and probably carbon monoxide. Oxides of nitrogen may also be present.
—Hazardous Polymerization: Will not occur.

SECTION VI. TOXICITY AND HEALTH HAZARD DATA

A. Exposure Limits:

Not established.

B. Exposure Effects:

Possible cancer hazard. May cause cancer based on animal data.

Carcinogenicity Status: Acetamide has been identified as a carcinogen or potential carcinogen for hazard communication purposes by:

International Agency for Research on Cancer (IARC) Monographs

California: WARNING: Acetamide is known to the State of California to cause cancer.

Inhalation: Harmful if inhaled.
Skin: Harmful if absorbed through the skin.
Eye: No specific hazard known. May cause transient irritation.
Ingestion: Harmful if swalled.

C. **First Aid:**

Inhalation: If inhaled, remove to fresh air. If not breathing, give artificial respiration. If breathing is difficult, give oxygen. Get medical attention.

Skin: In case of contact, immediately flush skin with plenty of water for at least 15 min while removing contaminated clothing and shoes. Wash contaminated clothing before reuse. Destroy or thoroughly clean contaminated shoes. If symptoms are present after washing, get medical attention.

Eye: In case of eye contact, immediately flush eyes with plenty of water for at least 15 min. Get medical attention if symptoms occur.

Ingestion: If swallowed, induce vomiting immediately as directed by medical personnel. Never give anything by mouth to an unconscious person. Call a physician or poison control center immediately.

SECTION VII.   VENTILATION AND PERSONAL PROTECTION

A. **Ventilation and Respiratory Protection:**

Use process enclosures, local exhaust ventilation or other engineering controls to reduce dust concentrations to an acceptable level.

B. **Respiratory Protection:**

If engineering controls are inadequate to control dust concentrations to an acceptable level, a NIOSH-approved dust respirator should be worn. If respirators are used, a program should be instituted to assure compliance with OSHA Standard 29 CFR 1910.134.

C. **Skin and Eye Protection:**

Impervious gloves and clothing should be worn. Safety glasses with side shields, goggles, or a face shield should be worn.

SECTION VIII.   SPECIAL STORAGE AND HANDLING PRECAUTIONS

Keep from contact with oxidizing materials.

SECTION IX.   SPILL, LEAK, AND DISPOSAL PROCEDURES

Sweep up material and package for safe feed to an incinerator. Dispose by incineration or contract with licensed chemical waste disposal agency. Discharge, treatment, or disposal may be subject to federal, state, or local laws.

The information contained herein is furnished without warranty of any kind. Users should consider these data only as a supplement to other information gathered by them and must make independent determinations of the suitability and completeness of information from all sources to assure proper use and disposal of these materials and the safety and health of employees and customers.

Chemical Group
Hoechst Celanese Corporation
P.O. Box 569320 / Dallas, Texas 75356-9320
Information phone 214 689 4000
Emergency phone: 800 424 9300 (CHEMTREC)

Acrylic acid, glacial

Issued April 19, 1991

#10

Identification

Product name: Acrylic acid, glacial

Chemical name: Acrylic acid

Inhibitor: Monomethyl ether of hydro-quinone (MEHQ, CAS No. 150-76-5), 200 + 20 ppm DISSOLVED AIR MUST BE PRESENT IN ORDER FOR INHIBITOR TO FUNCTION EFFECTIVELY

Chemical family: Acrylic monomer (carboxylic acid)

Formula: $CH_2CHCOOH$

Molecular weight: 72

CAS name: 79-10-7

Synonyms: Acroleic acid, ethylenecarboxylic acid; propene acid, 2-propenoic acid; vinylformic acid.

Department of Transportation information
Shipping name: Acrylic Acid
Hazard classification: Corrosive Material
United Nations number: UN2218
Emergency Response Guide no.: 29
Reportable Quantity: 5000 lb/2270 kg

Physical data

Boiling point (760 mm Hg.): 141 0°C (286.0°F)

Freezing point: 13 0°C (55°F)

Specific gravity (H_2O = 1 @ 20/20°C): 1 05

Vapor pressure (20°C): 4 mm Hg

Vapor density (Air = 1 @ 20°C): 2 5

Solubility in water (% by WT @ 20°C): Complete

Percent Volatiles by volume: 100

Appearance and odor: Clear, colorless, mobile liquid, strong, acrid odor

Fire and explosion hazard data

Flammable limits in air, % by volume
Upper: 8 0
Lower: 2 0

Flash point (test method):
Tag open cup (ASTM D1310): 129°F (54°C)
Tag closed cup (ASTM D56): 122°F (50°C)

Extinguishing media:
Use CO_2 or dry chemical for small fires, alcohol-type aqueous film-forming foam or water spray for large fires.

Special fire-fighting procedures:
If potential for exposure to vapors or products of combustion exists, wear complete personal protective equipment

Component information (See Glossary at end of MSDS for definitions)

Component, wt. % (CAS Number)	Exposure levels			Subject to SARA §313 reporting?
	OSHA PEL TWA	ACGIH TLV' TWA	IDLH	
* Acrylic acid 99 7%c (79-10-7)	10 ppm (skin)	2 ppm (skin)	NVE '	Yes

(1) All components listed as required by federal California New Jersey and Pennsylvania regulations
(2) No value established

and respirator approved by both NIOSH and MSHA and within the working limits of the respirator

Self-contained breathing apparatus with full facepiece operated in pressure-demand or other positive-pressure mode

Water spray can be used to reduce intensity of flames and to dilute spills to nonflammable mixture Use water spray to cool fire-exposed structures and vessels

Unusual fire and explosion hazards:
Rapid, uncontrolled polymerization can cause explosion.

Special hazard designations

	HMIS	NFPA	Key
Health:	3	3	0 - Minimal
Flammability:	2	2	1 - Slight
Reactivity:	2	2	2 - Moderate
Personal protective			3 - Serious
equipment:	G	—	4 - Severe

SARA §311 hazard categories
Acute health:	Yes
Chronic health:	Yes
Fire:	Yes
Sudden release of pressure:	No
Reactive:	Yes

Reactivity data

Stability:
Potentially unstable

Hazardous polymerization:
Can occur. UNCONTROLLED POLYMERIZATION CAN CAUSE RAPID EVOLUTION OF HEAT AND INCREASED PRESSURE WHICH CAN RESULT IN VIOLENT RUPTURE OF STORAGE VESSELS OR CONTAINERS

* Conditions to avoid:
Storage at temperatures above 25°C (77°F) and below 15°C (59°F), sunlight, x-ray or ultra-violet radiation; sparks and flame. Dissolved oxygen comes

from equilibration of the product with air Depletion of this dissolved oxygen severely reduces the effectiveness of the inhibitor and can lead to polymerization

Materials to avoid:
Peroxides, for example, t-butyl peroxide and hydrogen peroxide, other polymerization initiators: alkalis such as sodium hydroxide (caustic soda), oxidizing agents, amines

Hazardous combustion or decomposition products:
Oxides of carbon.

Health data

Effects of exposure/toxicity data
Acute
Ingestion (swallowing): Can severely irritate mouth, throat and stomach. Moderately toxic to animals (oral LD_{50}, rats 0 34 g/kg)
Inhalation (breathing): Can irritate nasal passages, throat and lungs Can cause pulmonary edema (accumulation of fluid in the lungs), signs and symptoms can be delayed for several hours Slightly toxic to animals (inhalation LC_{50}, rats. 4 hrs: 4000 ppm)
Skin contact: Can cause chemical burn Sensitization (allergic reaction) can occur Moderately toxic to animals by absorption (dermal LD_{50}, rabbits 0.29 g/kg)
Eye contact: Can cause chemical burn – damage irreversible Vapors are extremely irritating

Chronic

* Mutagenicity: Approximately 12 standard mutagenicity studies have been conducted with acrylic acid All in vivo studies and the majority of in vitro studies (including the Ames test) have been negative.
* Carcinogenicity: Acrylic acid was not carcinogenic to male and female rats in a lifetime drinking water study with concentrations up to 1200 ppm. It was not carcinogenic to male mice in several dermal carcinogenicity studies Equivocal

(continued)

* New or revised information; previous version dated February 28, 1991.

Acrylic acid, glacial

findings were noted for female mice in several dermal studies (i.e., some negative; some marginally increased for skin or lymph tumors). Results obtained in the latter studies are considered inconclusive at this time due to confusion in reporting, questionable study conduct or other considerations

Reproduction: No evidence of teratogenicity in rats; inhalation study. Reported to not adversely affect reproduction in rats. oral exposure (Intercompany Acrylate Study Group)

Other: Inflammation and alteration of nasal mucosa in rats and mice exposed to acrylic acid vapors for 90 days (Intercompany Acrylate Study Group)

Medical conditions aggravated by exposure:
Significant exposure to this chemical may adversely affect people with chronic disease of the respiratory system, skin and/or eyes.

Emergency and first aid procedures
Ingestion (swallowing): Patient should be made to drink large quantities of water. Do not induce vomiting. Contact a physician immediately.

Inhalation (breathing): Remove patient from contaminated area. If breathing has stopped, give artificial respiration, then oxygen if needed. Contact a physician immediately.

Skin contact: Remove contaminated clothing and wash contaminated skin with large amounts of water. If irritation persists, contact a physician.

Eye contact: Flush eyes with water for at least 15 minutes. Contact a physician immediately.

Spill or leak procedures

Steps to be taken if material is released or spilled:
Eliminate ignition sources. Caution: Spontaneous polymerization can occur. Avoid eye or skin contact. Place leaking containers in well-ventilated area. If fire potential exists, blanket spill with foam or use water spray to disperse vapors. Contain spill to minimize contaminated area and facilitate salvage or disposal

To clean up spill, flush area sparingly with water or use an absorbent. Avoid runoff into storm sewers and ditches which lead to natural waterways. If an odor or acidity problem exists, add lime or sodium bicarbonate. Call the National Response Center (800 424 8802) if spill is equal to or greater than reportable quantity (5000 lb/day) under "Superfund" All clean-up and disposal should be carried out in accordance with federal, state and local regulations. If required, state and local authorities should be notified

Waste disposal method:
This product when spilled or disposed is a hazardous solid waste as defined in Resource Conservation Recovery Act regulations (40CFR261). Preferred method is incineration or biological treatment in federal/state approved facility

Special protection information

Respiratory protection
Based on contamination level and working limits of the respirator, use a respirator approved by both NIOSH and MSHA (respirators are listed in order from minimum to maximum respiratory protection)

Up to 20 ppm – Chemical cartridge respirator with an organic vapor cartridge(s) and a full facepiece

Gas mask with organic vapor canister (chin-style or front- or back-mounted canister), with a full facepiece, providing protection against acid gases

Up to 50 ppm – Powered air-purifying respirator with loose-fitting full facepiece

Up to 500 ppm – Powered air-purifying respirator with tight-fitting full facepiece.

Up to 2000 ppm – Type 'C' supplied-air respirator with a full facepiece operated in pressure-demand or other positive-pressure mode.

Above 2000 ppm – Self-contained breathing apparatus with a full facepiece operated in pressure-demand or other positive-pressure mode

Ventilation
Local exhaust: Recommended when appropriate to control employee exposure
Mechanical (general): Not recommended as the sole means of controlling employee exposure

Protective gloves:
Neoprene or rubber

Eye protection:
Chemical safety goggles

Other protective equipment:
For operations where spills or splashing can occur, use chemical splash suit and neoprene or rubber boots. A safety shower and eye bath should be available

Special precautions

Precautions to be taken in handling and storing:
* Store in a cool, well-ventilated area at temperatures between 15°C and 25°C (59°F and 77°F). If product freezes, melt only in a temperature-controlled environment. Use only tempered water. 45°C (113°F) maximum temperature, to thaw bulk containers. Drums may be placed in a heated room at temperatures between 20°C and 33°C (68°F and 91°F). Product being melted, particularly in 55-gallon drums, should be agitated at regular intervals by rolling to assure thorough mixing and distribution of the polymerization inhibitor. NEVER USE STEAM OR ELECTRICAL HEATING SYSTEMS (SUCH AS TAPES, MANTLES OR JACKETS) TO THAW THIS PRODUCT. As soon as material is thawed, normal storage temperatures (15°C to 25°C, 59°F to 77°F) should be established. Keep away from heat, sparks and flame. Keep containers closed. Samples should be stored in opaque or amber glass containers. Use only DOT-approved containers. Use spark-resistant tools. Do not load into compartments adjacent to heated cargo. When transferring follow proper grounding procedures. Use with adequate ventilation. Avoid breathing vapor. Avoid contact with eyes, skin and clothing. Wash thoroughly with soap and water after handling. Wash contaminated clothing thoroughly before re-use. Discard contaminated leather clothing. AIR SHOULD BE USED TO BLANKET AND SPARGE STORAGE VESSELS IN ORDER FOR INHIBITOR TO FUNCTION. OXYGEN-FREE ATMOSPHERES SHOULD NEVER BE USED.

---

* New or revised information, previous version dated February 28, 1991.

Glossary for Components information table

ACGIH	- American Conference of Governmental Industrial Hygienists	**SARA**	– Superfund Amendments and Reauthorization Act
CAS	– Chemical Abstracts Service	**Skin**	– Potential contribution to overall exposure possible via skin absorption
IDLH	– Immediately Dangerous to Life or Health	**STEL**	– Short-term exposure limit 15-min time-weighted average
OSHA	– Occupational Safety and Health Administration	**TLV**	– Threshold limit value
PEL	– Permissible exposure limit	**TWA**	– 8-hour time-weighted average

Chemical Group
Hoechst Celanese Corporation
PO Box 569320/Dallas, Texas 75356-9320
Information phone 214 689 4000
Emergency phone: 800 424 9300 (CHEMTREC)

Reprinted with permission from Hoechst Celanese Corporation, Dallas, Texas.

Chemical Group
Hoechst Celanese Corporation
P.O. Box 569320 / Dallas, Texas 75356-9320
Information phone 214 689 4000
* Emergency phone: 800 424 9300 (CHEMTREC)

VINYL ACETATE

Issued February 16, 1990

#94

Identification

Product name: Vinyl acetate

Chemical name: Vinyl acetate

Inhibitor: Hydroquinone
(HQ, CAS no. 123-31-9), 3-5 ppm

Chemical family: Vinyl monomer
(acetate ester)

Formula: $CH_3COOCHCH_2$

Molecular weight: 86

CAS number: 108-05-4

CAS name: Acetic acid, ethenyl ester

Synonyms: VAM; VA; acetic acid vinyl
ester; acetic acid ethenyl ester,
1-acetoxyethylene; ethenyl acetate;
ethenyl ethanoate.

Department of Transportation information
Shipping name: Vinyl Acetate
Hazard classification: Flammable Liquid
United Nations number: UN1301
Emergency Response Guide no.: 26
* **Reportable Quantity:** 5000 lb/2270 kg

Physical data

Boiling point (760 mm Hg): 72.7°C
(163°F)

Freezing point: −100°C (−148°F)

Specific gravity (H₂O = 1 @ 20/20°C):
0.9338

Vapor pressure (20°C): 88 mm Hg

Vapor density (Air = 1 @ 20°C): 3 0

Solubility in water (% by WT @ 20°C): 2 3

Percent volatiles by volume: 100

Evaporation rate (BuAc = 1): 8 9

Appearance and odor: Clear, colorless,
mobile liquid, acrid, ether-like odor

Fire and explosion hazard data

Flammable limits in air, % by volume
Upper: 13 4
Lower: 2 6

Flash point (test method):
Tag closed cup (ASTM D56): 18°F (−8°C)

Extinguishing media:
Use CO_2 or dry chemical for small fires.

Component information (See Glossary at end of MSDS for definitions)[1]

Component, wt. % (CAS number)	OSHA PEL TWA; STEL (15)	ACGIH TLV* TWA; STEL (15)	IDLH	Subject to SARA §313 reporting?
• Vinyl acetate 99 9% (108-05-4)	10 ppm 20 ppm	10 ppm 20 ppm	NVE[2]	Yes

(Exposure levels)

(1) All components listed as required by federal California New Jersey and Pennsylvania regulations
(2) No value established

alcohol-type aqueous film-forming foam
or water spray for large fires Water may
be ineffective but should be used to
cool fire-exposed structures and vessels

Special fire-fighting procedures:
* If potential for exposure to vapors
or products of combustion exists,
wear complete personal protective
equipment and respirator approved
by both NIOSH and MSHA:

Self-contained breathing apparatus
with full facepiece operated in pressure
demand or other positive pressure
mode

Supplied-air respirator with full
facepiece and operated in pressure-
demand or other positive pressure
mode in combination with an auxiliary
self-contained breathing apparatus
operated in pressure-demand or other
positive pressure mode

Unusual fire and explosion hazards:
Rapid, uncontrolled polymerization
can cause explosion Vapor is heavier
than air and can travel considerable
distance to a source of ignition and
flashback Material creates a special
hazard because it floats on water

Special hazard designations

	HMIS	NFPA	Key
Health:	2	2	0 - Minimal
Flammability:	3	3	1 - Slight
Reactivity:	2	2	2 - Moderate
Personal protective			3 - Serious
equipment:	G	—	4 - Severe

SARA §311 hazard categories
Acute health:	Yes
Chronic health:	Yes
Fire:	Yes
Sudden release of pressure:	No
Reactive:	Yes

Reactivity data

Stability:
Potentially unstable

Hazardous polymerization:
Can occur. UNCONTROLLED
POLYMERIZATION CAN CAUSE
RAPID EVOLUTION OF HEAT AND
INCREASED PRESSURE WHICH CAN
RESULT IN VIOLENT RUPTURE OF
STORAGE VESSELS OR CONTAINERS

Conditions to avoid:
Temperatures above 38°C (100°F),
sunlight; x-ray or ultra-violet radiation.
sparks and flame

Materials to avoid:
Peroxides, for example, t-butyl
peroxide and hydrogen peroxide.
other polymerization initiators.
oxidizing agents

**Hazardous combustion or
decomposition products:**
Carbon monoxide

Health data

Effects of exposure/toxicity data

Acute
Ingestion (swallowing): Can cause
headache, drowsiness and
unconsciousness Slightly toxic to
animals (oral LD_{50}, rats 2 9 g/kg)
Inhalation (breathing): Can irritate nasal
passages, throat and lungs Practically
non-toxic to animals (inhalation LCLo.
rats, 8 hrs. 4000 ppm)
Skin contact: Can cause severe injury
(reddening and swelling) Prolonged
contact can cause blisters Sensitization
(allergic reaction) can occur Slightly
toxic to animals by absorption (dermal
LD_{50}, rabbits 2 4 g/kg)
Eye contact: Can cause severe injury −
damage reversible

(continued)

---

* New or revised information, previous version dated March 20, 1989

Hoechst

VINYL ACETATE

#94   Issued February 16, 1990

Chronic

Mutagenicity: *In vitro*, suggestive evidence of mutagenicity (Hoechst Celanese data) *In vivo*, not mutagenic (SPI Vinyl Acetate Study Group, 90-day study, micronucleus test).

Carcinogenicity: Oral – A two-year drinking water study showed no evidence of treatment-related carcinogenicity (rats, SPI), a second limited drinking water study showed suggestive evidence of treatment-related tumors in the uterus and thyroid (rats, Lijinski). Inhalation – evidence of treatment-related nasal tumors in a two-year study of rats exposed to 600 ppm VA (SPI); suggestion of a treatment-related lung tumor in mice exposed to VA by inhalation at 600 ppm for two years (SPI)

Reproduction: No evidence of effects on fetus in VA study by oral or inhalation routes A suggestion of slight effects on male reproduction in oral study (drinking water, SPI).

* Medical conditions aggravated by exposure:
Significant exposure to this chemical may adversely affect people with chronic disease of the respiratory system, skin and/or eyes

Emergency and first aid procedures

Ingestion (swallowing): Induce vomiting of conscious patient immediately by giving two glasses of water and pressing finger down throat Contact a physician immediately

Inhalation (breathing): Remove patient from contaminated area. If breathing has stopped, give artificial respiration, then oxygen if needed Contact a physician immediately

Skin contact: Remove contaminated clothing and wash contaminated skin with large amounts of water If irritation persists, contact a physician

Eye contact: Flush eyes with water for at least 15 minutes Contact a physician immediately

Spill or leak procedures

Steps to be taken if material is released or spilled:
Eliminate ignition sources. Caution. Spontaneous polymerization can occur. Avoid eye or skin contact. Place leaking containers in well-ventilated area. If fire potential exists, blanket spill with foam or use water spray to disperse vapors Contain spill to minimize contaminated area and facilitate salvage or disposal. To clean up spill, flush area sparingly with water or use an absorbent. Avoid runoff into storm sewers and ditches which lead to natural waterways Call the National Response Center (800-424-8802) if spill is equal to or greater than reportable quantity (5000 lb/day) under "Superfund" All clean-up and disposal should be carried out in accordance with federal, state and local regulations. If required, state and local authorities should be notified.

Waste disposal method:
This product when spilled or disposed is a hazardous solid waste as defined in Resource Conservation Recovery Act regulations (40CFR261). Preferred method is incineration or biological treatment in federal/state approved facility

Special protection information

* Respiratory protection

Based on contamination level, use a respirator approved by both NIOSH and MSHA:
≤140 mg/m3 – Type C supplied-air respirator with half-mask facepiece operated in pressure-demand mode
≤1400 mg/m3 – Gas mask with full facepiece and chin-type organic vapor canister (maximum service life, 2 hour). Gas mask with full facepiece and chest- or back-mounted organic vapor canister Type C supplied-air respirator with full facepiece operated in positive pressure mode Self-contained breathing apparatus with full facepiece operated in positive pressure mode
< 14,000 mg/m3 – Type C supplied-air respirator with half-mask or full facepiece operated in continuous-flow, pressure-demand, or other positive pressure mode Type C supplied-air

respirator with hood, helmet, or suit operated in continuous-flow mode
>14,000 mg/m3 – Self-contained breathing apparatus with full facepiece operated in pressure-demand or other positive pressure mode. Combination Type C supplied-air respirator with full facepiece operated in pressure-demand mode and with auxiliary self-contained air supply.

Unknown concentration – Self-contained breathing apparatus with full facepiece operated in pressure-demand or other positive pressure mode

Ventilation
Local exhaust: Recommended when appropriate to control employee exposure.
Mechanical (general): Not recommended as the sole means of controlling employee exposure.

Protective gloves:
Neoprene or rubber

Eye protection:
Chemical safety goggles

Other protective equipment:
For operations where spills or splashing can occur, use impervious body covering and boots A safety shower and eye bath should be available

Special precautions

Precautions to be taken in handling and storing:
Store in a cool, well-ventilated area Keep away from heat, sparks and flame Keep containers closed Use only DOT-approved containers Use spark-resistant tools Do not load into compartments adjacent to heated cargo When transferring follow proper grounding procedures Use with adequate ventilation Avoid breathing vapor Avoid contact with eyes, skin and clothing Wash thoroughly with soap and water after handling Wash contaminated clothing thoroughly before re-use Discard contaminated leather clothing

* New or revised information; previous version dated March 20, 1989

***** Glossary for Components information table

ACGIH	American Conference of Governmental Industrial Hygienists	**SARA**	Superfund Amendments and Reauthorization Act
CAS	Chemical Abstract Service	**Skin**	Potential contribution to overall exposure possible via skin absorption
IDLH	Immediately Dangerous to Life or Health	**STEL**	Short-term exposure level 15-min TWA
OSHA	Occupational Safety and Health Administration	**TLV**	Threshold limit value
PEL	Permissible exposure limit	**TWA**	8-hour time-weighted average

Chemical Group
Hoechst Celanese Corporation
PO Box 569320/Dallas, Texas 75356-9320
Information phone 214 689 4000
* Emergency phone: 800 424 9300 (CHEMTREC)

Reprinted with permission from Hoechst Celanese Corporation, Dallas, Texas.

REFERENCES

1. Fenlon, W. J., "The vapor phase flammability limits of acetic acid–air–nitrous oxide–water mixtures at 110°C and atmospheric pressure," Health and Safety Laboratory, Eastman Kodak Company, Rochester, NY 14650, KHSL Report No. TS-73-01, Feb. 2, 1973.
2. Shraer, B. I., *Soviet Chem. Ind.*, No. 10, pp. 28–31 (October 1970).
3. Lees, F. P., *Loss Prevention in the Process Industries*, Butterworths, London, 1980.
4. *Guidelines for Safe Storage and Handling of High Toxic Hazard Materials*, AIChE Center for Chemical Process Safety, New York, NY, 1988.
5. *Guidelines for Vapor Release Mitigation*, AIChE Center for Chemical Process Safety, New York, NY, 1988.
6. *Guidelines for the Use of Vapor Cloud Dispersion Models*, AIChE Center for Chemical Process Safety, New York, NY, 1987.
7. Special Issue on the Loss of Containment Conference, London, 1989. *J. Loss Prevent. Process Ind.*, 3(1) (1990).
8. Sax, N., *Dangerous Properties of Industrial Materials*, 9th ed., Van Nostrand Reinhold Co., New York, NY, 1988.
9. Reynolds, J. W., Tennessee Eastman Company Design Data Research Laboratory Report Number 7726-54, Kingsport, TN, Nov. 15, 1983.
10. Hitchcock, C. H., internal Eastman Chemical Company communication, Kingsport, TN, (1988).
11. *DIERS Project Manual*, AIChE Design Institute for Emergency Relief Systems, New York, NY, 1989.
12. *Emergency Relief Systems for Runaway Chemical Reactions and Storage Vessels*, SAFIRE Code, Fauske and Associates, Burr Ridge, IL, 1983.
13. Scott, I. G., Tennessee Eastman Company Technical Report TEAD-R-DC-706-2254, Kingsport, TN, April 24, 1986.
14. Jeffers, R. L., Eastman Chemical Company Notebook x21621, Kingsport, TN, 1991, p. 88.
15. *Design and Installation of Pressure Relieving Systems in Refineries*, Parts I and II, RP 520, American Petroleum Institute, 1976.

Index